Topics in
Current Physics

15

Topics in Current Physics   Founded by Helmut K. V. Lotsch

Topics in
Current Physics

15

Professor Myron B. Salamon, PhD

Department of Physics, University of Illinois at Urbana-Champaign,
Urbana, IL 61801, USA

ISBN-13: 978-3-642-81330-6   e-ISBN-13: 978-3-642-81328-3
DOI: 10.1007/ 978-3-642-81328-3

Library of Congress Cataloging in Publication Data. Main entry under title: Physics of superionic
conductors. (Topics in current physics ; v. 15). Includes bibliographies and index.
1. Superionic conductors. I. Salamon, Myron Ben, 1939-. II. Series. QC717.P49 537.6'2 79-9843

© by Springer-Verlag Berlin Heidelberg 1979

2153/3130-543210

Preface

Superionic conductors are solids whose ionic conductivities approach, and in some
cases exceed, those of molten salts and electrolyte solutions. This implies an unusual state of matter in which some atoms have nearly liquidlike mobility while
others retain their regular crystalline arrangement. This liquid-solid duality
has much appeal to condensed matter physicists, and the coincident development of
powerful new methods for studying disordered solids and interest in superionic
conductors for technical applications has resulted in a new surge of activity in
this venerable field. It is the purpose of this book to summarize the current research in the physics of superionic conduction, with special emphasis on those
aspects which set these materials apart from other solids. The volume is aimed towards the materials community and will, we expect, stimulate further research on
these potentially useful substances.

The usual characterization of the superionic phase lists high ionic conductivity;
low activation energy; and the open structure of the crystal, with its interconne-
ted network of vacant sites available to one ionic species. To these, as we demonstrate in this volume, should be added important dynamic and collective effects:
the absence of well-defined optical lattice modes, the presence of a pervasive,
low-energy excitation, an infrared peak in the frequency-dependent conductivity,
unusual NMR prefactors, phase transitions, and a strong tendency for the mobile
ion to be found between allowed sites. Thus, a study of the physics of these materials uncovers a wealth of new phenomena which justifies treating superionic conductors as a separate phase of matter. In examining these phenomena, we focus on
a few "classic" superionic conductors and have not attempted to survey the extensive literature in the chemistry and materials science of these materials. For
these, the books and reviews referred to in the various chapters will, we hope,
suffice.

The contributors to this volume, being currently active in research on superionic conductors, have written not reviews of the field but expositions of the
current state of understanding. In many cases, previously unpublished results are
presented. From the work presented here, a more complete picture of the nature of
the superionic phase emerges. These materials are not just solids with many vacancies nor liquids flowing through a rigid cage. The interactions between mobile

and cage ions are significant, destroying the usual phonon spectrum of the solid and mitigating the strong ion-ion repulsion. We are still far from a complete understanding of these materials; this volume will have served its purpose if it stimulates new research and helps to focus that research in fruitful directions.

In assembling a book such as this, the editor has placed great demands on the authors to produce contributions which are not only timely, but contain new material. In every case, the contributors have exceeded the editor's expectations, and done so very quickly. I would like to thank each of the authors for his excellent co-operation; special thanks to M.J. Delaney who, since he was at Illinois, was frequently pressed into service, reading and discussing the chapters as they developed. My thanks, too, to the series editor Dr. H. Lotsch for his encouragement and for the efficient handling of the many phases of producing this volume.

Urbana, Illinois
December 1978   *M.B. Salamon*

Contents

List of Contributors

BEYELER, HANS ULRICH
   Brown Boveri Research Center, CH-5405 Baden, Switzerland

BOYCE, JAMES BUCKLEY
   Xerox Palo Alto Research Center, 3333 Coyote Hill Road, Palo Alto, CA 94304, USA

BROESCH, PETER
   Brown Boveri Research Center, CH-5405 Baden, Switzerland

DELANEY, MICHAEL JOSEPH
   Department of Physics, University of Illinois at Urbana-Champaign,
   Urbana, IL 61801, USA

GEISEL, THEO
   Fachbereich Physik, Universität Regensburg,
   D-8400 Regensburg, Fed. Rep. of Germany, and
   Xerox Palo Alto Research Center, 3333 Coyote Hill Road, Palo Alto, CA 94304, USA

HAYES, TIMOTHY MITCHELL
   Xerox Palo Alto Research Center, 3333 Coyote Hill Road, Palo Alto, CA 94304, USA

PIETRONERO, LUCIANO
   Brown Boveri Research Center, CH-5405 Baden, Switzerland

REIDINGER, FRANZ
   Brookhaven National Laboratory, Upton, NY 11973, USA

RICHARDS, PETER MICHAEL
   Sandia Research Laboratory, Albuquerque, NM 87185, USA

SALAMON, MYRON B.
   Department of Physics, University of Illinois at Urbana-Champaign,
   Urbana, IL 61801, USA

SCHNEIDER, WALTER RUDOLF
   Brown Boveri Research Center, CH-5405 Baden, Switzerland

SHAPIRO, STEPHEN M.
   Brookhaven National Laboratory, Upton, NY 11973, USA

STRÄSSLER, SIEGFRIED
   Brown Boveri Research Center, CH-5405 Baden, Switzerland

USHIODA, SUKEKATSU
   Department of Physics, University of California, Irvine, CA 92714, USA

ZELLER, HANS RUDOLF
   Brown Boveri Research Center, CH-5405 Baden, Switzerland

1. Introduction

M. B. Salamon

Superionic conductors are unusual solids in which the regularity of the crystalline lattice is significantly disrupted. As a consequence, the basic concepts of solid-state physics - phonons, energy bands, Brillouin zones - must be, to some extent, modified or abandoned. The recent renewal of interest in the physical properties of superionic conductors is part of the general development of the physics of non-crystalline phases - abetted, in this case, by technological applications of these materials as solid electrolytes.

Dating from Faraday's observations in 1834, a considerable body of data on the basic phenomenology of high ionic conductivity in the solid state has been accumulated [1.1-3]. Examining these data, one finds no sharp demarcation between superionic conductors and other ionic crystals, all of which exhibit some ionic conductivity. Generally speaking, the combination of meltlike ionic conductivity in the solid phase ($\sigma \gtrsim 10^{-1}\Omega^{-1}cm^{-1}$) with a low activation energy ($\sim 10^{-1}$ eV) are the accepted hallmarks of the superionic phase. To these, the evidence in this volume will enable us to add: the absence of sharp features in Raman scattering, a peak in the frequency-dependent conductivity in the far infrared, and the presence of significant mobile ion density along the diffusion path or, equivalently, a transit time which is significant compared with the time spent at a lattice or interstitial site.

The basic properties and phenomenology of high ionic conductivity have been amply discussed in several recent reviews and conference proceedings [1.2,3] and will not be recounted here. Neither shall we attempt to catalog the growing number of superionic conductors nor to discuss their technological applications. Rather, our purpose is to examine the physics of superionic phases - both experimental manifestations and model calculations which reproduce these results with a minimum of unjustified assumptions. Both experimentally and theoretically, the main focus is on several "classic" superionic conductors - α-AgI, α-RbAg$_4$I$_5$, β-alumina, the fluorites, and the copper halides - plus a few quasi-one-dimensional materials.

Throughout this volume the reader is assumed to be familiar with the classical concepts of atomic diffusion in solids which are outlined here. In its simplest form, a diffusing ion is said to traverse a barrier of height Δ (the *activation energy*) at a rate

$$\nu = \nu_0 \exp(-\Delta/k_B T) \quad , \tag{1.1}$$

called the *jump frequency*. The *attempt frequency* ν_0 is typically comparable to a phonon frequency and is often taken to be the frequency of oscillation of the ion within the cell associated with its lattice site. In superionic conductors, where one species is more mobile than the others, the less mobile ions are said to constitute the "cage" while the attempt frequency measures the vibration of the mobile ion within the cage. From (1.1), the elementary theory leads to a diffusion constant

$$D = \nu a_0^2 = \nu_0 a_0^2 \exp(-\Delta/k_B T) \quad , \tag{1.2}$$

where a_0 is the *jump distance*. The *ionic conductivity* σ can be calculated from the diffusion constant by using the Einstein relation

$$\sigma = ne^2 D/k_B T \tag{1.3}$$

where n is the density of diffusing ions. A more complete description of diffusion involves the 3N normal coordinates of the N-body system in terms of which the crystal potential is written [1.4]. In this model, the jump frequency is computed from the rate at which the representative point of the system moves from one potential minimum to the next in configuration space. The minima are separated by a saddle point in the 3N-dimensional potential surface, over which the representative point must pass. The activation energy, in this approach, measures the difference in potential between the bottom of the well and the saddle point.

Two characteristic times are frequently employed to characterize ionic conductivity: the mean residence (or dwell) time τ_{res} (or τ_d), which is essentially ν_0^{-1}, and the transit (or flight) time τ_{trans} (or τ_f) which measures the time spent between sites. To describe ionic diffusion in solids, a jump diffusion model is usually employed which makes the assumption of instantaneous jumps between sites, i.e., $\tau_{trans} \ll \tau_{res}$. The theme of this volume, which will readily become apparent, is the inadequacy of the simple jump model for the understanding of superionic conductors. Translated into probabilities, the premise of instantaneous jumps means that the mobile ion should be found with much higher probability on a lattice or interstitial site than at the saddle point position.

In Chapter 2, Boyce and Hayes examine EXAFS data for a number of superionic conductors and conclude, in each case, that the best fits to the data result when the mobile ion is permitted to occupy the volume outside the cage-ion radii with essentially uniform density. Even more pictorial evidence of this is presented in Chapter 3, in which Shapiro and Reidinger employ Fourier difference maps on their neutron diffraction data to demonstrate the high probability of finding mobile ions at saddle-point positions. This casts doubt on the applicability of the standard jump-diffusion model to superionic conductors. Rather than an isolated case, the presence of substantial saddle-point density (or equivalently, considerable time spent at the saddle point) is demonstrated to be characteristic of the superionic phase. Their inelastic neutron scattering results display a dispersionless low-energy mode which also appears to be characteristic of these materials.

In Chapter 4, Beyeler et al. focus on the frequency-dependent conductivity in attempting to construct a model which goes beyond simple hopping. They concentrate on the inability of the crystal to come into equilibrium during the short time the ion resides on the lattice site, calculating correction terms to the usual many-body diffusion model [1.4]. That these additional terms are required is seen by examining the infrared peaks in the conductivity of several of these materials. Such broad, low-frequency features seem to be characteristic of the superionic phase. A summary of the optical data is given by Delaney and Ushioda in Chapter 5. The absence of sharp Raman features, discussed in that chapter, demonstrates clearly the strong coupling of the phonon modes of the crystal, cage as well as mobile species, resulting in the absence of any clear optical phonon lines. The same paucity of structure is evident from inelastic neutron diffraction (Chapter 3) and appears to be a fundamental characteristic of the superionic phase.

Nuclear resonance has been frequently used to study atomic mobility. Richards in Chapter 6 deals with resonance results on superionic conductors, and concludes that restricted dimensionality frequently plays a dominant role in the NMR properties. Nothing definitive has yet resulted from these measurements. The major problem appears to be competing effects from various sources of relaxation in different frequency regimes.

The strongest evidence for coherent mobile-ion effects comes from the ordering transitions which bound the superionic phase of most of these materials. The various models for phase transitions are examined by Salamon in Chapter 7 in the light of the modern theory of critical phenomena, and a survey of experimental results is presented. A close connection between ionic conduction and short-range order is found, and the mediation of ion-ion interactions by distortions of the cage lattice is shown to be crucial.

Cognizant of the breakdown of simple hopping, Geisel presents an exposition of continuous diffusion models in Chapter 8. These models, in which the detailed coupling to the cage lattice is treated as a frictional drag, are shown to be quite successful in explaining the optical and inelastic neutron diffraction spectra. In the large friction limit, qualitative agreement with experiment is obtained.

Finally, some comment on the title of this volume is in order. The use or avoidance of the term "superionic" appears to divide researchers in this field into two groups. The assertion is made that the term, despite its widespread use, is a misnomer [1.3]. However, it must be kept in mind that the word "ion" is derived from *iōn*, the present participle of the Greek "to go," and thus implies motion.

4

The problem with the present combination is that "super-" is of Latin derivation. A better choice would have been "hyperionic" perhaps, but we now have a term of general acceptance for, as we demonstrate in this volume, a phase of matter with a variety of characteristic and physically interesting properties.

References

1.1 W. van Gool (ed.): *Fast Ion Transport in Solids* (North-Holland, Amsterdam 1973)
1.2 J. Hladik: *Physics of Electrolytes* (Academic Press, London, New York 1972)
1.3 S. Geller (ed.): *Solid Electrolytes*, Top. Appl. Phys. Vol.21(Springer, Berlin, Heidelberg, New York 1977)
1.4 C.P. Flynn: *Point Defects and Diffusion* (Oxford University Press, London 1972)

2. Structure and Its Influence on Superionic Conduction: EXAFS Studies

J. B. Boyce and T. M. Hayes

With 17 Figures

Superionic conductors are a class of materials which achieve ionic conductivities comparable with those of molten salts while still in the solid phase. As discussed in Chap.7, the transition from insulating to conducting behavior may take place sharply at a specific temperature, as in the case of AgI, or gradually over a large temperature range, as in the case of PbF_2 [2.1]. In either case, it involves a disordering of one of the ion sublattices of the material. At low temperature, all the ions are situated on well-defined lattice sites and have a very low mobility. As the temperature is increased, the mobile ions begin populating interstitial sites. In the superionic phase, these ions are distributed over a large number of available sites [2.2]. It follows that this superionic phase is quite naturally characterized structurally in terms of the mobile and immobile sublattices [2.3]. The immobile ions form a complex structure through which the mobile ions move. This structure is not rigid since these ions execute large vibrations about their lattice sites; nonetheless, they do not leave those sites and so do not contribute to the ionic conductivity. The positions of the immobile ions define characteristic voids which are populated to varying degrees by the mobile ions and through which these ions move. In AgI, for example, the iodine forms a bcc lattice, while the Ag ions are located in and move among the tetrahedral voids, with negligible occupation of the octahedral locations [2.4,5].

Since many of the fundamental questions about superionic conductors are structural in nature, the structural probes have been essential in understanding these materials. The transition to the superionic state involves the disordering of the arrangement of the mobile ions, and so must be characterized structurally. In the superionic phase itself, elementary considerations of mass and temperature suggest that the mobile ions spend a significant fraction of time in flight from one location to another [2.4,6,7]. Therefore, structural studies yield substantial information not only on the sites that are occupied, but also on the flight path from one site to another. These measurements can thus yield insight into the conduction process as well as the order-disorder transformation.

In this chapter, we discuss the structural information probed through the extended X-ray absorption fine structure (EXAFS), and the relationship between this information and the high ionic conductivity of these materials. The EXAFS consists of the oscillations, as a function of photon energy, in the absorption cross section for the photoexcitation of an electron from a deep core state to a continuum state. These oscillations are a final state electron effect, arising from the interference between the outgoing wavefunction and that small fraction of itself which is scattered back from the near-neighbor atoms. This interference reflects directly the net phase shift of the backscattered electron, which is predominantly proportional to the product of the momentum of the electron, k, and the distance traveled. The atomic identity of both the excited and backscattering atoms has a more subtle but nonetheless significant effect on the interference. As a consequence, analysis of the EXAFS can yield not only the distance but also the type and number of the nearest neighbors of the excited atoms.

Since EXAFS arises from scattering by the near neighbors of the excited atom species *only*, a given measurement involves a subset of those pair correlation functions which are probed in a single diffraction measurement. This is a substantial simplification. In X-ray and neutron diffraction studies of superionic conductors, the complicated diffraction pattern contains both Bragg peaks (or Debye lines) from long-range order and a liquid-like diffuse scattering pattern from short-range correlations [2.8,9]. These two contributions are often dominated by the stationary lattice and the mobile-ion/mobile-ion correlations, respectively. EXAFS, on the other hand, measures primarily the pair correlation function of the mobile species with respect to the immobile ions. Thus, the EXAFS technique is especially well suited to determining the path taken by the conducting ions.

In Sect.2.1 we discuss the EXAFS technique and compare it with x-ray and neutron scattering. In Sect.2.2 various structural models for superionic conductors are discussed. In Sects.2.3,4 the experimental results on AgI and the cuprous halides, respectively, are presented and analyzed in terms of these structural models. We show that an excluded volume model not only explains the measured pair distribution function well, but can also yield important insight into the conduction process.

2.1 Technique of EXAFS

2.1.1 Theory

The microscopic origin of the EXAFS is very well understood, having been discussed at length in the literature [2.10-14]. Of particular importance is the demonstration by LEE and PENDRY [2.12] that, except in unusual circumstances, electrons which have been scattered by more than one neighboring atom make a negligible contribution to the measured EXAFS. This leads to an enormous simplification in the interpretation of the EXAFS, and is used in the following treatment. From general considerations [2.13,14], the absorption cross section for the photoexcitation of an electron from the K shell of atom species α by an X-ray photon of energy E can be expressed as

$$\sigma_\alpha(E) = \sigma_\alpha^0(E)[1 + \Delta_\alpha(k)] \quad , \tag{2.1}$$

where the EXAFS for randomly oriented local environments or for powders is given by

$$k\Delta_\alpha(k) = \sum_\beta \int_0^\infty dr/r^2 \, p_{\alpha\beta}(r) \, 2\text{Re}\{\exp(2ikr)\Lambda_{\alpha\beta}(k,r)\} \tag{2.2}$$

and

$$\Lambda_{\alpha\beta}(k,r) \sim (-2i\pi^2)t_\beta^+(-k,k) \, \exp[-2\mu(k)r + 2i\delta_\alpha(k)] \quad . \tag{2.3}$$

Here, $k^2/2 = E_{1s} + E$ is the final state electron energy with E_{1s} the binding energy of the K-shell electron. The $\sigma^0(E)$ factor in (2.1) contributes a broad, atomlike background to $\sigma_\alpha(E)$, essentially featureless except for the K edge. The term which gives rise to the EXAFS, Δ, has been divided into two distinct contributions: direct structural information in $p(r)$ and complicated energy dependences in Λ. $p_{\alpha\beta}(r)$ is the radial distribution of atom species β about the excited species α, defined so that $\int_0^\infty dr \, p_{\alpha\beta}(r)$ equals the total number of β atoms in the sample. Each atom species in the system is represented in the sum over β in (2.2). The angular distribution of neighbors does not enter except for a single crystal or an oriented polycrystalline sample for which (2.2) is modified by a factor $3(\hat{r}_{\alpha\beta}\cdot\hat{e})^2$, with $\hat{e}$ the photon polarization and $\hat{r}_{\alpha\beta}$ the unit vector from atom α to atom β [2.10,15]. In the factor $\Lambda_{\alpha\beta}(k,r)$ are included the t-matrix of the scattering atom (t_β^+), the electron mean free path (μ), and the $\ell = 1$ phase shift due to the potential of the excited atom (δ_α). Note that (2.2) does not involve the correlations between all atom pairs, as does the analogous expression for a diffraction study. This represents a significant advantage and will be discussed later.

Consider the special case where $p_{\alpha\beta}(r)$ contains a single Gaussian peak,

$$p_{\alpha\beta}(r) = \frac{N_\beta}{(2\pi\epsilon_{\alpha\beta}^2)^{\frac{1}{2}}} \exp[-(r - R_{\alpha\beta})^2/2\epsilon_{\alpha\beta}^2] \quad . \tag{2.4}$$

Here $R_{\alpha\beta}$ is the mean separation of atoms α and β, $\epsilon_{\alpha\beta}^2$ is the mean square deviation from $R_{\alpha\beta}$, and N_β is the number of β neighbors at a distance $R_{\alpha\beta}$ from atom α. A Gaussian is often not the appropriate model distribution function, especially in the superionic phase, but it qualitatively reveals the various functional dependences of the EXAFS. Using (2.4) in (2.2) and (2.3) gives [2.16]

$$k\Delta_\alpha(k) \sim \sum_\beta \frac{N_\beta}{R_{\alpha\beta}^2} |t_\beta^+(-k,k)| \exp[-2\mu(k)R_{\alpha\beta}]$$

$$\exp(-2k^2\epsilon_{\alpha\beta}^2) \sin[2kR_{\alpha\beta} + 2\delta_\alpha(k) + \gamma_\beta(k)] \quad , \tag{2.5}$$

where γ_β is the phase of t_β^+. It is worth commenting on three aspects of the EXAFS apparent in (2.5). First, the EXAFS amplitude decreases with increasing k due to a Debye-Waller-like factor, $\exp(-2k^2\epsilon^2)$, and the k dependence of the t matrix. This effect is more important at elevated temperatures or in disordered materials where ϵ is large, precisely the case for superionic conductors. As will be seen below, however, sufficient signal is available even in the superionic phase to yield meaningful results. Second, note that ϵ is substantially larger for more distant shells. The $\exp(-2k^2\epsilon^2)$ factor makes the EXAFS sensitive mainly to near-neighbor correlations. But this is the quantity of interest in superionic conductors, since it reveals the location of the mobile ions relative to the immobile ions. Third, the argument of the sine contains a k-dependent phase shift in addition to 2kR. If one expands the phase shifts in a Taylor series as $\delta = \delta_0 + \delta_1 k + \ldots$ and $\gamma = \gamma_0 + \gamma_1 k + \ldots$, then it is seen that the EXAFS oscillates in k space with a frequency $(2R + 2\delta_1 + \gamma_1)$, not just 2R. This means that the phase shifts have to be known or determined from a standard in order to obtain the distance R from the measurement.

The objective of the experimental studies can be stated quite simply — the extraction of the $p_{\alpha\beta}(r)$. This can be complicated in practice. Each shell of near neighbors of radius r_j in $p_{\alpha\beta}$ introduces a sinusoidal term of argument ~$2kr_j$ in $k\Delta$. The backscattering of more distant atoms makes, therefore, a higher frequency contribution to the oscillations in k space. The k-space interpretation can be hampered by the superposition of the contributions from many shells of near neighbors. The situation is improved by a Fourier transform into coordinate space using the complementary variables 2k and r. High frequency becomes large r, and the contributions of the various shells are often separated spatially to a large extent. Representing the limited k range of the experiment by a window function, W(k), the experimental quantity of interest corresponds to the Fourier transform of $W(k) \cdot k\Delta$ [2.14]

$$\phi_\alpha(r) = \sum_\beta \int_0^\infty \frac{dr'}{r'^2} p_{\alpha\beta}(r')\xi_{\alpha\beta}(r - r') \tag{2.6}$$

for r > 0. As represented by (2.6), the Fourier transform of the EXAFS, ϕ, is the sum of ξ peaks located at {r}. ξ is analogous to the peak function of X-ray or neutron diffraction and the {r} are determined by the peaks in $p_{\alpha\beta}(r)$. Explicitly, $\xi_{\alpha\beta}(r,r'')$ is the Fourier transform of $W(k) \cdot \Lambda_{\alpha\beta}(k,r'')$. The exact form of the window function is discussed in Sect.2.1.3. The r" variable in ξ, which arises from the mean-free-path factor in Λ, has been suppressed in (2.6). The inelastic electron scattering processes which underlie this factor are not a significant consideration in the principal use to which (2.6) will be put in this paper — comparisons among the EXAFS signals from similarly positioned nearest neighbor shells in similar systems. These processes must be taken into account, however, in comparisons among shells at substantially different radii and in comparisons between data and calculated spectra [2.11,12].

The discussion of K-shell EXAFS data analysis which follows is based upon (2.1-3), and (2.6). It is also possible to measure the EXAFS on the absorption cross section for the photoexcitation of less tightly bound electrons, such as those in the L or M shells. Unlike electrons from the K shell, these electrons are characterized by a mixture of final state anuglar momenta. As a consequence, the single term in (2.1) is replaced by a sum of three terms, and the interpretation of experiment is more complicated [2.17]. Although strictly applicable only to K-shell photoexcitation, many of the principles discussed in the following treatment can be generalized to apply to the other shells as well.

2.1.2 Experiment

The most direct and commonly used method to acquire EXAFS data is a transmission experiment. The experimental setup used at the Stanford Synchrotron Radiation Laboratory (SSRL) is shown in Fig.2.1. The X-ray source for the experiments to be discussed below is synchrotron radiation from SPEAR, the Stanford storage ring; conventional X-ray sources have also been used but with much longer counting times due to the reduced photon intensity [2.18]. The X-ray beam is monochromatized by two diffractions from a silicon crystal and its incident intensity, I_0, is measured by a gas proportional counter. The transmitted intensity, I, is measured in a similar manner, yielding the absorptance, $\mu x = \ln(I_0/I)$, where x is the sample thickness. The absorptance is proportional to the sample's absorption cross section.

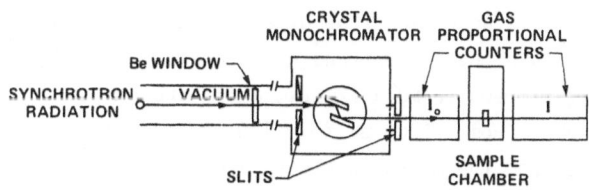

Fig.2.1 Experimental configuration for a transmission EXAFS experiment

$\sigma_\alpha(E)$ is thereby measured as a function of X-ray energy. A typical spectrum is shown in Fig.2.2 for the case of the Cu K edge in CuBr at 77 K. The EXAFS is manifest as sharp oscillations in the absorption cross section extending for about 800 eV above the Cu K edge at 8.98 keV. The prominent spike in the absorption cross section just

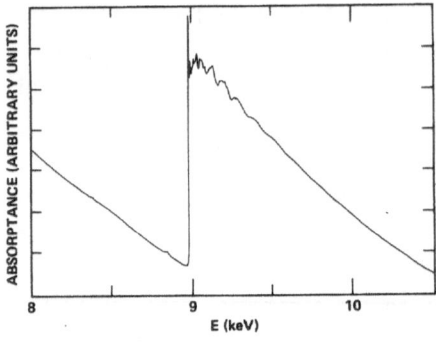

ABSORPTANCE (ARBITRARY UNITS)

8   9   10
E (keV)

Fig.2.2 The absorptance of CuBr at 77 K
as a function of X-ray photon energy, including the onset of the Cu K-shell absorption at 8.98 keV

above the K edge is not part of the EXAFS. It is the "white line" and is more properly
viewed as arising from either an unusually high density of final states or the substantial overlap of an unusually localized final state wavefunction with the K shell
[2.19].

There are other methods by which EXAFS data can be acquired. In general, these
methods involve measuring the intensity of a characteristic by-product of the
K-shell absorption event as a function of incident X-ray photon energy. For example,
in studying a system which is very dilute in the atom species of interest, the
method of preference involves monitoring the intensity of the characteristic
fluorescence produced by radiative decay of the hole in the K shell [2.20]. Similarly, surface atoms can be probed in preference to bulk atoms by monitoring the
characteristic Auger spectrum [2.15]. In each case, the absorption cross section
can be obtained and analyzed as described below. Transmission experiments were used
for the work discussed here.

2.1.3 Data Reduction and Analysis

The experimentally determined absorptance, illustrated in Fig.2.2, represents the
sum of two distinct contributions: the K-shell absorption which contains the EXAFS
information, $\sigma_\alpha(E)$, and a slowly varying background absorption, σ_{bg}. This background
absorption is due to the photoexcitation of those electrons of all atom species in
the system which are less tightly bound than the K electrons of interest. σ_{bg} is approximated by a polynomial, such as a Victoreen formula [2.18], and fitted to the
absorptance over an energy range extending from ~ 1000 eV to ~ 100 eV below the
K edge. The resulting σ_{bg} is subtracted from the measured absorptance over its complete range, both above and below the K edge, yielding a quantity proportional to
the $\sigma_\alpha(E)$ defined in (2.1).

Next, $\sigma_\alpha^0(E)$ is approximated and $\Delta_\alpha(E)$ extracted using (2.1). In order to accommodate its rapid variation in the region of the K edge, σ_α^0 is approximated as
a polynomial of 4th or 6th order in $(E - E_{th})^{\frac{1}{2}}$. E_{th} is chosen as that value of E
which corresponds to one-half the step height in absorption at the K-shell thres-
hold, excluding the contribution due to the threshold spike or white line. The

parameters in the polynomial are chosen by a least squares fit to the K-shell absorptance, weighted by $(E - E_{th})$, over an energy range extending from just above the threshold spike to the upper limit of the data range. The sinusoidal variations in Δ as a function of $(E - E_{th})^{\frac{1}{2}}$ are of sufficiently high frequency relative to variations in σ_α^0 that fitting the absorptance to a polynomial of low order will ignore Δ_α and approximate σ_α^0.

The remaining step in obtaining $k \cdot \Delta_\alpha(k)$ as a function of electron momentum, k, is the identification of the zero of conduction electron energy, that value of E which corresponds to k = 0 for the final state electrons. This is difficult to accomplish in an absolute sense, as it differs from the threshold for continuum excitation by an unknown self-energy correction and a Fermi energy (which is zero for an insulator). In the technique under discussion here, the analysis will involve a comparison among reduced data sets. Accordingly, the zero of conduction electron energy is chosen arbitrarily as E_{th}, making k proportional to $(E - E_{th})^{\frac{1}{2}}$. The systematic k-scale distortion introduced in this way will not affect the comparisons in most cases. Note, however, that this approach could be incorrect in comparing EXAFS from substantially different systems, such as a metal and an insulator, where the self-energy correction could be different.

An alternative technique for analyzing the EXAFS data has been put forward by LEE and BENI [2.21]. The structural information is obtained by comparing the data in k space with calculations of the EXAFS based on theoretically derived phase shifts and backscattering t matrices [2.22,23]. In this technique, it is essential to adjust the zero of conduction electron energy used in reducing the experimental data in order to obtain correspondence with the calculations. This adjustment corrects for errors in assigning the "muffin-tin zero" in the calculation as well as for the self-energy.

If E_{th} is chosen as the zero of conduction electron energy, the resulting $k \cdot \Delta(k)$ for CuBr is shown in Fig.2.3. The procedure for extracting Δ has normalized it properly to the number of Cu atoms in the sample, allowing it to be

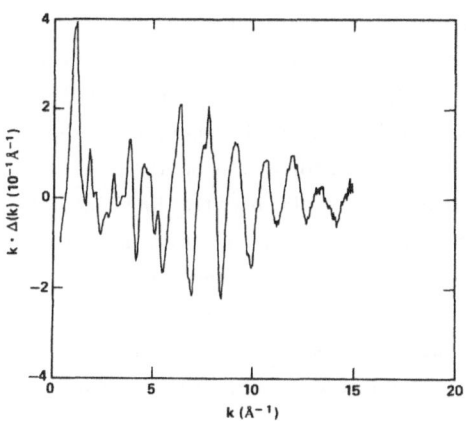

Fig.2.3 The EXAFS oscillations, $k \cdot \Delta(k)$, on the Cu K-shell absorption in CuBr at 77 K as a function of photoelectron momentum k

compared directly with (2.2). The peak at k ~ 1 Å^{-1} corresponds to the threshold
spike in Fig.2.2. The EXAFS starts above that feature, and extends to about
15 Å^{-1}. The influence of many shells of near neighbors is manifest in the presence of many frequencies in the EXAFS.

The comparison among EXAFS data sets, or between data and a calculated spectrum, can be accomplished either in k space or in r space following a Fourier
transform. There are advantages to be associated with each approach. The k space
data contain more information in the sense that they have not been "degraded" by
the use of a window function, a necessary step in the Fourier transform approach.
On the other hand, as initially observed by SAYERS et al. [2.24], the extraction
of structural information is often simplified substantially in r space due to the
spatial separation of the near neighbor shells. A combined approach is also used
[2.25,26], in which the signal arising from a single isolated shell in r space is
transformed back into k space for comparison purposes. This appears to offer no
advantage over a direct comparison in r space except that possibly the k space
expression, (2.2), may, in some cases, be simpler than the r-space expression,
(2.6). It has the disadvantage associated with an additional window function and
Fourier transform. In any case, it is likely that these relative advantages and
disadvantages are unimportant compared with the advantages accruing from the use
of a consistent, rigorous prescription for comparing two data sets. The r-space
approach is used in the following discussion.

In transforming the EXAFS into r space, one faces the problem of minimizing
the undesirable effects of the limited data range available for transformation.
In EXAFS, the low k cutoff is the most significant problem, arising from the necessity of avoiding the white line. In the absence of a reliable technique for re-
moving that peak, the EXAFS data set must be terminated in the region of 2 to
3 Å^{-1}. The large k cutoff is much less of a problem since the EXAFS signal is
typically very small for k larger than 15 to 20 Å^{-1}. In any case, the inacces-
sibility of a portion of the k-space range of kΔ results in the shape and extent
of the peak function, ξ, being quite sensitive to the manner in which the data is
terminated in the transform. The problem of selecting an appropriate window function occurs analogously in the Fourier transformation of X-ray or neutron diffraction results, and has been treated extensively [2.27]. The optimum window function
is chosen to compromise between a long-range or oscillating peak function on the
one hand, and an intolerable broadening of the r-space information on the other.
The customary choice in diffraction has been to eliminate the oscillations in
r space at the expense of structural resolution. In the following analysis, a different approach is taken and a square window, which has been broadened by con-
volution with a Gaussian, is used.

The effect of this broadening in k space is to impose an exponential localization in r space, lessening substantially the interference of neighboring peaks

and the accompanying distortion in apparent peak position and magnitude [2.14].
This choice minimizes the loss of structural resolution, but necessitates the
use of analysis techniques which are unaffected by the residual oscillations in
the peak function.

The data in Fig.2.3 have been transformed using a square window $k = (2,8, 13.6)$
$Å^{-1}$, with a Gaussian broadening of 0.7 $Å^{-1}$. The absence of data for negative k
has resulted in a ϕ which is complex. The real part and the magnitude of $\phi(r)$ for
the K edge of Cu in CuBr at 77 K are shown in Fig.2.4a. The signals from the first
few shells of atoms around each Cu are clearly evident to the right, while the
small peaks near the origin result from incomplete removal of σ_α^0. The first peak
at 2.2 Å is due to the Br first neighbors, the second peak at 3.7 Å is due to the
Cu second neighbors, and so on. The actual position of the first and second neighbors is 2.46 Å and 4.01 Å, respectively, different from the position of the peaks
in the EXAFS due to the phase shift in (2.3). There is some interference between
the peaks due to the second and third neighbors but the first neighbor peak is
well resolved. The $\phi(r)$ for CuBr at 370°C, just below the γ the β phase transition,
is shown in Fig.2.4b and is seen to be remarkably different. The first neighbor
peak height has decreased and the peak has broadened. In addition, it has shifted
inward by about 0.1 Å despite the fact that lattice expansion has caused an increase
of 0.03 Å in the Cu-Br near-neighbor spacing. These changes are explained by the
excluded volume model discussed in Sect.2.2. First, the procedure by which detailed
structural information is extracted from the EXAFS is considered.

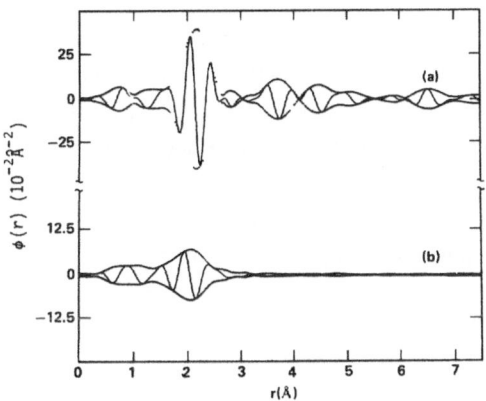

Fig.2.4a and b The real part (solid
line) and the magnitude of the Fourier
transform $\phi(r)$ of the EXAFS on the Cu
K-shell absorption in CuBr at (a) 77 K
and (b) 370°C. Note the factor of 2
scale change. The date were transformed
using a square window with k between
2.8 and 13.6 $Å^{-1}$, broadened by a Gaussian of width 0.7 $Å^{-1}$

The procedure used is based on (2.6). The variations which are observed among
the $\phi(r)$'s obtained from different systems or at different temperatures are assumed
to be due solely to changes in $p_{\alpha\beta}(r)$. The peak function $\xi_{\alpha\beta}(r)$ is insensitive to
changes in crystal structure, local bonding, thermal effects, etc. [2.22]. Thus, the
shift or broadening of $\phi(r)$ is due to a shift or broadening of a peak in p(r). The

unique peak function, $\xi_{\alpha\beta}$, associated with scattering from species β about excited atom species α is extracted from the ϕ measured using a structurally known sample, or standard. Then the ϕ obtained from a structurally unknown system can be analyzed to yield its $p_{\alpha\beta}(r)$. In the specific case of the superionic conductors, the standard is the material at low temperatures (77 K) where the structure is well known and $p_{\alpha\beta}(r)$ for near neighbors is a narrow Gaussian. Knowing $\phi(r)$ and $p(r)$, one obtains $\xi(r)$ from (2.6). Using the ξ obtained from the 77 K data, the unknown $p(r)$ at elevated temperatures, including the superionic phase, can be obtained. The near-neighbor peak dominates in $\phi(r)$, so what is obtained is the anion-cation pair distribution function versus temperature. Since the position of the immobile ions is well known (Br in CuBr), the location of the mobile ion (Cu in CuBr) versus temperature can be obtained.

The application of this procedure for structural analysis is straightforward. $\xi_{\alpha\beta}$ is extracted from the standard (77 K data) with a known $p_{\alpha\beta}(r)$ using (2.6). A simulated $\phi_m(r)$ for the unknown (high-temperature data) can be calculated from this $\xi_{\alpha\beta}$ and a model calculation for the $p_{\alpha\beta}(r)$, again using (2.6). The model parameters in $p_{\alpha\beta}(r)$ are varied until the best fit to the measured ϕ is found, corresponding to a minimum in the reliability-of-fit parameter, R:

$$R^2 = (2N)^{-1} \sum_{}^{N} [Re\{\phi - \phi_m\}]^2 / [(Re\{\phi\})^2 + (Re\{d\phi/dr\})^2]$$

$$+ [Im\{\phi - \phi_m\}]^2 / [(Im\{\phi\})^2 + (Im\{d\phi/dr\})^2]$$

(2.7)

where the sum extends over all N points in the range of the structural feature being fit. This measure of the quality of fit is a standard fractional difference least squares for the real and imaginary parts individually except for the $d\phi/dr$ terms in the denominators. These terms have been introduced to enhance the sensitivity of R to variations in the shape of the peaks in $p(r)$, at the expense of some of its extraordinary sensitivity to position.

The theory of EXAFS and the analysis of the data has been described in some detail in the last three sections. Before applying these ideas to superionic conductors, a comparison of EXAFS with X-ray and neutron scattering is made.

2.1.4 Contrast with Diffraction Studies

X-ray [2.28] and neutron [2.29] diffraction are the structural probes of choice for many solids. Since their application to superionic conductors is discussed in detail in Chap.3, we present here only a brief comparison between the diffraction and EXAFS measurements, emphasizing those areas where the techniques differ. As will be seen, the two types of experiments often complement one another in the study of superionic conductors.

In the ideal case of a single crystal sample, X-ray and neutron diffraction have the potential of yielding the complete atomic distribution. The basic quantity determined in a diffraction study is the diffracted intensity as a function of momentum transfer, i(q), where q = 4π sinθ/λ. When properly normalized and corrected for self-scattering, i(q) is analogous to the EXAFS Δ(k). In order to contrast these two quantities, it is convenient to reformulate the usual expression for i(q) for a single crystal sample so as to resemble (2.2):

$$i(\underline{q}) \propto \sum_{\alpha} c_{\alpha} \sum_{\beta} \int d\underline{r} \, \rho_{\alpha\beta}(\underline{r}) \, \exp(i\underline{q}\cdot\underline{r}) \, \frac{f_{\alpha}(q)f_{\beta}^{*}(q)}{<|f(q)|^{2}>} \quad , \qquad (2.8)$$

where c_{α} is the concentration of atom species α and $\rho_{\alpha\beta}(\underline{r})$ is the angle-dependent distribution of atom species β about species α. The $\rho_{\alpha\beta}(r)$ defined in connection with (2.2) is the angular average of $4\pi r^{2}\rho_{\alpha\beta}(\underline{r})$ (see [2.17,18]). $f_{\alpha}(q)$ is the scattering factor for atom species α, and is effectively independent of q in the case of neutron scattering. When the integral over $\underline{r}$ in (2.8) is evaluated, the long-range order in a single crystal sample contributes only to Kronecker delta functions at the reciprocal lattice vectors, the usual Bragg peaks. For a randomly oriented polycrystalline sample, these become the Debye lines, independent of the angle of $\underline{q}$. If there is short-range order in the sample, then it will contribute to a diffuse background, oscillating slowly with q. The long- and short-range order contributions to i(q) are illustrated in Fig.2.5 for a polycrystalline sample of Ag_2S in the superionic phase [2.9].

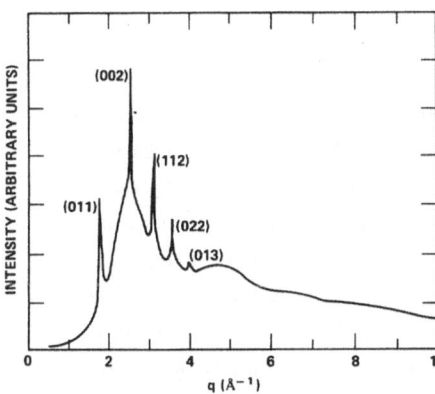

Fig.2.5 The X-ray scattering intensity of superionic Ag_2S at $205^{\circ}C$, showing the combination of both Debye peaks and diffuse background observed in diffraction studies [2.9]

Apart from the added angular information which is contained in (2.8) for a single crystal sample, the most important difference between (2.2) and (2.8) is the additional sum over atom species present in (2.8). Each diffraction measurement is a function of *all* of the atom pair correlations. For a multicomponent

system, this complicates the interpretation of the diffraction data. Consider, for
example, a two-component system. In principle, the unique extraction of the three
independent pair correlations requires three independent measurements in which the
peak functions are varied. This has been accomplished occasionally for the diffuse
portion of i(q) by combining three diffraction measurements (taking advantage of
the isotopic dependence of the neutron scattering factors) [2.30]. It is much more
common to analyze the Bragg peaks and the diffuse background from a single diffraction measurement with some simple model [2.5]. This procedure is most successful
when applied to diffraction data from a single crystal sample. The uniqueness of
the results is difficult to establish given the similar shapes of the various peak
functions. The neutron scattering peak functions are, in fact, identical in shape.
This situation becomes more complicated as the number of atom species increases,
since i(q) is a sum of $N(N + 1)/2$ independent terms for an N-component system. For
these reasons, the amount of local structural information which has been extracted
from the diffraction studies declines strongly as the chemical and structural complexity of the system increases. In contrast, the EXAFS data from the K-shell absorption of atom species α is a function of only those pair correlations involving
that species, numbering N for an N-component system. On a more subtle level, the
peak functions for different scattering atoms tend to be more distinct in EXAFS than
in diffraction studies. Taken in conjunction, these two differences simplify the
structural analysis of EXAFS data for a multicomponent system.

In addition to simplyfying the analysis, the absence of some correlations
from the EXAFS measurement can be of particular importance when this technique is
applied to system such as a superionic conductor, wherein one of the atom species
can be broadly distributed throughout the structure. Consider, for example, the
diffraction pattern for Ag_2S shown in Fig.2.5. In principle, the Debye lines contain
contributions from S-S, S-Ag, and Ag-Ag correlations. The analysis of this data
[2.9] shows, however, that only S-S correlations contribute significantly to those
lines. Accordingly, an analysis of the Debye lines will yield definitive results
for the S-S correlations but less information concerning the S-Ag and Ag-Ag correlations, due to the broad distribution of Ag ions throughout the structure. On
the other hand, the diffuse background in Fig.2.5 will yield the short range Ag-Ag
correlations. Only short-range contributions to the S-Ag correlations will be revealed, such as possible distortions of the S nearest neighbor positions about an
occupied Ag site. This is very significant since it is just the long-range S-Ag
correlations which will give the information needed concerning the path taken by
the conducting Ag ions. In contrast, the EXAFS probes both the long- and short-range
correlations, but only as they affect the near neighbor environment. Thus, the
EXAFS measured on the Ag K-shell absorption cross section will be dominated by the
important S-Ag correlations and have no contribution from the S-S correlations. Thus,
the EXAFS technique can be particularly useful in determining the conduction path
in these superionic conductors.

There is an additional, more subtle distinction to be drawn between the results of the two techniques, based on the respective regions of momentum space which contribute strongly to the Fourier transform. In the transform of typical EXAFS data, the quantity complementary to r, 2k, ranges approximately from 6 to 30 or 40 $\mathring{A}^{-1}$, with the largest signal in the 10 to 26 $\mathring{A}^{-1}$ region. In a diffraction measurement, the analogous range of q is typically 0 to 18 $\mathring{A}^{-1}$. The information is uniformly distributed over this range for neutrons, but is strongest at q = 0 for X-rays. The EXAFS transform contains relatively little information for small momentum transfer, and, accordingly, is particularly sensitive to the broadening of a peak in the pair correlation. The EXAFS signal can disappear relatively quickly with substantial broadening. It is often true, however, that the absence of a contribution from the sharp distribution (e.g., S-S in Ag_2S) in the EXAFS compensates, in part, for this relative disadvantage in a study of a superionic conductor.

In addition, the EXAFS measurement is especially sensitive to the shape of a peak in a correlation function due to the ready accessibility of very large momentum values. This sensitivity to shape will prove to be important in the EXAFS studies of superionic conductors discussed in detail in this chapter.

2.2   Structural Considerations for Superionic Conduction

2.2.1   General Considerations

Some qualitative observations can be made concerning the nature of superionic conduction and the special characteristics of those materials which exhibit it (see, for example, [2.3,31]). An ion can hop in a solid, and thereby contribute to the ionic conductivity, only if its immediate environment includes a vacant site which is energetically accessible. This general condition can be divided into two, both of which must hold:

(I) The lattice must include more sites than there are ions to fill them. It is implicit in this that the sites be energetically similar.

(II) The lattice must further be such that the ions can hop from their sites to nearby vacant ones. This implies only modest barriers between the nearly equivalent sites.

A third condition also has to be added, since the above two criteria can be satisfied without appreciable long-range motion of the ions; namely,

(III) There must be connected paths through the lattice; otherwise there is a large hopping rate without significant dc conduction.

Consider the expression for the hopping ionic conductivity as the product of the fractional number of defects, n_d, and the hopping rate for a defect, ν_d:

$$\sigma \propto n_d v_d \propto [\exp(-U_F/k_B T)] \cdot [\exp(-U_m/k_B T)] \tag{2.9}$$

where U_F is the formation energy per defect and U_m is the barrier height for motion of the defect. The above statements then imply that U_F and U_m have to be small: U_F, so that the interstitial and the lattice sites are nearly equally populated ($n_d \sim 1$); and U_m, so that the interstitial ion can move easily.

A consideration of the structure can be useful in determining whether these criteria can be satisfied. Consider the first condition, that there be more sites than ions to occupy them. An ion is allowed to move if its neighboring site is unoccupied, but the energy difference may be too large to allow the hop unless the vacant site is nearly equivalent to the occupied site. If the sites are structurally equivalent, then they will be energetically equivalent ($U_F = 0$). If the sites are structurally inequivalent, then they may or may not be energetically equivalent depending on other considerations.

The second criterion is most likely satisfied by face-shared polyhedra rather than edge-shared polyhedra since the opening is larger in the face-shared case. In other words, the energy of the ion in the face site must not be too different from that in the centers of the polyhedra, i.e., $U_m \sim k_B T$. ARMSTRONG et al. [2.32], have pointed out various chemical requirements for U_m to be small. One such requirement is that the mobile cation be stable in both four-fold and three-fold coordinated sites. The coordination number in the face of both an octahedron and tetrahedron is three. So for U_m to be small, the energy of the mobile ions in a three-fold coordinated face site must not be very different from that in the four-coordinated lattice site. Such is the case for Ag and Cu ions thereby accounting, in part, for the predominence of Ag and Cu compounds among superionic conductors.

2.2.2 Pair Potentials

The above considerations are useful for a qualitative understanding of superionic conductivity. They are too vague, however, for a microscopic understanding of this phenomenon. Both the structure and the conductivity are derived from the potential energy of the mobile ion as a function of temperature and position in the crystal. The pair potential for any two ions can be written as

$$V_{ij}(r) = A_{ij} \exp[(r_i + r_j - r)/\rho] + \frac{q_i q_j e^2}{r} - \frac{1}{2}(\alpha_i q_j + \alpha_j q_i)\frac{e^2}{r^4} \quad . \tag{2.10}$$

The first term is the overlap repulsion between closed shell ions with r_i the ionic radius of the ith ion and r the distance between ions i and j. The second term is the Coulomb potential with q_i the fraction of charge on the ith ion. The third term is the polarization self-energy of the ions, with α_i the polarizability of the ith ion.

This potential is quite complicated but has served as the starting point for various theoretical calculations. An early example is the work of FLYGARE and HUGGINS [2.33], who calculated the potential energy of ions of varying size in the α-AgI lattice as a function of position using (2.10). They simplified the calculation by setting the polarizability of the mobile Ag ion to zero and ignoring the cation-cation contribution in (2.10). They considered only motion along a (100) "tunnel" connecting octahedral sites in the iodine bcc lattice and adjusted the ionic radii of Ag and iodine to minimize the apparent activation energy for the path. This required rather small radii: r_{Ag} = 0.83 Å and r_I = 1.76 Å. If the ionic radii are increased, however, the considerations of FLYGARE and HUGGINS suggest that the ion path deviates from the octahedral sites toward the tetrahedral sites. This would result in a very different conduction path than the (100) tunnel, one through the tetrahedral sites as required by the recent structural studies discussed below.

Other theoretical calculations have been performed using molecular dynamic simulations with potentials similar to (2.10) as the starting point. This is a powerful technique which has the capability of yielding all the pair correlation functions, as well as information on the ionic motion, such as the diffusion coefficient. The various parameters of the potential must be known, however, and this is usually not the case. Two different molecular-dynamics calculations have been performed on CaF_2 [2.34,35] using a rigid ion potential. The results differ in that one calculation yields half the fluorine ions occupying the octahedral sites in the fluorite lattice [2.34], whereas the other yields no such octahedral occupation [2.35]. The more complicated case of α-AgI has been treated by SCHOMMERS [2.36] using (2.10) with the polarizability neglected (α_i = 0) and with a harmonic term added for the iodine ions. The Ag diffusion coefficient is obtained, as well as the Ag current-current correlation function, the results agreeing reasonably well with experiment.

2.2.3 Anharmonic Model

The use of a realistic potential, such as (2.10), to analyze most structural data is too complex due to the large number of parameters and the many-body nature of (2.10). Even the molecular dynamics calculations have simplified this potential, ignoring either the Coulomb term [2.34,35] or the polarizability term [2.36], and only a few such calculations have been performed. Approximations are essential. The most widely accepted one is the independent oscillator model with anharmonic terms. The fact that the mobile ions actually move from their lattice sites implies a strong anharmonicity in the potential felt by these ions, whereas anharmonicity may be small for the immobile ions. The use of an Einstein oscillator model is appropriate since the superionic phase is well above the Debye temperature in these materials. Consider, for example, materials with an fcc immobile-ion lattice, such as the low temperature γ-phase of CuCl [2.37], CuBr [2.38] and CuI, as well as superionic α-CuI [2.39,40]. The potential for these materials can be written as

$$V_{Cu} = V_0 + \frac{1}{2} \alpha u^2 + \beta xyz \quad ,$$

$$V_{halogen} = V_0' + \frac{1}{2} \alpha' u^2 \quad , \tag{2.11}$$

where $\underline{u} = (x,y,z)$ is the displacement of the ions from their equilibrium sites, the tetrahedral sites for the Cu ions. The cubic term has the effect of increasing the Cu-ion potential for displacements toward a corner of the tetrahedron and decreasing it for displacements toward the face. Equation (2.11) is also appropriate for the fluorite materials, such as CaF_2 [2.41,42], SrF_2 [2.42], and BaF_2 [2.43], where the metal ion is fixed in an fcc lattice and the F ions are mobile. Equation (2.11) has been used to fit the X-ray and neutron diffraction data on the cuprous halides and the fluorites and has yielded reasonably good fits. This is also the situation for the materials with bcc lattices, such as AgI, where (2.11) is modified somewhat to reflect the different symmetry. For example, in AgI, HOSHINO et al. [2.44] use

$$V_{Ag} = V_0 + \frac{\alpha_1}{2} (x^2 + y^2) + \frac{\alpha_2}{2} z^2 + \frac{\beta}{2} (x^2 z - y^2 z) \quad , \tag{2.12}$$

and obtain a good fit to their X-ray and neutron diffraction data. Extracting the harmonic and anharmonic coefficients in (2.11) and (2.12) from diffraction data specifies the position of the mobile and immobile ions, and yields some insight into the conduction path.

Equations (2.11) and (2.12) are not a good approximation to (2.10) if the ions approach closely to one another since it underestimates the hard-core repulsion term in (2.10). The ions in superionic conductors do come close together since the mobile ions must squeeze through the faces of polyhedra to move from one site to another. An indication that (2.12) is not adequate is found by CAVA et al. [2.5], in analyzing their neutron diffraction data on α-AgI. They find that even fourth-order terms in the anharmonic potential have to be included and are significant. The importance of these higher order terms has also been pointed out by HARADA et al. [2.38], in the case of CuBr. So the anharmonic model with only cubic terms does not appear to be sufficient. In fact, such a model does not yield a good fit to the EXAFS data. So a different approach has been proposed [2.6]: the excluded volume model.

2.2.4 Excluded Volume Model and Cation-Anion Correlations

It is well established [2.45] that the pair correlation functions in a liquid can
be explained using a model system of hard spheres. In this approximation, both the
attractive potential arising from atomic polarizability and the repulsive core-core
interaction [the third and first terms in (2.10)] are replaced by an effective hard-
sphere pair potential

$$
V_{ij}(r) = \begin{cases} 0, \ r > r_i + r_j = r_{excluded} \\ \infty, \ r \leq r_i + r_j = r_{excluded} \end{cases} \ , \tag{2.13}
$$

where r_i is the effective hard-sphere radius of atom i, usually somewhat less than
the atomic radius. It has been shown [2.45] that the pair correlations which result
from this model are nearly identical to those resulting from a much more sophisticated interaction such as a Lennard-Jones potential. An analogy has been drawn
[2.6] between this model of liquids and the cation-anion correlations in superionic
conductors, based upon the observation that high ionic conductivity implies a relatively weak *effective* coupling between the mobile ions and the fixed ion lattice.
An obvious complication is the long-range Coulomb interaction, the second term in
(2.10). One expects that these interactions do not have a large effect on the cation-
anion correlations beyond fixing the crystal structure, since the substantial can-
cellation of opposing forces in such ordered arrays means that the Coulomb and dipole forces are probably slowly varying along the conduction path. The Coulomb
forces should have a large effect on the mobile ion correlations, however, and
therefore possibly on the conductivity.

An important distinction between the hard-sphere liquid and superionic conductors
is that the immobile ions are fixed to certain sites. The cation-anion interactions
are treated as hard spheres, placing the mobile ions randomly in space except that
they may not approach closer to an immobile ion site than $r_{excluded}$, which is
roughly the sum of the effective hard sphere radii of the cation and anion. The
cation-anion pair correlation function may then be calculated quite simply from
geometrical considerations, and compared with the results of experiment. This comparison with experiment is quite simple since only one parameter need be specified,
$r_{excluded} = r_i + r_j$.

Ionic conduction in this model will depend on the interconnection of the interstitial volumes formed by the immobile-ion lattice, a percolation-type problem. It
will also depend on the likelihood that a mobile ion will move between neighboring
volumes, given the assumed uniform distribution of the mobile ions in the interstitial voids of the immobile-ion lattice. That portion of the temperature dependence
of the conductivity which arises from these simple geometrical arguments can be

calculated as that of a dilute Boltzmann gas in a complex interconnected space. The neglect of Coulomb and dipole interactions, as well as the mobile-ion correlations, can prevent a determination of the magnitude of the conductivity. These structural considerations, however, can go a long way toward explaining quantitatively the observed temperature dependence of the conductivity. This suggests that the temperature dependence due to the other factors may be small. This model will be discussed further in connection with the analysis of the EXAFS data in the next section.

2.3 EXAFS Investigations of bcc Superionic Conductors: AgI

In this section the structural information probed by EXAFS and the relationship of this information to the high ionic conductivity in these materials is discussed. This information includes the positions of the ions, both mobile and immobile, as well as the motion of the mobile ions from one site to another. The data to be discussed is that on AgI and on the cuprous halides since these are the materials that have been extensively studied using EXAFS. AgI will be discussed in detail since it is the archtypical superionic conductor, and the most studied structurally.

This section is the first of two sections, divided according to the structure of the immobile-ion lattice. This concerns the bcc materials represented by AgI. The next treats the fcc materials represented by CuI. The nature of the conduction path in these two structures is different. In the bcc structure, the largest voids are tetrahedra which share faces with one another. So the Ag ions can move among equivalent sites by passing through the tetrahedral faces. The fcc structure contains tetrahedra which share corners with one another and which share faces with octahedra and not with other tetrahedra. Similarly, the octahedra share faces only with tetrahedra. So the conduction path of the Cu ions consists of alternating tetrahedral and octahedral sites [2.46]. This fact is probably partially responsible for the fcc structures having higher activation energies and smaller ionic conductivities than the bcc structures [2.47].

Materials with the bcc structure comprise a large fraction of the Ag salts which are superionic conductors. Examples include α-AgI, α-Ag$_2$S, α-Ag$_2$Se, α-Ag$_3$SBr, and α-Ag$_3$SI. AgI is one of the most studied of these, and the one considered in detail here. In the low temperature β phase, the ionic conductivity is normal [2.47-49], being small and highly activated ($U \simeq 0.96$ eV). It achieves a value of only $\sigma \simeq 10^{-4} \, \Omega^{-1} cm^{-1}$ just below the $\beta \rightarrow \alpha$ phase transition at $T_c = 147^{\circ}C$, as shown in Fig.7.1. At T_c it transforms to the α phase and the ionic conductivity increases by four orders of magnitude to $\sigma \simeq 1 \, \Omega^{-1} cm^{-1}$. This value of σ is comparable with that of molten salts even though T_c is a factor of 2 below the melting point of AgI, $T_m = 555^{\circ}C$. In this superionic α phase, the ionic conductivity is large and only slightly temperature dependent: $U \simeq 0.05$ eV $\simeq k_B T$ [2.49,50]. It increases by only a factor of approximately 2 from T_c to T_m and actually decreases by 12 percent on melting [2.50].

2.3.1 Early Structural Studies

Many structural studies of AgI have been reported. In 1934, STROCK [2.2], from an analysis of the Bragg X-ray peaks obtained from a powdered sample, determined the following structure. Below $T_c = 147°C$, the Ag ions reside inside the regular tetrahedra of iodine in the hexagonal wurtzite structure, the β phase. (In this low temperature region, AgI also has a metastable γ phase that has the face-centered-cubic structure, sphalerite type. This phase can be formed by applying pressure, ≈ 20 bars at room temperature, to the stable β phase [2.51].) At T_c, AgI transforms to the α phase in which the iodine ions form a body-centered cubic lattice and the Ag sublattice is disordered. The nature of this disorder, according to STROCK, consists of the two cations per unit cell equally distributed over 42 crystallographic sites. Although the Ag ions are fixed in their wurtzite lattice sites below T_c, their distribution becomes liquid-like above T_c. This could then qualitatively account for both the high Ag ionic conductivity of $\approx 1\ \Omega^{-1}cm^{-1}$ in the α phase and the four orders of magnitude increase in conductivity at T_c.

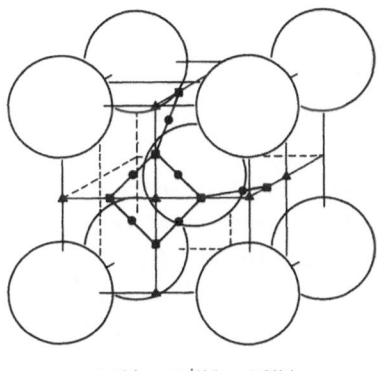

▲ 6(b) ■ 12(d) ● 24(h)

Fig.2.6 STROCK model [2.2] for superionic AgI. The large ions are the iodine in a bcc array. Two Ag ions are randomly distributed over 42 crystallographic locations in the unit cube: the 6(b) sites, the 12(d) sites, and the 24(h) sites

STROCK's proposed structure for α-AgI is shown in Fig.2.6. The 42 sites available for the two Ag ions in the unit cell ($a_0 = 5.07$ Å) consist of 12 equivalent tetrahedrally coordinated positions, 12(d), 24 three-fold coordinated sites, 24(h), and 6 octahedrally coordinated sites, 6(b). The largest volume for the Ag ions is inside the distorted tetrahedra at the 12(d) sites with the smallest inside the distorted octahedra at the 6(b) positions. The octahedra are very distorted and may more correctly be considered as two-fold coordinated sites since two iodine ions are 2.53 Å away with four iodine at 3.57 Å distance. The 24(h) sites are at the center of the shared faces of the distorted iodine tetrahedra.

In 1957, HOSHINO [2.52] confirmed the liquid-like Strock model using both the Bragg peaks and the diffuse X-ray background. But to explain the absolute intensities of the peaks, large Debye-Waller factors, B, had to be introduced. They yield a root-mean-square deviation for both the Ag and I ions of $(<u^2>)^{\frac{1}{2}} = 0.3$ Å at 250°C, using $B = 8\pi^2<u^2>$. So, in addition to the equal distribution of the two Ag ions over 42 sites, there are, according to HOSHINO, large thermal vibrations and possibly large iodine lattice distortions due to the disorder.

This picture of the α phase was modified somewhat by BURLEY [2.51] who proposed a model for the memory effect. This effect is the observation that if γ- or β-AgI is heated into the α phase above 147°C, but not exceeding about 170°C, then the original low temperature phase (γ or β) is regenerated upon cooling below 147°C. BURLEY's explanation is that α-AgI derived from the γ phase has the tetrahedral sites preferentially occupied, whereas the α-AgI derived from the β phase has the trigonal sites preferentially occupied. This preference retains the memory of the type of symmetry of the starting low temperature phase, cubic for γ and hexagonal for β so that this starting phase is regained on cooling. However, above 170°C this memory is lost and the site occupation is presumably random as given by STROCK.

More recently, neutron diffraction experiments were performed on powder samples by BÜHRER and HÄLG [2.53]. The diffuse background scattering showed oscillations, indicating a liquidlike Ag lattice with short-range order. However, the Bragg peaks did not fit the Strock model. Rather they are best fit by placing the Ag ions on the 24(g) sites, which are positions displaced in (100) directions about the 12(d) tetrahedral sites. In addition, large Debye-Waller factors are needed, corresponding to rms displacements of $(<u^2>)^{\frac{1}{2}} = 0.3$ Å for iodine at 195°C, in rough agreement with HOSHINO.

The neutron diffraction investigation of WRIGHT and FENDER [2.54] is in basic agreement with the results of BÜHRER and HÄLG. WRIGHT and FENDER also point out that the large Debye-Waller factors imply a mean thermal displacement for Ag of approximately $(<u^2>)^{\frac{1}{2}} = 0.4$ Å at 255°C which argues against a static displacement of this order of magnitude from the tetrahedral site. Rather, they argue that the displacement of the Ag ions is dynamic and can be accounted for by anharmonic vibrations with reduced energy barriers in certain directions.

2.3.2 EXAFS Study

The structural picture of the superionic phase of AgI was refined further by the EXAFS study [2.4,6] which we now discuss in detail.

The data of this study were collected at SSRL using a transmission experiment as discussed in Sect.2.1. The samples used were polycrystalline disks about two absorption lengths thick ($\approx$ 100 μm) at the Ag K edge (25.5 keV) for optimum signal-to-noise ratio [2.55]. For the high-temperature measurements, this sample was contained in a boron nitride cell with thin windows ($\approx$ 120 μm thick) for minimal holder absorption at the Ag K edge. The data were collected in the insulating β phase at 77 K, 20°C and 98°C and in the superionic α phase at 198°C and 302°C. Using the procedures discussed in Sect.2.1, the EXAFS on the Ag K edge was extracted as a function of final-state electron momentum from the measured absorption and Fourier-transformed into real space. The resultant complex transform, φ(r), is shown in Fig.2.7 for the 77 K and the 471 K data. In order to enable a direct comparison among the data at different temperatures, all data were reduced in an identical manner. For example, in this case, a square window k = 3.1 to 11 Å^{-1}, broadened with a Gaussian of width 0.7 Å^{-1}, was used in the Fourier transform to real space.

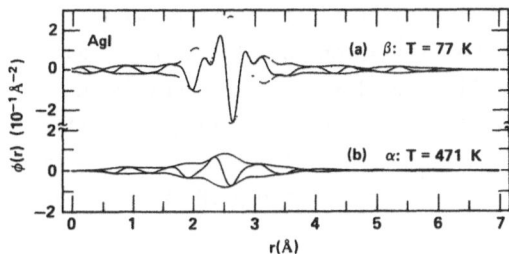

Fig.2.7a and b The real part (solid line) and the magnitude of the Fourier transform φ(r) of the EXAFS on the Ag K-shell absorption in AgI: (a) at 77 K well below the normal-superionic transition temperature of 420 K and (b) at 471 K well above the transition

Before discussing the change of the EXAFS of AgI with temperature, first consider the 77 K data, Fig.2.7a. At this temperature AgI has the hexagonal wurtzite structure (a = 4.59 Å, c = 7.51 Å) [2.51,56] in which each Ag atom has four iodine nearest neighbors at 2.82 Å and 12 Ag next-nearest neighbors at 4.59 Å. Examining Fig.2.7a, one observes a large peak near 2.82 Å and much smaller peaks on both sides of it. The large peak with extensive structure between 1.6 and 3.8 Å is therefore associated with the four iodine near neighbors to the Ag. It peaks at 2.56 Å, 0.26 Å below the actual peak in the Ag-I pair distribution function. This shift is due to the k dependence of the initial phase shift and the phase of the t matrix as discussed in connection with (2.3) and (2.5). The structure in this first neighbor peak is not due to structure in $p_{Ag-I}(r)$ since this function is a narrow Gaussian at this low temperature. Rather, it is due to resonances in the backscattering t matrix of the heavy iodine backscatterer. Similar signatures are observed in other materials with heavy ion backscatterers, such as Ag metal [2.57], and are predicted theoretically [2.12,21,22]. So the structure of this peak in Fig.2.7a is due to the peak function, ξ_{Ag-I}.

The peaks below 1.5 Å are fully understood to be artifacts of the data-reduction process, due to incomplete removal of σ_{bg} and σ^0 in extracting the EXAFS oscillation from the total absorption. The peaks beyond 3.8 Å are due to more distant neighbors. For example, the small peak due to the 12 Ag second neighbors is at 4.41 Å, 0.18 Å below the actual Ag-Ag distance due again to the phase shift. These peaks are well separated from the near-neighbor peak in real space and so do

not interfere with it. This is a major advantage of using the real space data since
in k space these various peaks beat with one another [2.24]. Only the first neighbor
peak was considered since the further neighbors do not contribute significantly to
the EXAFS at elevated temperatures. As discussed in Sect.2.1, from this spectrum one
obtains ξ_{Ag-I} for the four iodine near neighbors since the first neighbor $p_{Ag-I}(r)$
is a narrow Gaussian centered at 2.82 Å. This ξ_{Ag-I} was used to obtain the unknown
first neighbor $p_{Ag-I}(r)$ at elevated temperatures.

From a detailed numerical analysis of the ϕ's in Fig.2.7, it was clear that
significant changes occur on heating to 198°C into the superionic α phase: the
main peak shifts slightly to lower r and becomes asymmetric, and the amplitude decreases. Since the shifts are small, the Ag ions must retain their tetrahedral
positions in the high temperature phase. The data are inconsistent with any significant number of Ag ions in octahedral sites. This conclusion of tetrahedral occupation is consistent with the fact that the 12(d) sites have the largest available space for the Ag ions, thus yielding the lowest potential energy for the cation.
But, because the peak shifts inward and a shoulder develops, the pair distribution
function is asymmetric.

Consider in detail the various possibilities for p(r) at 198°C. In each case
p(r) refers to the near-neighbor Ag-I pair correlation function and so the subscripts are dropped. For each model, a p(r) was constructed and a model EXAFS
function, $\phi_m(r)$ was obtained using the ξ_{Ag-I} obtained at 77 K. This model function
was compared with the data in the interval r = 1.75 to 3.4 Å and the quality of
fit calculated using (2.7). The following possibilities were considered [2.4,6]:

(I) *Strock Model* [2.2,52]
In this model, two Ag ions are distributed over the 42 crystallographic sites of
Fig.2.6. These are 12 four-fold coordinated sites with near-neighbor spacing
r_0 = 2.83 Å at 200°C, 24 three-fold coordinated sites with r_0 = 2.69 Å, and six
octahedral sites with four iodine ions at r_0 = 3.57 Å and two at r_0 = 2.53 Å. The
simulated $\phi_m(r)$ for this model is shown in Fig.2.8b and is quite different from
the experimental result of Fig.2.7b, as is evidenced by the fact that R = 11.9%.

(II) *Bührer and Hälg Model* [2.53]
This model displaces the Ag ions from the 12(d) sites of Fig.2.6 at (0.25, 0, 0.5)
in the (100) directions by 0.29 Å. This then places the Ag ions at the 24(g) sites
at (0.193, 0, 0.5). The two Ag ions presumably randomly occupy these 24 sites. The
near-neighbor environment consists of two iodine ions at r_0 = 2.72 Å and two at
r_0 = 2.97 Å for 195°C. The $\phi_m(r)$ for this pair distribution function is shown in
Fig.2.8c. It does not fit the experimental data of Fig.2.7b and has a reliability
of fit of 13.1%.

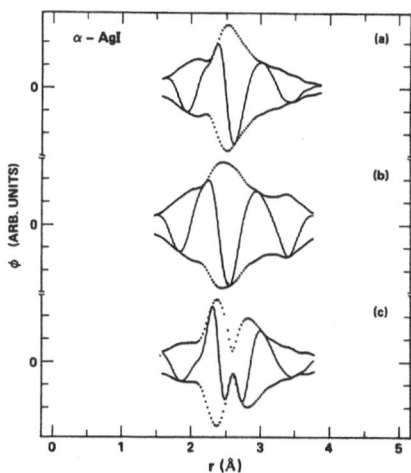

Fig.2.8a-c Synthesis of the EXAFS $\phi_m(r)$ for
the superionic phase of AgI at 198°C, using
three different structural models for $p_{Ag-I}(r)$:
(a) Ag in sites displaced ≈0.1 Å from the
tetrahedral center toward the tetrahedral
face [2.4]; (b) STROCK model [2.2] of Figure
2.6; and (c) BÜHRER and HÄLG model [2.53]

(III) *Single Gaussian* [2.4]

The simplest case is a single Gaussian for p(r), (2.4). The number of iodine near
neighbors, N, the Ag-I near-neighbor spacing, r_0, and the relative Ag-I mean square
displacement, ε, were adjusted to fit the data at 198 C and the reliability-of-fit
parameter, R, (2.7), was minimized. The best fit yielded R = 4.5%. The obtained
parameters are r_0 = 2.81 Å, N = 2.72 neighbors and ε = 0.11 A. This value of the
near-neighbor spacing is to be compared with the center of the tetrahedral location
of 2.83 Å at this temperature. Also, note that the fit yielded a reduced amplitude
with the number of near neighbors being 2.72 and not 4 as would be expected. This
fit is not exceptionally good and, in addition, systematic deviations between the
data and the simulation were observed.

(IV) *Generalized Bührer and Hälg Model* [2.4]

This consists of a general (100) displacement from the tetrahedral sites. The p(r)
for such a model is two Gaussian peaks of equal amplitude and width, displaced in
opposite directions from the center of the tetrahedral location at r_0 = 2.83 Å. The
Gaussian width, amplitude, and positions were varied for a best fit, yielding
R = 4.2% and N = 1.28 neighbors at r_0 = 2.74 Å and at r_0 = 2.88 Å with ε = 0.07 Å.
Note that the best such fit yielded a reduced amplitude of 2.56 neighbors instead
of 4.

(V) (110) *Displacement* [2.4]

Instead of the (100) displacement of BÜHRER and HÄLG [2.53], the Ag ions can be
displaced in the (110) direction from the tetrahedral center toward the tetrahedral
face. This is a likely displacement for Ag motion between tetrahedral locations
through the shared faces. In this case, the pair distribution function would consist of two Gaussian peaks with amplitudes in the ratio of 3:1 since this displacement would move the Ag ion closer to the three iodines of the face and further from

the one iodine at the corner of the tetrahedron. The total amplitude, Gaussian width, and positions were varied for a best fit. This yielded R = 3.1% with 1.95 neighbors at 2.77 Å and 0.65 at 2.93 Å, both with ε = 0.08 Å. Again note that the amplitudes are reduced giving 1.95 + 0.65 = 2.60 neighbors instead of 3 + 1 = 4. The obtained $\phi(r)$ for this model is shown in Fig.2.8a and compares favorably with the data.

Examining the above models, it is clear that the Ag-I correlations peak near the radius which corresponds to Ag in the tetrahedral site, r_0 = 2.83 Å, far from the trigonal and octahedral distances. In addition, there are a number of systematics evident in the relative quality of these fits to the 198°C data. The most prominent is that the data can be fitted well only if the total number of iodine neighbors of Ag is allowed to be less than four. The first two models have four neighbors, while the last three have between 2.57 and 2.72 neighbors. It is clear, therefore, that any model used to fit the data must explain this *apparent* decrease in the number of nearest neighbors with increasing temperature. The second significant systematic is illustrated by the improvement in fit when the pair distribution is allowed to be asymmetric, as in going from a single Gaussian to the 3:1 two-Gaussian peak model. This asymmetry is evident in the ϕ's, and becomes more pronounced with increasing temperature. In the excluded volume model, these two features are shown to be different aspects of the same effect [2.6]. We now consider the model in detail.

As discussed in Sect.2.2, the excluded volume model approximates the anion-cation interaction with a hard-sphere potential, (2.13). The iodine ions are fixed to bcc lattice sites with lattice constant a_0 = 5.07 Å at 200°C, and the Ag ions are uniformly distributed over the remaining allowed volume. Thus, the location of the Ag ions in this bcc lattice is determined primarily by the one parameter, $r_{excluded}$, which is roughly the sum of the effective hard-sphere radii of Ag and iodine.

In fitting the EXAFS data using this model, two principal parameters were varied: the radius of the excluded volume, $r_{excluded}$, and the width of a Gaussian with which the pair function, exp $[-V_{ij}(r)/k_B T]$, is convoluted, ε. The radius is the central feature of the model, and is expected to decrease slowly with temperature as atoms with higher kinetic energy approach slightly closer together. The Gaussian is intended to accommodate the rms displacements of the iodine atoms as well as the slight softness of the actual core-core interactions, and so is expected to broaden slowly with temperature. Finally, a modest dilation of the iodine "cage", Δr_{I-I}, was allowed to account for local distortions of those cages actually occupied by Ag atoms. No variations in the amplitude of the peak was allowed, corresponding to the conservation of four nearest-neighbor iodines. In applying this model to the wurtzite phase, an additional constraint was added, namely, that the ions do not occupy the octahedral sites. This is undoubtedly a valid assumption due to the small conductivity and low disorder in this phase. No such constraint was applied to the superionic phase where the Ag ions are permitted to occupy all the voids in the bcc iodine lattice allowed by the hard-core repulsion of (2.13).

The parameter values obtained for each data set are shown in Table 2.1. Also included is r_{face}, that value of $r_{excluded}$ which corresponds to the allowed volume just touching the face of the tetrahedral cage. The obtained values of $r_{excluded}$ are such that only the trigonal and tetrahedral locations are occupied in the superionic phase. The distorted octahedral locations are not occupied since they are too small, having $r_0 = 2.53$ Å $< r_{excluded} = 2.68$ Å at 198°C. The temperature dependence of the parameters are as expected, $r_{excluded}$ decreasing and ε increasing with increasing temperature. The slight dilation of the iodine cage about a resident Ag ion partially explains the large Debye-Waller factors [2.5,52-54]. The larger dilation in the bcc phase is probably a result of the substantial distortion of the iodine cage from a regular tetrahedron. The R values range from 1.54% to 2.6%, a substantial improvement over the previous models. An example of the fit to the AgI data at 198°C is shown in Fig.2.9. The data is shown in Fig.2.9a with the model simulation in Fig.2.9b over the range used to calculate R. The difference between the data and excluded volume fit is presented in Fig.2.9c. This difference corresponds to R = 2.0%.

Table 2.1. Parameters obtained in fitting the excluded volume model to the EXAFS data on AgI. Included are the quality of fit, R, for each set of parameters and the value of the near-neighbor spacing for an Ag ion in the tetrahedral face, r_{face}

Temperature	20	98	198	302
		[°C]		
$r_{excluded}$[Å ± 0.01]	2.723	2.694	2.682	2.671
ε[Å ± 0.01]	0.045	0.050	0.055	0.068
Δr_{I-I}[Å ± 0.02]	0.03	0.02	0.08	0.08
R	1.5%	1.8%	2.0%	2.6%
r_{face}[Å]	2.676	2.666	2.736	2.751

It is very significant that this model completely accounts for two trends in the EXAFS data with increasing temperature: the apparent decrease in the number of nearest neighbors[1], and the increasing asymmetry of the first peak in the pair correlation function. The connection between these effects is evident upon examination of the Ag-I pair correlation functions deduced from the EXAFS data, shown in Fig.2.10. As $r_{excluded}$ decreases from the tetrahedral center values [corresponding

---

[1]In the discussion of the EXAFS data on AgI in [2.4], the amplitude decrease was attributed to the smeared-out distribution of iodine ions seen by an Ag ion in flight. This explanation is generally consistent with the results from the excluded model (see, for example, Fig.2.10), but is not quantitative.

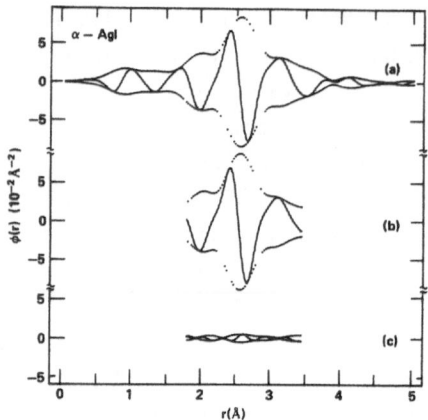

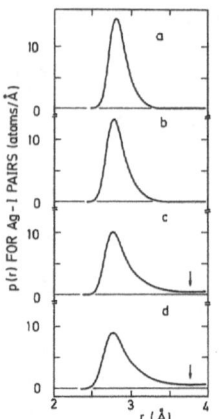

Fig.2.9a-c The data, (a),and simulation, (b), for the EXAFS on superionic AgI at 198°C. The simulation was obtained using the excluded volume model. The difference between the data and simulation, (c), corresponds to an R=2.0%

Fig.2.10a-d The radial distribution function of iodine ions about each Ag in AgI in the β phase at (a) 20°C and (b) 98°C, and in the α phase at (c) 198°C and (d) 302°C. Each distribution function is normalized to contain four iodine ions in the first peak. The arrows in (c) and (d) locate the contribution from a silver ion occupying the center of a face between adjacent tetrahedra

to a delta-function p(r)], the peak in the pair correlation function moves to lower r and a long, slowly varying tail develops beyond the tetrahedral center. The tail makes only a very small contribution to the EXAFS due to the absence of low k components in the transform. The excluded volume model works because it has just the proper relationship between peak asymmetry and tail growth in the pair correlation, seen as a phase shift and amplitude decrease in the data. It is this attribute, even more than the improved R, which recommends this model.

It is interesting to note that the peaks which fit the EXAFS in all the above models are relatively narrow compared with the large Debye-Waller factors obtained from X-ray and neutron diffraction studies as discussed above. The broadening in the nearest neighbor position which is observed in EXAFS, however, should not be confused with the much larger rms deviation from the average lattice position deduced from these studies (≈ 0.3 Å for iodine and 0.4 Å for Ag at 195°C [2.53]). This broadening includes effects which do not broaden the EXAFS. Included in this category are long-wavelength acoustic phonons and certain types of static disorder, such as the distortion of the iodine cage, Δr_{I-I}.

2.3.3 Other Recent Structural Studies

Other recent structural studies on AgI have been reported. HOSHINO et al. [2.44], showed that the Ag ions are located near the tetrahedral locations, consistent with the EXAFS results. Their X-ray and neutron data could be fit well using a (110) displacement from the tetrahedral center or the anharmonic potential of (2.12). CAVA et al. [2.5] reported neutron diffraction on a single crystal of both β- and α-AgI. At 23°C, in the β phase, the Ag ions were found to occupy the tetrahedral sites, with no more than 0.04 Ag ions in the octahedral sites. This result was as- sumed in the analysis of the EXAFS data in the β phase. The best fit to the α-phase data called for large and highly anharmonic displacements about the tetrahedral sites in the direction of the cage faces, and avoiding the octahedral locations. These results are qualitatively consistent with the results of the excluded volume model. The Ag charge density obtained by CAVA et al. [2.5] is consistent with that deduced from the EXAFS data, Fig.2.10. A quantitative comparison would be of inter- est.

Both the diffuse background and the Debye peaks in the X-ray diffraction pat- tern of AgI have been analyzed by SUZUKI and OKAZAKI [2.8]. To analyze the Debye peaks, a small divided cell model was used in which the Ag ions are distributed on the corners of small cubic cells in the voids of the iodine bcc lattice. Since this method yields the intensities of the Debye peaks quite well, they conclude that the Ag-ion distribution is liquid-like. In addition, they analyzed the diffuse halos in the diffraction profile to obtain the Ag-Ag radial distribution function. It yields a mean distance between silver ions of 2.75 Å and a coordination number of 5 Ag near neighbors to each Ag ion. This large number of Ag near neighbors is surprising, as is the small near-neighbor spacing of 2.75 Å, closer than the four iodine ions. These results may be connected with the authors having neglected the exclusion of the Ag ions from the space occupied by the iodine ions.

2.3.4 Structural Model for Superionic Conduction in bcc Conductors

Consider the relationship between the structural information and superionic conduc- tion in AgI from the viewpoint of the excluded volume model. First consider the relationship between $r_{excluded}$ and r_{face}. Below T_c, $r_{excluded}$ is greater than r_{face}. The allowed volume, therefore, does not touch the face, confining each Ag to the central region of its tetrahedral cage, as represented schematically in Fig.2.11a. Furthermore, the tetrahedra share faces only in pairs in this wurtzite structure, and so an Ag ion would have to move into an octahedral site to move any distance through the hexagonal iodine lattice. For either reason, cation conductivity re- quires an activated hop between allowed volumes. One would then expect normal ionic conductivity with a normal activation energy, as observed (U = 0.96 eV) [2.48]. Above T_c, however, $r_{excluded}$ is less than r_{face}. Furthermore, each cage shares faces with four other cages in this bcc structure. There is a continuous path through the solid by means of these multiply connected tetrahedral regions, as represented schemati- cally in Fig.2.11b. One would then expect a high ionic conductivity with no ap- preciable activation energy, as observed (U ≃ 0.05 eV ≃ $k_B T$) [2.50]. It is inter- esting to note that the sharp transition from normal to superionic behavior is

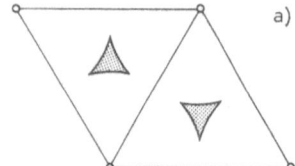

a)

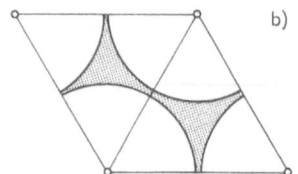

b)

<u>Fig.2.11a and b</u> A schematic two-dimensional represen-
tation of the allowed (shaded) and excluded regions for
cation centers in the excluded volume model in the (a)
insulating and (b) conducting phases. The anion centers
are indicated by open circles. In the insulating phase
the allowed regions are isolated, whereas in the superionic phase they are connected, allowing for high conductivity

driven by the increase in r_{face} resulting from the structural change in the iodine
cage, rather than by an unusual decrease in $r_{excluded}$, as seen in Table 2.1.

An estimate of the temperature dependence of the conductivity expected for
α-AgI on the basis of this model was obtained by considering the geometrical factors [2.6]. The conductivity of a dilute Boltzmann gas with a fixed scattering
length is proportional to $T^{-\frac{1}{2}}$. In the excluded volume model, this conductivity
must be multiplied by the probability that a cation will locate and pass through
the opening in one of the faces of its tetrahedral cage. To lowest order, this
geometrical factor is proportional to the area of the opening divided by the allowed volume. Both are functions of T specified by $r_{excluded}$ and determined by
the EXAFS data. The net effect of these three temperature-dependent factors is to
increase the conductivity by a factor of 1.27 ± 0.06 between 198°C and 302°C.
The reported conductivity [2.50] increases by a factor of 1.26. This is a remarkable agreement considering that presumably important effects have been neglected,
including the details of the Ag-I interaction, Ag-Ag correlations, etc. One must
conclude that these other effects do not dominate the temperature dependence of
the conductivity.

Previously proposed models for superionic conductivity have been based upon
either hypothetical conduction paths or general considerations like residence and
hopping times or Brownian motion in one dimension. In the second class of models,
the mobile ion is viewed as vibrating about its lattice site for a residence
time. Then the ion hops or diffuses to a neighboring site during a hopping time.
These times are treated as independent parameters in the lattice gas models (see
Chap.4 for a discussion). More unified approaches have been applied to describe
motion of the ions by assuming that the ions undergo Brownian motion in a one-
dimensional sinusoidal or harmonic potential (for a discussion of these models
see Chap.8). Similarly, in the excluded volume model, the cation would bounce about
the tetrahedral void for a time in the allowed volume before it finds the opening
in the cage face and "jumps" to the next allowed volume. The models are similar in
this general sense.

The only conductivity model which discussed the conduction path in any detail is the "tunnel" model of FLYGARE and HUGGINS [2.33] mentioned in Sect.2.2. They assumed conduction along the (100) line connecting octahedral sites, and adjusted the ionic radii of the Ag and iodine to minimize the apparent activation energy for the path. This required rather small radii. EXAFS results [2.4] and the neutron diffraction study of CAVA et al. [2.5] have shown that the trigonal sites are much more likely than the octahedral sites to be part of the conduction path, ruling out this model in detail. But, as mentioned earlier, if the ionic radii are increased to larger values than were used, the considerations of FLYGARE and HUGGINS suggest that the ion path moves from the octahedral sites toward the tetrahedral sites. This would result in a very different conduction path, quite similar to the one proposed in the excluded volume model.

So in this model for bcc materials, the cations move as a gas in the allowed regions of the bcc anion lattice. Those allowed regions are centered at the tetrahedral sites where the cations are confined by the hard-sphere anion walls. They remain in this cavity until their motion is directed toward the opening in the tetrahedral face and they travel to the neighboring tetrahedral region. Since the faces do not lie in a straight line, the cations upon entering the neighboring tetrahedron collide with an anion, bouncing back and forth off the spherical anion walls until another face opening is found and the process is repeated.

2.4 EXAFS Investigations of fcc Superionic Conductors: Cuprous Halides

The fcc superionic conductors provide two inequivalent sites for the mobile ions, the tetrahedral and octahedral locations shown in Fig.2.12. In contrast with the bcc structure, these tetrahedra share faces only with octahedra. If the mobile ions pass through the faces in this structure, their conduction path will consist of alternating tetrahedral and octahedral sites [2.46]. Examples of this type of ionic conductor with cations mobile include α-Ag_2Te, α-CuI, α-Cu_2S and α-Cu_2Se. The

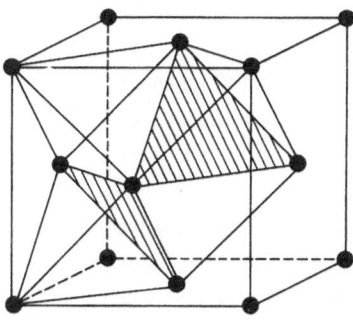

Fig.2.12 Face-centered-cubic structure showing the tetrahedral and octahedral voids. When motion is through the faces of the polyhedra in this structure, the path of the mobile ions consists of alternating tetrahedral and octahedral sites

Table 2.2 Selected structural·and conductivity data on the cuprous halides for the
various phases

Material	γ Phase		β Phase		α Phase	
	Structure	Maximum Conductivity $[\Omega^{-1}cm^{-1}]$	Structure	Temperature Range $[°C]$	Structure	Temperature Range $[°C]$
CuCl	Zincblende	≈ 0.1	Wurtzite	407-422	-	-
CuBr	Zincblende	≈ 0.1	Wurtzite	385-469	Br bcc	469-488
CuI	Zincblende	≈ 0.1	Wurtzite	369-407	I fcc	407-600

fluorite-structured ionic conductors, such as PbF_2, CaF_2, $SrCl_2$, and UO_2, also
fall into this category with mobile anions. In addition, the low-temperature γ phase
of CuCl, CuBr, and CuI has such a structure. The cuprous halides are discussed in
this section.

Selected structural and conductivity data for the cuprous halides are shown in
Table 2.2. All three have a low-temperature zincblende γ phase in which the ionic
conductivity achieves a high value, about 0.1 $\Omega^{-1}cm^{-1}$ [2.58,59]. The conductivity
at the high temperature limit of the γ phase is predominantly due to the Cu ions
[2.58,59]. with the halogen ions forming an immobile fcc lattice. All three materials transform to the hexagonal wurtzite phase at elevated temperature with the
ionic conductivity σ increasing discontinuously by a small factor [2.58]. At even
higher temperatures, CuCl melts while both CuBr and CuI transform to an α phase. The
α phase of these two halides differ in that CuBr has a bcc bromine lattice similar
to AgI, whereas CuI has an fcc iodine lattice. In the α phase both materials approach a conductivity of 1 $\Omega^{-1}cm^{-1}$. This situation is to be contrasted with that of
AgI, where the conductivity is low in the low-temperature phase but has a four-order-
of-magnitude discontinuity at the phase transition.

In this section, CuI will be discussed in detail since it has an fcc α phase
that exists over a wide temperature range. CuCl and CuBr will be treated less
thoroughly since they are similar to CuI both in structure and in the behavior of
the conductivity, except for the α phase.

2.4.1 CuI Structural Studies

In 1952 MIYAKE et al. [2.40], observed , in addition to the structural changes
listed in Table 2.2, a large loss of intensity in the X-ray Bragg peaks as CuI
is heated to elevated temperatures. The measured Debye-Waller factors are large
above 200°C in the γ phase. For the α phase, a displaced-ion model was proposed
in which the Cu ions are displaced in the (111) directions from the tetrahedral
center toward the tetrahedral faces. Only half the tetrahedral sites, the ones
on the zincblende lattice, were determined to be so occupied (T_d^2). The obtained

displacement is too large, however, to be consistent with the model. It is
≈ 0.96 Å, whereas the distance to the face is only 0.89 Å. This places the Cu ions
through the tetrahedral face into the octahedral region. Also in 1952, KRUG and
SIEG [2.60] observed a similarly large reduction in peak intensity and attributed
it to a 25 percent occupation of the octahedral sites, with the remaining three-
quarters of the Cu ions occupying one-half of the tetrahedral sites of the zinc-
blende lattice (T_d^2). MATSUBARA [2.39] proposed an anharmonic oscillator model and,
using the data of [2.40], obtained the following coefficients of the potential
(2.11) for the Cu ions:

$$\alpha_{Cu} = 1.8 \times 10^{-12} erg/Å^2$$
$$\beta_{Cu} = 2.9 \times 10^{-12} erg/Å^3 \text{ at } 470°C .$$

(2.14)

Recently, BÜHRER and HÄLG [2.61] explained their neutron diffraction data on α-CuI
by assuming that all the tetrahedral locations were statistically occupied (O_h^5).
In addition, they proposed that the Cu ions are displaced in the (111) directions
toward the faces of the tetrahedra. The Cu ions are then randomly distributed on
the 32(f) sites at a distance 0.54 Å from the center of the tetrahedron. They also
questioned the β phase having the wurtzite structure (C_{6v}^4), since a satisfactory
explanation of this data was obtained with space group D_{3h}^1.

2.4.2 EXAFS and Structural Models for CuI

The EXAFS data [2.62] were collected on the Cu K edge (8.98 keV) using a sample
of approximately three absorption lengths thick (≈ 20 μm) contained in a boron
nitride cell. Data were collected in all three phases from 77 K to 470°C. The experimental arrangement and the data reduction were the same as that for AgI. The
resultant complex transform of the EXAFS into real space, $\phi(r)$, is shown in
Fig.2.13 for the γ-phase data at low (77 K) and high (200°C) temperatures and for
the superionic α-phase data at 470°C. In all cases, a square window, k = 2.8 to
10.4 Å^{-1}, with a Gaussian broadening of width 0.7 Å^{-1}, was used for the Fourier
transform to real space.

The structure between 1.4 and 3.2 Å in $\phi(r)$ at 77 K (Fig.2.13a) is due to the
four iodine first neighbors to the Cu in the zincblende lattice. The peaks beyond
3.5 Å are due to more distant neighbors. As in the case of AgI, this $\phi(r)$ at 77 K
yields ξ_{Cu-I} since the first neighbor $p_{Cu-I}(r)$ is a narrow Gaussian centered at
2.61 Å. Using this ξ_{Cu-I} the unknown near-neighbor $p_{Cu-I}(r)$ at elevated temperatures
was obtained. From Figs.2.13b,c it is seen that p(r) broadens asymmetrically and
shifts inward with increasing temperature.

Consider the various models for p(r). For each, a model EXAFS function, $\phi_m(r)$
was obtained as for AgI. This model function was compared with the data in the

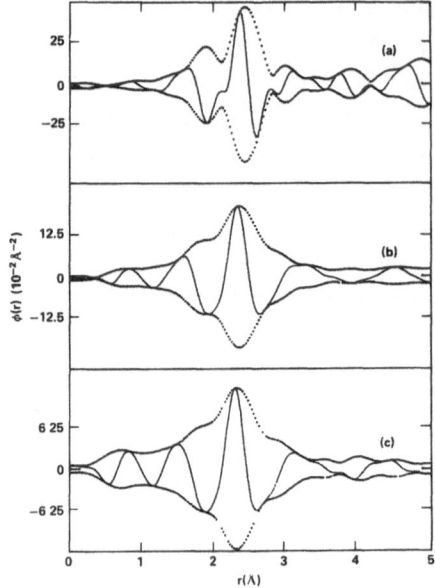

Fig.2.13a-c The real part (solid line) and the magnitude of the EXAFS in real space on the Cu K edge in CuI: (a) at 77 K where the Cu ions are located at the tetrahedral sites of the zincblende lattice, (b) at $200^{\circ}C$ where $\sigma \sim 10^{-4} \ \Omega^{-1}cm^{-1}$, and (c) at $470^{\circ}C$ in the superionic α phase. Note the difference in vertical scales

interval r = 1.4 to 3.2 Å and the R value, (2.7), evaluated. The following three possibilities were considered:

(I) *Displaced Site Model*

BÜHRER and HÄLG [2.61] proposed a model similar to that of MIYAKE et al. [2.40], in which the Cu ions are displaced from the tetrahedral centers at (1/4, 1/4, 1/4), the 8(c) sites of O_h^5, to the 32(f) sites at (x, x, x). At $445^{\circ}C$, BÜHRER and HÄLG obtained x = 0.30 and a lattice constant a_0 = 6.16 Å. The near-neighbor environment for this model consists of three iodine atoms at a distance r_0 = 2.54 Å and one iodine atom at r_0 = 3.21 Å. This specific model does not fit the EXAFS data well, yielding a reliability-of-fit parameter of 12% at $450^{\circ}C$ and 10% at $470^{\circ}C$.

(II) *Anharmonic Oscillator Model*

The anharmonic potential of (2.11) with the parameters determined by MATSUBARA [2.39], (2.14), was compared with the EXAFS data for α-CuI. The probability that the Cu ion is displaced to a position $\underline{r}_{Cu}$ from the tetrahedral center as origin is

$$\mathscr{P}_{Cu}(\underline{r}_{Cu}) \sim \exp[-V_{Cu}(\underline{r}_{Cu})/k_B T] \quad , \tag{2.15}$$

where V_{Cu} is the Cu potential, (2.11). Similarly, for the iodine ions, one has

$$\mathscr{P}_I(\underline{r}_i) \sim \exp[-V_I(\underline{r}_i - \underline{r}_i^0)/k_B T] \tag{2.16}$$

where $\underline{r}_i^0$ is the equilibrium position of the ith iodine atom among the four iodine ions of the tetrahedron. The Cu-I near-neighbor pair correlation function is given by

$$p_{Cu-I}(r) = 4\pi r^2 <\rho_{Cu-I}(\underline{r})> \quad , \tag{2.17}$$

where

$$\rho_{Cu-I}(\underline{r}) = 4 \int d\underline{r}' \mathscr{P}_I(\underline{r}') \mathscr{P}_{Cu}(\underline{r} + \underline{r}') \quad . \tag{2.18}$$

The factor of 4 in (2.18) comes from the number of iodine near neighbors. This model does not fit the EXAFS data well either, yielding R = 13.5% for the 470°C $\phi(r)$. The main reason that the anharmonic model fits so poorly is that it does not allow for an adequate inward shift of p(r) or for a large asymmetry.

(III) *Excluded Volume Model*
This model does account for the observed inward shift and asymmetry in p(r), and enabled a good fit of the CuI data at all temperatures. The fitting procedure was as described for AgI with the addition of one parameter, c_{oct}, the concentration of Cu ions in the octahedral sites. Since the near-neighbor environment is different for Cu ions in the octahedral and tetrahedral sites, the relative occupation of these sites will be determined by considerations in addition to the hard-sphere repulsion of (2.13). This is accounted for by allowing the two sites to differ in occupation. The octahedral site energy is higher than that for the tetrahedra, so that $c_{oct} = 0$ at low temperatures and increases with increasing temperature. This increase is expected to be augmented by the increase in ionic conductivity, since the anticipated conduction path consists of alternating tetrahedral and octahedral locations.

This model yielded good fits with R ranging from 1% to 3% for the data at all temperatures. The p(r)'s determined by these fits are shown in Fig.2.14. At 20°C where the conductivity is small, p(r) is close to Gaussian and centered on the tetrahedral site. As the temperature increases, p(r) becomes broader and more asymmetric. In addition, some density builds up at larger distances, specifically near the arrows in Fig.2.14 corresponding to Cu ions in the faces of the tetrahedra. Above 300°C there is a large density at this face site, corresponding to the Cu ions moving through the tetrahedral faces into the octahedral regions. The increase in p(r) at large r could not be fully explained with the Cu ions only in the tetrahedral locations but could be accounted for by some octahedral occupation. Also, a substantial improvement in the quality of fit was obtained at elevated temperatures with $c_{oct} > 0$. For example, in α-CuI at 470°C, the excluded volume model gives R = 1.81% with $c_{oct} = 30\%$ and R = 2.48% with $c_{oct} = 0$, an increase by a factor of 1.37. In addition, the c_{oct} determined from the EXAFS data is zero at low temperatures, as expected, and increases in a systematic way with increasing temperature. So an increase in c_{oct} is a very likely explanation for the obtained p(r).

The fitted parameters, $r_{excluded}$ and c_{oct}, are shown in Fig.2.15 along with r_{face}, the near-neighbor Cu-I spacing of a Cu ion in the face shared by a

38

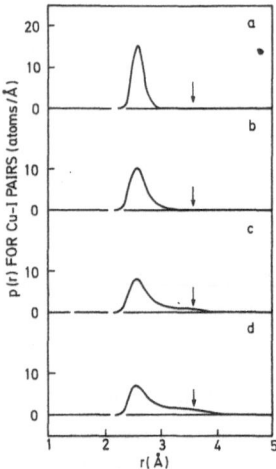

Fig.2.14a-d The radial distribution function of iodine ions about each Cu in CuI in the γ phase at (a) 22°C, (b) 300°C, (c) 350°C, just before the γ→β phase transition, and in the α phase at (d) 470°C. The arrows indicate the location for Cu ions in the center of a tetrahedral face, the bridging site between the tetrahedra and octahedra

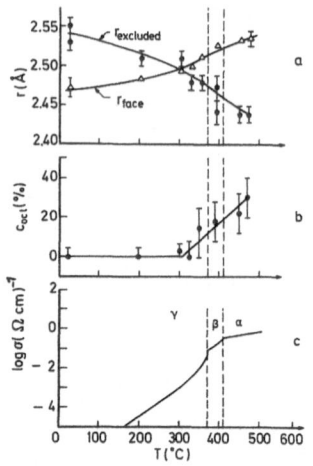

Fig.2.15a-c The parameters (a) $r_{excluded}$ and (b) c_{oct}, the concentration of Cu ions in the octahedral sites, obtained from the EXAFS data on CuI using the excluded volume model. Also included are the near-neighbor spacing for Cu ions in the tetrahedral face, r_{face}, and (c) the ionic conductivity [2.59]

tetrahedron and octahedron. It is seen that when $r_{excluded}$ becomes less than r_{face}, the concentration of Cu ions in the octahedral sites begins to increase. It is also the same temperature region where the ionic conductivity becomes large, as shown in Fig.2.15c. Due to the more complex conduction path than for bcc AgI, a quantitative comparison between the conductivity and the parameters obtained from the excluded volume model has not been made. The qualitative correlations between the ionic conductivity and both $r_{excluded}$ and c_{oct} are, however, quite suggestive.

In the superionic α phase, the obtained octahedral occupation is approximately 30%. Equal occupation of all sites would yield c_{oct} = 33%, near the obtained value. The large octahedral occupation results from the very large allowed volume in an octahedron, V_{oct}, which is approximately 10 times larger than the allowed tetrahedral volume, V_{tet}. The Cu-ion density of the sites, however, is not uniform. It is determined by the potential energy difference between the sites, $\Delta E = E_{oct} - E_{tet}$, given by $c_{oct}/(1 - c_{oct}) = (4V_{oct}/8V_{tet})\exp(-\Delta E/k_B T)$. This expression with c_{oct} = 30% yields a small energy difference, ≈ 0.16 eV, between the tetrahedral and octahedral sites. This compares favorably with the total activation energy for motion of 0.2 eV determined from NMR [2.63] and conductivity [2.64]. This energy difference between the sites along the conduction path accounts in part for the lower conductivities and higher activation energies for the fcc materials, compared with the bcc materials. For example, the activation energy is 0.2 eV in α-CuI but only 0.05 eV in α-AgI.

Similar behavior has been observed in PbF_2 using neutron diffraction [2.65]. In this fluorite-structured material there are two tetrahedral and one octahedral locations in the fcc Pb lattice for two F ions. An equal occupation of these sites would yield an occupation of 0.67 per site. An occupation of about 0.4 is observed [2.65] in the octahedral location, showing that, even in the high temperature super-ionic phase of PbF_2, there is a preference for the tetrahedral locations.

2.4.3 CuBr

In 1952, HOSHINO [2.66], and KRUG and SIEG [2.60] observed the structural changes for CuBr listed in Table 2.2 as well as an anomalous loss in intensity of the Bragg peaks with increasing temperature. More recently, BÜHRER and HÄLG [2.61] obtained similar results, although they place the Cu ions on the tetrahedral 12(d) sites in the bcc α phase, whereas HOSHINO [2.66] supported the STROCK model, Fig.2.6. HARADA et al. [2.38] have proposed an anharmonic model to fit their neutron data on single crystal γ-CuBr up to about 300°C. The potential is that appropriate for the fcc structure, (2.11), and yielded the following parameters:

$$\left.\begin{array}{l} \alpha_{Cu} = 0.94 \times 10^{-12} erg/Å^2 \\[2ex] \beta_{Cu} = 1.0 \pm 0.3 \times 10^{-12} erg/Å^3 \\[2ex] \alpha_{Br} = 1.49 \times 10^{-12} erg/Å^2 \\[2ex] \beta_{Br} \simeq 0 \quad . \end{array}\right\} \tag{2.19}$$

The EXAFS for CuBr [2.62] on the Cu K edge were collected and reduced in a manner similar to that on CuI. Only the γ phase was studied. Some of the spectra are shown in Figs.2.2-4 and were discussed in connection with the data reduction procedure. The anharmonic oscillator model with the parameters of [2.38], (2.19), was compared with the EXAFS data using (2.15-18) for $p_{Cu-Br}(r)$ and the ξ_{Cu-Br} extracted from the 77 K data, as for CuI. This model did not fit well, yielding, for example, R = 23% at 225°C and R = 26% at 370°C. The excluded volume model did fit well, giving R = 0.9% at 225°C and R = 1.2% at 370°C. It yielded a good fit to the data since it accounts properly for the observed inward shift and increasing asymmetry in p(r) with increasing temperature, whereas the anharmonic oscillator model does not. The resulting parameter values for this model as a function of temperature in the γ phase are shown in Fig.2.16. Again, it is seen that $r_{excluded}$ and c_{oct} correlate strongly with the increase in the ionic conductivity: $r_{excluded}$ becomes less than r_{face} and c_{oct} increases when the conductivity becomes large.

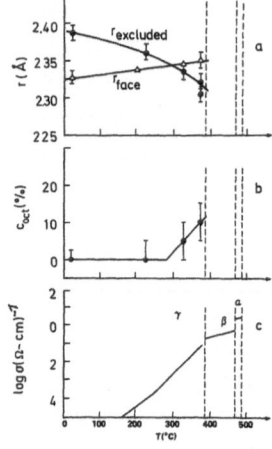

Fig.2.16a-c The parameters, (a) $r_{excluded}$ and (b) c_{oct}, the concentration of Cu ions in the octahedral sites obtained from the EXAFS data on CuBr using the excluded volume model. Also included is the near-neighbor spacing for Cu ions in the tetrahedral face, r_{face}, and (c) the ionic conductivity [2.58]

2.4.4 CuCl

SAKATA et al. [2.37] compared their neutron diffraction data on single crystal γ-CuCl with three models: the harmonic oscillator model, the anharmonic oscillator model, and the disordered model in which the Cu ions occupy the four positions displaced from the normal tetrahedral positions in (111) directions toward the tetrahedral faces. The latter two models fit the data best but were essentially indistinguishable from one another. They favored the anharmonic model since it fit all the data from room temperature to about 300°C with no additional parameters. The parameters obtained for the potential of (2.11) are

$$
\left.
\begin{aligned}
\alpha_{Cu} &= 0.74 \times 10^{-12} \text{erg/\AA}^2 \\[6pt]
\beta_{Cu} &= 1.15 \pm 0.66 \times 10^{-12} \text{erg/\AA}^3 \\[6pt]
\alpha_{Cl} &= 1.35 \times 10^{-12} \text{erg/\AA}^2 \\[6pt]
\beta_{Cl} &\simeq 0 \quad .
\end{aligned}
\right\}
\tag{2.20}
$$

Using an integrated intensity analysis of their neutron diffraction data, SCHREURS et al. [2.30] obtained results that agree quite well with the above. In addition, by varying the Cu isotopic abundance, they obtained the Cu-Cu, Cu-Cl, and Cl-Cl pair correlation functions. These results showed that the Cu-Cu distribution is broader and flatter than the Cl-Cl distribution, consistent with the Cu-sublattice disorder.

The EXAFS on the Cu K edge were compared with the predictions of the anharmonic model and the excluded volume model in the manner describe above. The anharmonic model, with the parameters of (2.20), did not fit well. For example, R = 26% at

250°C and R = 25% at 350°C. The excluded volume model did fit well yielding R = 1.2% at 250°C and R = 1.7% at 350°C. It accounts for the observed inward shift and increasing asymmetry with increasing temperature while the anharmonic model does not. The excluded volume model parameters are shown in Fig.2.17.

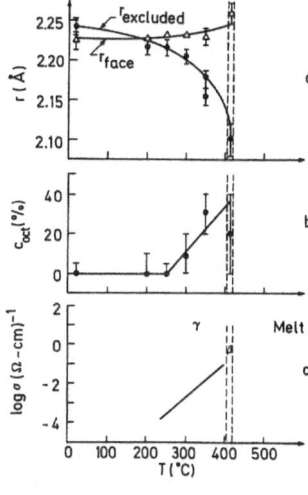

Fig.2.17a-c The parameters, (a) $r_{excluded}$ and (b) c_{oct}, the concentration of Cu ions in the octahedral sites obtained from the EXAFS data on CuCl using the excluded volume model. Also included is the near-neighbor spacing for Cu ions in the tetrahedral face, r_{face}, and (c) the ionic conductivity [2.58]

2.4.5 Discussion

In the analysis of the EXAFS data on the fcc phases of the cuprous halides, the excluded volume model fits well since it accounts for the two prominent features of the data with increasing temperature: (I) the increasingly inward, rather than outward, shift in p(r); and (II) the increase in the asymmetry of p(r). The anharmonic oscillator model does not fit the data well, since it cannot account for the inward shift and asymmetry of p(r), and, at the same time, the sharp cutoff at low r in p(r) due to the hard-core repulsion. The displaced-site models can be viewed as crude approximations to the anharmonic oscillator and excluded volume models, and the specific ones proposed did not fit the EXAFS data.

In addition, the excluded volume model yields a finite occupation of the octahedral sites in all three copper halides just before the γ → β phase transition. In superionic α-CuI, this occupation reaches about 30% and corresponds to an octahedral-site energy that is 0.16 eV higher than that of the tetrahedral site.

The situation on the conductivity in these fcc materials is more complicated than for that for the bcc materials. Two qualitative correlations, however, can be made between the conductivity and the parameters: $r_{excluded}$ becomes less than r_{face} and c_{oct} increases in the region where the conductivity becomes large. The fact that $r_{excluded}$ becomes less than r_{face} implies that the portion of the barrier for motion due to an ion squeezing through the tetrahedral face decreases to zero, i.e., $U_m = 0$. The remaining portion of the barrier, namely, the difference in site energy between the tetrahedral and octahedral locations, does not, however, go to zero, i.e., $U_F > 0$. As a result, the transition from the low-temperature ionic-in-sulating to the high-temperature ion-conducting phase is less sharp and more complicated than that for AgI.

2.5 Summary

The EXAFS data on superionic conductors strongly support an excluded volume model for the cation-anion pair correlations. In this model, the cation-anion potential interaction is approximated as a hard-sphere repulsion. This neglects the Coulomb and dipole interactions, which are most likely slowly varying along the conduction path. AgI is a bcc material in which only equivalent tetrahedral locations form the conduction path. The resulting pair correlations from the excluded volume model yield an excellent fit to the EXAFS data on AgI. For the fcc cuprous halides, the model is generalized since the conduction path necessarily includes octahedral sites which have a higher energy than the tetrahedral sites. This slightly more complicated model yields pair correlations in excellent agreement with EXAFS data on the cuprous halides. In each case, the excluded volume model yields substantially better agreement than proposed displaced-site and anharmonic models. This is because the excluded volume model is the ultimate in anharmonicity; that is, it offers no resistance to motion in some directions and infinite resistance in others.

In addition, the conductivity is easily examined in the excluded volume model. In the simple case of AgI, from the EXAFS data alone, the model can predict both the discontinuous increase in conductivity at the phase transition and the slight temperature dependence above it. Even in the case of the more complicated conduction path in the cuprous halides, this model yields suggestive correlations with the conductivity data. Judging by the success of its applications to date, the excluded volume model has the potential to bring together, in a new way, important and diverse aspects of superionic conduction.

Acknowledgement. We wish to thank J.L. Beeby, J.C. Mikkelsen, Jr., and W. Stutius for their contributions to the development of the ideas presented in this chapter.

References

2.1 For a discussion of the various types of transitions observed in ionic conductors see, for example, the reviews by: M.O'Keeffe: In *Superionic Conductors,* ed. by G.D. Mahan, W.L. Roth (Plenum Press, New York 1976) p.101
      G.D. Mahan: Ibid., p.115
      B.A. Huberman: Ibid., p.151
2.2 L.W. Strock: Z. Phy. Chem., Abt. B *25*, 411 (1934); *31*, 132 (1936)
2.3 For recent reviews of the structures of many superionic conductors see, for example: K. Funke: Prog. Solid State Chem. *11*, 345 (1976)
      S. Geller: In *Solid Electrolytes,* ed. by S. Geller, Top. Appl. Phys., Vol.21 (Springer, Berlin, Heidelberg, New York 1977) p.41

2.4 J.B. Boyce, T.M. Hayes, W. Stutius, J.C. Mikkelsen, Jr.: Phys. Rev. Lett. *38*, 1362 (1977)
2.5 R.J. Cava, F. Reidinger, B.J. Wuensch: Solid State Comm. *24*, 411 (1977)
2.6 T.M. Hayes, J.B. Boyce, J.L. Beeby: J. Phys. C *11*, 2931 (1978)
2.7 See, for example: G. Eckold, K. Funke, J. Kalus, R.E. Lechner: J. Phys. Chem. Solids *37*, 1097 (1976), and references contained therein
2.8 M. Suzuki, H. Okazaki: Phys. Stat. Sol. (a) *42*, 133 (1977)
2.9 Y. Tsuchiya, S. Tamaki, Y. Waseda, J.M. Toguri: J. Phys. C *11*, 651 (1978)
2.10 E.A. Stern: Phys. Rev. B *10*, 3027 (1974)
2.11 C.A. Ashley, S. Doniach: Phys. Rev. B *11*, 1279 (1975)
2.12 P.A. Lee, J.B. Pendry: Phys. Rev. B *11*, 2795 (1975)
2.13 T.M. Hayes, P.N. Sen: Phys. Rev. Lett. *34*, 956 (1975)
2.14 T.M. Hayes, P.N. Sen, S.H. Hunter: J. Phys. C *9*, 4357 (1976)
2.15 P.A. Lee: Phys. Rev. B *13*, 5261 (1976)
2.16 E.A. Stern, D.E. Sayers, F.W. Lytle: Phys. Rev. B *11*, 4836 (1975)
2.17 See, for a discussion: F.W. Lytle, D.E. Sayers, E.A. Stern: Phys. Rev. B *15*, 2426 (1977)
2.18 F.W. Lytle, D.E. Sayers, E.A. Stern: Phys. Rev. B *11*, 4825 (1975)
2.19 See, for a discussion: M. Brown, R.E. Peierls, E.A. Stern: Phys. Rev. B *15*, 738 (1977)
2.20 J. Jaklevic, J.A. Kirby, M.P. Klein, A.S. Robertson, G.S. Brown, P. Eisenberger: Solid State Comm. *23*, 679 (1977)
2.21 P.A. Lee, G. Beni: Phys. Rev. B *15*, 2862 (1977)
2.22 B.-K. Teo, P.A. Lee, A.L. Simons, P. Eisenberger, B.M. Kincaid: J. Am. Chem. Soc. *99*, 3854 (1977)
     P.A. Lee, B.-K. Teo, A.L. Simons: J. Am. Chem. Soc. *99*, 3856 (1977)
2.23 S.J. Gurman, J.B. Pendry: Solid State Comm. *20*, 287 (1976)
2.24 D.E. Sayers, E.A. Stern, F.W. Lytle: Phys. Rev. Lett. *27*, 1204 (1971)
2.25 G.S. Brown, P. Eisenberger, P. Schmidt: Solid State Comm. *24*, 201 (1977)
2.26 G.S. Brown, L.R. Testardi, J.H. Wernick, A.B. Hallak, T.H. Geballe: Solid State Comm. *23*, 875 (1977)
2.27 See, for a discussion: J. Waser, V. Schomaker: Rev. Mod. Phys. *25*, 671 (1953)
2.28 A. Guinier: *X-Ray Diffraction* (Freeman, San Francisco 1963)
2.29 G.E. Bacon: *Neutron Diffraction* (Clarendon Press, Oxford 1975)
2.30 J. Schreurs, M.H. Mueller, L.H. Schwartz: Acta Cryst. A *32*, 618 (1976)
2.31 For a review of these structural criteria and references to other work, see: W. van Gool: Annu. Rev. Mater.Sci. *4*, 311 (1974)
     S. Geller: "Halogenide Solid Electrolytes" in *Solid Electrolytes*, ed. by S. Geller, Topics in Appl. Phys.Vol.21 (Springer, Berlin, Heidelberg, New York 1977) p.56
2.32 R.D. Armstrong, R.S. Bulmer, T. Dickinson: J. Sol. State Chem. *8*, 219 (1973)
2.33 W.H. Flygare, R.A. Huggins: J. Phys. Chem. Solids *34*, 1199 (1973)
2.34 A. Rahman: J. Chem. Phys. *65*, 4845 (1976)
2.35 M. Dixon, M.J. Gillan: J. Phys. C *11*, L165 (1978)
2.36 W. Schommers: Phys. Rev. Lett. *38*, 1563 (1977); Phys. Rev. B *17*, 2057 (1978)
2.37 M. Sakata, S. Hoshino, J. Harada: Acta Cryst. A *30*, 655 (1974)
2.38 J. Harada, H. Suzuki, S. Hoshino: J. Phys. Soc. Jpn. *41*, 1707 (1976)
2.39 T. Matsubara: J. Phys. Soc. Jpn. *38*, 1076 (1975)
2.40 S. Miyake, S. Hoshino, T. Takenaka: J. Phys. Soc. Jpn. *7*, 19 (1952)
2.41 B.T.M. Willis: Acta Cryst. *18*, 75 (1965)
     H.B. Strock, B.W. Batterman: Phys. Rev. B *5*, 2337 (1972)
2.42 M.J. Cooper, K.D. Rousse: Acta Cryst. A *27*, 622 (1971)
2.43 M.J. Cooper, K.D. Rousse, B.T.M. Willis: Acta Cryst. A *24*, 484 (1968)
2.44 S. Hoshino, T. Sakuma, Y. Fujii: Solid State Comm. *22*, 763 (1977)
2.45 See, for a discussion: J.L. Finney: Proc. R. Soc. London A *319*, 495 (1970)
2.46 L.V. Azaroff: J. Appl. Phys. *32*, 1658 (1961)
2.47 For a review of the experimental data on the conductivity, see K. Funke, Ref.3
2.48 R.N. Schock, E. Hinze: J. Phys. Chem. Sol. *36*, 713 (1975)
     H. Hoshino, M. Shimoji: J. Phys. Chem. Sol. *35*, 321 (1974)
2.49 P.C. Allen, D. Lazarus: Phys. Rev. B *17*, 1913 (1978)
2.50 A Kvist, A.-M. Josefson: Z. Naturforsch. *23*A, 625 (1968)

2.51 G. Burley: J. Chem. Phys. *38*, 2807 (1963)
2.52 S. Hoshino: J. Phys. Soc. Jpn. *12*, 315 (1957)
2.53 W. Bührer, W. Hälg: Helv. Phys. Acta *47*, 27 (1974)
2.54 A.F. Wright, B.E.F. Fender: J. Phys. C *10*, 2261 (1977)
2.55 For a discussion of signal-to-noise considerations, see: B.M. Kincaid: SSRP Report No. 75/03 (1975)
2.56 B.R. Lawn: Acta Cryst. *17*, 1341 (1964)
2.57 J.B. Boyce, T.M. Hayes, W. Stutius, J.C. Mikkelsen: Unpublished data
2.58 J.B. Wagner, C. Wagner: J. Chem. Phys. *26*, 1597 (1957); and references contained therein
2.59 T. Jow, J.B. Wagner: J. Electrochem. Soc. *125*, 613 (1978)
2.60 J. Krug, L. Sieg: Z. Naturforsch. *7*a, 369 (1952)
2.61 W. Bührer, W.Hälg: Electrochim. Acta *22*, 701 (1977)
2.62 J.B. Boyce, T.M. Hayes, W. Stutius, J.C. Mikkelsen, Jr.: Bull. Amer. Phys. Soc. *23*, 241 (1978); and to be published
2.63 J.B. Boyce, B.A. Huberman: Solid State Comm. *21*, 31 (1977)
2.64 W. Jost: *Diffusion in Solids, Liquids and Gases* (Academic Press, New York 1960) p.188
2.65 J.D. Axe, S.M. Shapiro, N. Wakabayashi: To be published
     S.M. Shapiro: In *Superionic Conductors*, ed. by G.D. Mahan, W.L. Roth (Plenum Press, New York 1976) p.261
2.66 S. Hoshino: J. Phys. Soc. Jpn. *7*, 560 (1952)

3. Neutron Scattering Studies of Superionic Conductors

S. M. Shapiro and F. Reidinger

With 22 Figures

Over the past 30 years, neutron diffraction and inelastic scattering have proved
to be invaluable techniques for the study of structure and dynamics of solids.
Although the neutron flux impinging upon a sample in a neutron instrument is
several orders of magnitude less than photon flux available in X-ray sources, the
favorable scattering properties of neutrons as compared to X-rays more than compensates for the lack of intensity. We shall not attempt to review the bases of
neutron scattering since there exist several excellent reviews on the properties
of neutrons and their use [3.1-4]. Descriptions of the instrumentation have also
been adequately discussed in these references. Suffice it to say that the irregular,
but limited, variation of scattering lengths within the periodic table and the
low kinetic energy of a thermal neutron beam make it possible to probe positions
and dynamics of almost all elements in the periodic table. Because X-rays have
extremely large energies and interact with the electrons of an atom, high-resolution
inelastic studies are presently impossible and the observation of light elements
in compounds containing heavy elements is extremely difficult. However, the combined
use of X-ray and neutron diffraction has proved invaluable in unraveling structures of complicated solids and alloys [3.5].

In this chapter we discuss the structural studies of superionic conductors (SI)
by means of neutron and X-ray diffraction and their dynamical properties by neutron-
scattering techniques. Because of the breakdown of long-range order on one or more
sublattices in the SI phase, the relative importance of the two techniques is
altered when compared to the study of ordered solids. The least affected is the
determination of the average structure from the study of the Bragg peaks because
the static coherence is maintained by the rigid sublattice. Unique interatomic distances of the disordered sublattice, however, can no longer be obtained from such
studies and we have to rely on the contribution of the diffuse scattering between
the Bragg peaks. Diffuse intensity close to the Bragg peaks, Huang scattering, contains information about the distortions of the rigid sublattice. Phonon studies
by means of neutron scattering are expected to exhibit anomalous line broadening
because the largely uncorrelated motion of the mobile atoms will destroy the coherence of the vibrational states. Additionally, energy analysis of the diffuse
scattering contains information about the time scale of the diffusing ion.

From the numerous studies on SI materials we have selected only a few to discuss below. Structural studies on AgI, the fluorites, and β-alumina will be reviewed and comparison with other disordered systems such as the hydrides will be discussed. There has been a more modest effort on the dynamical behavior of the SI conductors and the results on AgI, $RbAg_4I_5$, fluorites and β-alumina will be reviewed.

3.1 Neutron Scattering

3.1.1 Scattering Function

We refer the reader to the articles of LOVESEY [3.6] and DACHS [3.7] for detailed discussion of the formalism for writing the scattering functions for neutron diffraction and inelastic scattering. We shall present only a summary of the features of the scattering function relevant to the results which follow.

The differential scattering cross section for neutrons can be written as the sum of a coherent part and an incoherent part,

$$\frac{d^2\sigma}{d\Omega d\omega} = \frac{d^2\sigma}{d\Omega d\omega}\bigg|_c + \frac{d^2\sigma}{d\Omega d\omega}\bigg|_i \quad . \tag{3.1}$$

In general terms, the coherent part gives us information about the cooperative effects among different atoms, such as Bragg scattering in the structural sense, or phonon behavior in the dynamical sense. The incoherent part tells us about the correlation of an atom with itself and provides information about individual particle motion such as diffusion.

Dealing first with the coherent cross section we can write

$$\frac{d^2\sigma}{d\Omega d\omega}\bigg|_c = \frac{k_f}{k_i} \left| F(\underline{Q},\omega) \right|^2 \quad , \tag{3.2}$$

where the conservation of momentum and energy yield

$$\underline{Q} = \underline{k}_f - \underline{k}_i \quad ,$$

$$\hbar\omega = \frac{\hbar^2}{2m_N} (k_i^2 - k_f^2) \quad . \tag{3.3}$$

$\hbar = h/2\pi$ (normalized Planck's constant)

Here k_i and k_f are the critical and final momentum of the neutron and Q and $\hbar\omega$ are the momentum and energy transferred to the solid. $F(Q,\omega)$ is the generalized structure factor which is composed of an elastic part which measures the average structure and an inelastic part which probes the time-dependent fluctuations of the average structure.

In the harmonic approximation we have, for phonon creation,

$$|F(Q,\omega)|^2 = (2\pi)^3 \frac{N}{V} |F_0(Q)|^2 \delta(Q - G)\delta(\hbar\omega)$$

$$+ \sum_p |\ g_p(Q)|^2 [n(\omega) + 1]\delta(Q \pm q - G)\delta(\hbar\omega + \hbar\omega_s) \qquad . \qquad (3.4)$$

In (3.4) N is the number of unit cells in a crystal; V the volume of the unit cell; G is a reciprocal lattice vector; and p denotes the mode of wave vector q and branch j. The Bose-Einstein population factor is written $n(\omega) = [\exp(\hbar\omega/k) - 1]^{-1}$.

3.1.2 Elastic Scattering

The first term in (3.4) is the structure factor for elastic or Bragg scattering;

$$F_0(Q) = \sum_j b_j \exp(iQ\cdot R_j) \exp(- \tfrac{1}{2} Q\cdot g_j \cdot Q) \qquad (3.5)$$

where b_j is the scattering length of atom j, R_j is its position within the unit cell and g_j the variance-covariance matrix which contains the displacements of atom j. The exponent $\frac{1}{2} Q\cdot g_j \cdot Q$ is, in the physics community, usually referred to as the Debye-Waller factor; in the crystallographic community it is better known as $\beta_j^{\ell m} h_\ell h_m$ where $\beta^{\ell m} = 2\pi^2 \sigma^{\ell m}$ are the thermal parameters and $h_\ell = Q/2\pi$ is one of the Miller indices (h k ℓ).

For X-ray scattering b_j has to be replaced by the atomic scattering factor $f_j(Q)$ which depends upon the scattering angle because the diameter of the electron cloud of the atom is comparable with the wavelength of the X-ray. In structural studies of SI conductors, where the mobile species is a heavy atom with many electrons (e.g., Ag β-alumina, and Ag_2S), X-rays may be a more suitable tool than neutrons because the atom of interest has a large scattering power. On the other hand, the absence of strong absorption, the independence of the scattering length with angle, and the relative ease of data acquisition at temperatures other than room temperature can favor the use of neutrons.

Equation (3.5) is adequate for the majority of structural investigations whose major goal is the determination of coordinates R_j and the interpretation of the characteristic bonding distances and angles. These "well-behaved" structures exhibit long-range order which is only disturbed by harmonic vibrations of the atoms around their equilibrium positions. The mean square displacements, which constitute the variance-covariance matrix $\underline{\underline{\sigma}}_j$ of (3.5), are usually regarded as an annoyance, and are frequently reduced by performing the structure determination at very low temperatures. The major difficulty in solving such structures is the determination of suitable phases and once a reasonable starting model has been obtained, it is straightforward to find unique values of R_j and $\underline{\underline{\sigma}}_j$. This uniqueness is no longer guaranteed if the translational symmetry of one sublattice is violated on as large a scale as in the superionic conductors. A particular example of this difficulty is the abundance of models for the Ag distribution in the SI phase of AgI (see Sect.3.2). We want to emphasize that in the SI compounds, the problems arise not with the overall structure of the various compounds, but with physically signifi-cant features which cannot readily be presented by a set of positional and thermal parameters but require a detailed knowledge of the probability density (PD) of the total unit cell. We therefore require for a definite structure determination of a superionic conductor

   - that the PD of the unit cell can be qualitatively represented by Fourier techniques down to the noise level;
   - that the PD be quantitatively described by whatever means necessary such as partial atoms, off-center atoms or higher-order thermal tensors; and
   - that the merit of each parameter introduced be demonstrated by significance tests.

The most recent of these techniques, namely, the description of a complex PD by anharmonic parameters, higher order cumulants, or higher order tensors has al-ready been applied by COOPER et al. [3.8], who analyzed "well-behaved" structures such as CaF_2 at low temperatures. JOHNSON [3.9] also used these techniques in analyzing the structure of certain organic compounds. WILLIS's approach [3.8] is intuitively attractive as he proposes the following relation between the probability density, $\rho(\underline{r})$, and the site potential $V(\underline{r})$:

$$\rho(\underline{r}) \sim \exp[-V(\underline{r})/k_B T] \quad . \tag{3.6}$$

The structure factor, which is the Fourier transform of $\rho(\underline{r})$, is quite complex even for sites of cubic symmetry and becomes even more so when the site symmetry is re-duced. Johnson's formalism, on the other hand, is applicable to any site symmetry and stresses the need for a structure factor which allows the refinement of all relevant parameters with a minimum of correlations. In this method $F_0(\underline{Q})$ is re-placed by a more general structure factor $\phi(\underline{Q})$,

$$\phi(\underline{Q}) = F_0(\underline{Q})\left[1 + \frac{i^3}{3!} C^{jk\ell} H(Q_j Q_k Q_\ell) + \frac{i^4}{4!} D^{jk\ell m} H(Q_j Q_k Q_\ell Q_m)\right] \quad ,$$

where $H(Q_j \ldots)$ are orthogonalized Hermite polynomials and the tensor components of $C^{jk\ell}$ and $D^{jk\ell m}$ are restricted only by site symmetries [3.10]. The PD's, which can be obtained from $\phi(Q)$ by a Fourier transformation, retain all the potentially significant features of COOPER's formalism [3.8] but share none of its shortcomings. (To prevent divergence at large r, fourth- or higher-order terms with even parity are required in the potential $V(r)$ [3.11].

Structural analysis by a least squares refinement is relatively recent [3.12] when compared to the Fourier techniques. The presence of disorder in SI conductors complicates the least squares refinement and stresses the importance of Fourier methods for analysis and representation of a structure. As the average electron or nuclear density, $\rho(r)$, retains its periodicity, it can be represented as a three-dimensional series,

$$\rho(r) = \sum_{Q-\infty}^{\infty} F_0(Q)S(Q) \exp(iQ \cdot r) \quad , \tag{3.8}$$

where $F_0(Q)$, which is proportional to the square root of the observed intensity, is obtained from the experiment and $S(Q)$, the phase of $F_0(Q)$, is obtained from a model structure.

Shortcomings of a proposed model of the structure are revealed by Fourier synthesis of the difference between the observed structure factor $F_0(Q)$ and that calculated from a specific model $F_c(Q)$,

$$\Delta\rho(r) = \frac{1}{V} \sum_{Q} \left[F_0(Q) - F_c(Q) \right] \cdot S(Q) \exp(iQ \cdot r) \quad . \tag{3.9}$$

For noncentrosymmetric structures, $\Delta\rho(r)$ is less reliable because $S(Q)$ is, in general, a complex quantity; whereas in centrosymmetric structures $S(Q)$ is limited to either ± 1. This *difference Fourier method* may show regions of nuclear density at r_i, but they are only significant if they provide a statistical improvement in a least squares refinement of the structure with an atom or partial atom placed at positions r_i.

Fourier methods are known to suffer from termination effects which often increase the noise level and prevent the simultaneous representation of mobile and fixed ions. Partial Fourier synthesis can sometimes circumvent this problem. This involves replacing $F_c(Q)$ in (3.9) by $F_c'(Q)$ where F_c' is a calculated structure factor with the features of interest (such as partial atom at r_i) being omitted so they will appear in the $\Delta\rho(r)$ map. This will be illustrated in Sect.3.2.1 for AgI.

3.1.3 Inelastic Scattering

The second term in (3.4) is the dynamical structure factor for coherent inelastic scattering,

$$g_p(\underline{Q}) = \sum_j \frac{b_j(\underline{Q} \cdot \underline{\xi}_{j,p}^{\,\prime})}{(2NM_j\omega_p/\hbar)^{\frac{1}{2}}} \exp(i\underline{Q} \cdot \underline{R}_j) \exp(-\underline{Q} \cdot \underline{\underline{\sigma}}_j \cdot \underline{Q}) \qquad . \tag{3.10}$$

By satisfying the delta function in (3.4) with a judicious choice of incident and scattered neutron momentum, the phonon dispersion curves can be measured. With the assumption of a lattice dynamical model the measured dispersion curves can be re-produced by using the interatomic forces as adjustable parameters. Measurements of this sort performed on β-AgI and PbF_2 will be discussed in Sect.3.3. In an anharmonic solid, or a highly defective one, the response is no longer a delta function at ω_p, but a broadened Lorentzian whose width Γ is inversely proportional to the lifetime. The separation of the contributions of anharmonicity and defect scattering to the linewidth is one of the difficulties in interpreting the inelastic results. Below, we shall discuss one attempt at separating the various contributions in the case of PbF_2.

The incoherent part of the scattering function gives information on the indivi-dual motion of a particle. Incoherent scattering arises from differences in scat-tering lengths due to the natural abundance of isotopes with different scattering lengths, or due to nuclear spin where the scattering length is different for the spin of the neutron and the nucleus being parallel or antiparallel. The most famous example of incoherence is that of hydrogen.

If an atom is diffusing randomly through a solid (or liquid) its random motion will increase the energy width of the scattered neutron beam so that the scattering will be considered as quasielastic. For random diffusion, and small momentum trans-fers, the intensity of the quasielastic incoherent scattering is a broadened Lorentzian

$$\left| F_{inc}(Q,\omega) \right|^2 = \frac{DQ^2/\pi}{\omega^2 + DQ^2} \quad , \tag{3.11}$$

with full width at half maximum, $\Gamma = 2DQ^2$. D is the diffusion constant and in super-ionic conductors is $\sim 10^{-5}$ cm^2/s, comparable to the diffusion constant in a liquid. By a measurement of the Q dependence of the linewidth, information about the rest time and jump distance of the diffusion process can be determined . Equation (3.11) is the result for a simple random motion which could be translational or rotational. In practice the situation is more complicated and we shall see from examples of measurements in AgI and NH_4 β-alumina.

3.2 Structural Studies

In this section we shall demonstrate how the techniques described in Sect.3.1 have
been used to determine the disorder in several superionic conductors. The proto-
typical systems that will be discussed are AgI, PbF_2, and β-alumina. It will also
be shown that, even though diffraction methods yield a time-averaged results, the
observed density $\rho(\underline{r})$ deduced is useful in arriving at dynamical models for the
ionic motion.

3.2.1 AgI

Silver iodide is the most studied superionic conductors because it has one of the
highest ionic conductivities ($\sigma = 1.3 \ \Omega^{-1}cm^{-1}$ at T = 418 K) [3.13]. On cooling,
AgI undergoes an abrupt first-order phase transition at $T_c \simeq 147^{o}C$ where the symmetry changes from the highly conducting cubic α phase, where the Ag^{+} ions are
highly disordered, to the low-temperature normally conducting wurzite (β) phase.

The structure of β-AgI is hexagonal wurtzite as determined from early X-ray
studies on both powders and single crystals [3.14-16]. However, many uncertainties
about the details of the structure remained. HELMHOLTZ [3.16] concluded that the
Ag^{+} ion was displaced from the ideal wurtzite structure. From superior data BURLEY
[3.17] concluded that the β-AgI retained the ideal wurtzite structure. In the most
recent structural study using neutron diffraction CAVA et al. [3.18] were unable
to establish uniquely the structure despite the excellent quality of the data. They
do, however, obtain a structure consistent with Burley's determination.

In Table 3.1, the best parameters obtained from the refinement of CAVA et al.
[3.18] are shown. The R factors, which represent the quality of the fit to the data,
were expected to approach the 1% agreement obtained from intensities of symmetry-
related reflections, but clearly do not. The Fourier synthesis of the structure is
shown in Fig.3.1.

The Ag thermal parameters show that the mean square displacements (MSD) are isotropic at all temperatures. BÖHRER and BRÜESCH [3.19] propose that the MSD are
dominated by a low-lying optic mode they observed in their measurements of the phonon
dispersion curves. Because of the particular eigenvectors they assign to this mode,
U_{11} should be significantly larger than U_{33}, but as seen in Table 3.1, the U_{ii}'s
are nearly isotropic. BÖHRER and BRÜESCH also claim that the displacements of
this low-lying mode would favor the jump of an Ag ion into the large interstitial
volumes of β-AgI. However, the least squares refinement of the structure shows negligible occupation of the octahedral site at 000¼ which is the largest site in
β-AgI.

At T = $147^{o}C$, the first-order phase transition to the body-centered-cubic
(bcc) α phase occurs. The phase transition is strongly first order, and no precur-
sor effects have been observed. The mechanism of the transition is not well under-

Table 3.1. Lattice constants, positional and thermal parameters for β-AgI. Both Ag⁺ and I⁻ occupy position 2(b) 1/3 2/3 z of space group P6₃mc with $z_I = 0$. Estimated standard deviations in parenthesis. [3.18]

Temperature [°C]		23	80	140
Lattice constants a[Å]		4.599(3)	4.597(5)	4.598(3)
c[Å]		7.520(5)	7.520(10)	7.514(7)
z_{Ag}		0.628(1)	0.630(1)	0.630(2)
I⁻thermal parameters a	U_{11}	0.046(2)	0.056(3)	0.072(3)
	U_{33}	0.046(2)	0.051(3)	0.065(4)
Ag⁺thermal parameters a	U_{11}	0.082(4)	0.103(7)	0.130(8)
	U_{33}	0.068(5)	0.083(7)	0.112(7)
	R	6.5%	8.9%	8.8%

a For both atoms the factor for second-order thermal motion is
$$T = \exp\{-2\pi^2[(h^2 + 2hk + k^2)U_{11}a^{*2} + \ell^2 U_{33}c^{*2}]\},$$ where a^* and c^* are reciprocal lattice constants.

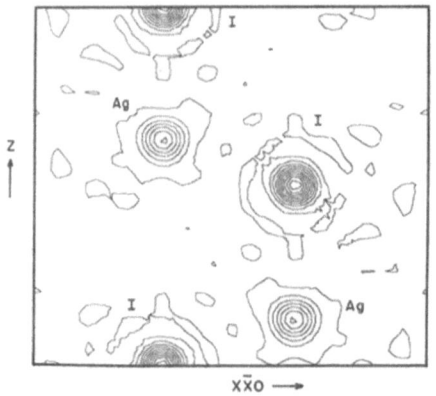

z ↑

x̄x̄o ⟶

Fig.3.1. Section ρ(xx̄z) of Fourier synthesis of the scattering density of β-AgI at room temperature

stood. BURLEY [3.17] showed that the (0001) plane of the β phase, the close-packed plane, transforms to the (110) plane, which is the most densely packed plane in the bcc α phase. The other (h k i 0) planes of the β phase were later shown to transform as [3.20]

β ↔ α
(0001) ↔ (011)
(10Ī0) ↔ (100)
(12̄30) ↔ (01Ī)

Quite rencently CAVA [3.21] even questioned the invariance of the close-packed plane (0001) ↔ (110). He showed that the rearrangement of the I atoms in AgI is similar to the rearrangement of atoms in a martensitic transformation of a metal undergoing a bcc → hcp transformation. In the latter system, the (110) plane of the bcc structure generally does not transform into the (0001) plane of the hcp structure, but into another plane whose orientation is close to it.

BOHRER and BROESCH [3.19], in their study of the lattice dynamics of the β phase, suggested that the low-lying mode at $\hbar\omega \sim 2.0$ meV plays an important role in the phase transformation. The atomic displacements of this mode contain the motions of the I ions which favor the reorganization of the iodine sublattice into the bcc lattice of the α phase. This mode could, therefore, be considered as the "soft" mode which nucleates the phase transition. No temperature dependence is seen because of the strongly first-order nature of the transition.

In the high-temperature α phase the Ag ions are disordered and the ionic conductivity approaches that of an ionic liquid. The first structural investigation of α-AgI was performed in 1934 by STROCK [3.22] using X-ray diffraction. He proposed that the mobile silver ions distribute themselves uniformly over the many sites shown in Fig.3.2. Numerous other X-ray studies on powders led to several different models of the disorder. Because of the difficulty in obtaining single crystals of α-AgI there were no diffraction studies on single crystals which could have provided data of sufficient quality to allow the unique determination of the disorder. Recently, techniques have been developed [3.21] which enabled single crystals to survive the transformation into the α phase. With these, a neutron diffraction study of α-AgI was undertaken and the precise measurements permitted a detailed representation of the structure and the Ag density [3.18]. This structure analysis basically confirmed the STROCK [3.22] model because significant Ag density was found at the accessible interstices of the bcc array of iodine ions. A direct Fourier analysis of the observed structure factors (3.8) gives little information about the Ag ion distribution because $\rho(\underline{r})$ is dominated by the iodine density as shown in Fig.3.3a. With a partial Fourier synthesis using (3.9), the silver density can be represented in detail as shown in Fig.3.3b for $T = 160^{\circ}$C. This section, which represents a face of the cube, contains all of the possible interstices among the bcc I⁻ array whose size allows for silver occupancy. The equilibrium site is clearly the tetrahedral site ¼ 0 ½ . It is evident that the equivalent octahedral sites at ½ 0 ½ and ½ 00 are local minima in the density. Appreciable density appears near 0.4 0 0.4 which is a trigonally coordinated site bridging two tetrahedral sites and displays the characteristics of a saddle point. Using a least squares refinement, the data are best represented by placing an Ag⁺ ion at the ¼ 0 ½ site with anisotropic MSD, $U_{11} = 2U_{22}$ and $U_{33} = U_{22}$ and anharmonic third- and fourth-order parameters. The results are given in Table 3.2.

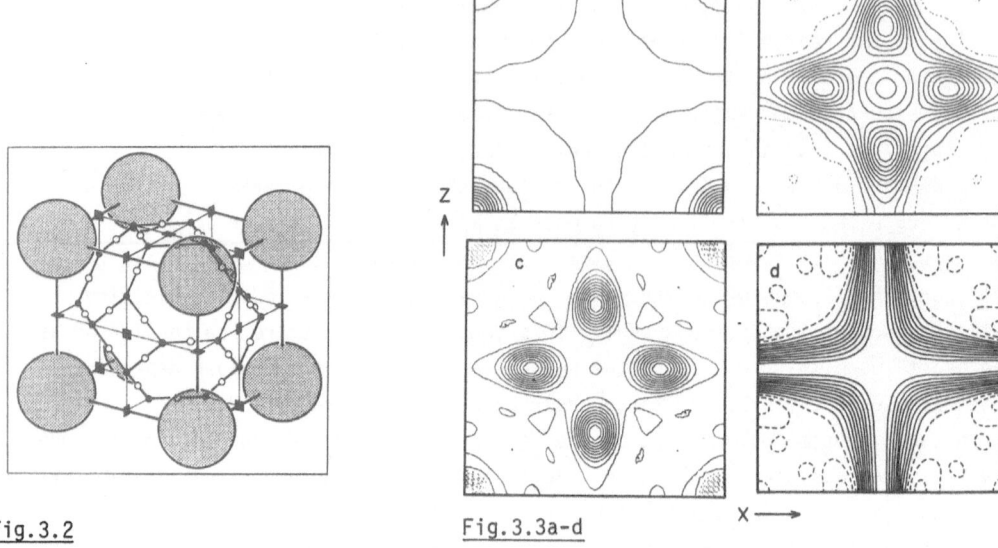

Fig.3.2   Fig.3.3a-d

Fig.3.2. Unit cell of cubic α-AgI. The large shaded circles are iodine atoms and the open and closed symbols are the positions of the silver atoms allowed by symmetry according to STROCK [3.22]. The solid circles are the tetrahedral sites occupied by the mobile Ag ions as shown in Fig.3.3b

Fig.3.3a-d. Sections $\rho(x0z)$ of Fourier synthesis of the scattering density in (a) α-AgI at T = 160°C showing mostly the iodine density [3.18]. (b) α-AgI at T = 160°C showing only the silver density [3.18]. (c) $VD_{0.75}$ at T = 127°C showing the deuterium density [3.23]. (d) in Ag_2S at T = 300°C showing only the silver density [3.24]

The interesting feature of the Ag densities is that there is a continuous density between the tetrahedral sites. This is in contrast to the isomorphic compound vanadium deuteride ($VD_{0.79}$) where the V atoms form the bcc lattice for the deuterium atoms. The $\rho(\underline{r})$ for $VD_{0.79}$ at 125°C is shown in Fig.3.3c [3.23]. Here the D atoms are strongly localized within the tetragonal site with very little density in between. On the opposite extreme the Ag^+ density map of Ag_2S (Fig.3.3d) shows that the silver density is not localized at all [3.24]. In this case, the Ag ions look like a "lattice liquid."

Let us now consider some models of the ionic motion arrived at from these structural studies. If we consider an ion as a free oscillator, the relation between the eigenfrequency ω_i and the U_{ii} for a classical oscillator is given by

$$m\omega_i^2 U_{ii} = k_B T \quad . \tag{3.12}$$

In the infrared and Raman spectra of AgI [see Chap.5] a broad peak is observed at $0.7 \cdot 10^{12}$ Hz (= 2.9 meV). If we identify this mode as the local oscillation of the

Table 3.2. Lattice constants and thermal parameters for α-AgI. I^- and Ag^+ ions occupy, respectively, positions 2(a) 0 0 0 and 12(d) ¼ 0 ½ of space group Im3m. Estimated standard deviations in parentheses. [3.18]

Temperature [$^{\circ}$C]	160	200	240	300
Lattice constant a[Å]	5.067(6)	5.069(4)	5.079(8)	5.079(4)
Scale factor	1.196(16)	0.766(25)	0.749(24)	1.186(25)
Extinction parameter, G	0.019(1)	0.019(4)	0.019(4)	0.029(4)
I^- thermal parameters [a] $B[Å^2]$	7.83(7)	8.49(20)	9.19(21)	10.32(13)
$D_{1111} \times 10^4$	0.075(7)	0.031(23)	0.082(25)	0.115(7)
$D_{1122} \times 10^4$	0.019(2)	0.22(8)	0.032(9)	0.027(6)
Ag^+ thermal parameters [b] U_{11}	0.287(4)	0.316(9)	0.342(12)	0.367(5)
U_{22}	0.152(1)	0.170(5)	0.192(7)	0.198(3)
$C_{122} \times 10^3$	-0.340(17)	-0.375(51)	-0.491(61)	-0.423(23)
$D_{1111} \times 10^4$	1.05(23)	1.16(83)	2.40(129)	-
$D_{2222} \times 10^4$	0.18(4)	-	-	-
$D_{1122} \times 10^4$	0.16(3)	-	-	-
R_w	1.44%	2.29%	2.70%	1.44%
R	1.22%	1.75%	2.30%	0.86%

[a] For I^-: $T = \exp\{-B \sin^2\theta/\lambda^2\}$; $c_{jk\ell} \equiv 0$; $D_{1111} = D_{2222} = D_{3333}$, $D_{1122} = D_{1133} = D_{2233}$, all other $D_{jk\ell m} \equiv 0$.

[b] For Ag^+: $T = \exp\{-2\pi^2[h^2U_{11} + (k^2 + \ell^2)U_{22}]a^{*2}\}$; $C_{122} = -C_{133}$; all other $C_{jk\ell} \equiv 0$; $D_{2222} = D_{3333}$, $D_{1122} = D_{1133}$. Except for D_{1111} and D_{2233} other $D_{jk\ell m} \equiv 0$. Omitted values were not significant and were excluded from the refinement.

Ag ions in AgI, then the Ag ions can be treated as classical oscillators since this energy is much less than k_BT (T = 420 K). The values of U_{ii} obtained from (3.12) are 0.17 $Å^2$ and 0.22 $Å^2$ at T = 160°C and 300°C, respectively. These are very close to the averaged values of 0.197 $Å^2$ and 0.254 $Å^2$ obtained from Table 3.2. This close agreement indicates that the Ag-Ag interaction and the general disorder do not affect the oscillatory behavior of the Ag ions. Additionally, we assume that during the diffusive motion, the Ag ions can be treated as free particles in thermal equilibrium with the lattice. We can then use (3.6) to relate the $\rho(\underline{r})$ at a saddle point ($\underline{r}_s$) to the energy V_s needed to promote an Ag^+ ion to a saddle point,

$$\rho(r_s) = \rho(0) \exp(-V_s/k_BT) \quad .$$

Noting that two tetrahedral sites and four octahedral sites contribute to the density at r_s, we can calculate the activation energies of the tetrahedral and octahedral sites to be $V_T = 0.06$ eV and $V_0 = 0.12$ eV, respectively. The activation energy obtained from diffusion constants measurements is $V_D = 0.06$ eV [3.25] which is in close agreement with V_T. From this we conclude that the Ag^+ motion between the tetrahedral sites is the basic step in the diffusion process. The Ag ions spend a significant time between the tetrahedral sites, so that the diffusion should be considered as continous. This model differs from the earlier interpretation of the EXAFS data, but, in some ways, is qualitatively similar to the latest interpretation (Chap.2). The present model also differs from the model deduced from the quasi-elastic neutron studies (Sect.3.3.1). In these models, jump diffusion is assumed and the EXAFS and neutron scattering obtain different values of the rest time and transit time. The EXAFS study did not find a difference between the MSD of the α and β phases as shown in Tables 3.1,2. The model obtained from the quasi-elastic neutron scattering studies incorrectly predicts the translatory ionic motion as being along the [100] directions. A more detailed discussion of this model will be given in Sect.3.3.1.

To see how a jump diffusion model would appear in a Fourier synthesis, it is interesting to study $\rho(\underline{r})$ for $VD_{0.79}$ of Fig.3.3c. The jump diffusion model appears valid for deuterium in $VD_{0.79}$ because the deuterium density is strongly localized at the tetrahedral sites even though the diffusion constant and activation energy are very similar to that of AgI. This implies that the ionic motion is a very rapid jump among the tetrahedral sites spending a negligible amount of time in between them [3.26].

3.2.2 Fluorites

Compounds with the fluorite crystal structure exhibit high ionic conductivity at elevated temperatures and provide one of the simplest systems in which to study ionic motion. Measurements of the heat capactiy in several fluorites revealed a broad anomaly below the melting temperature which was interpreted as a melting of the anion sublattice [3.27]. Figure 3.4 shows the unit cell of the fluorite structure XY_2. It consists of cubic arrays of anions at ¼ ¼ ¼ with the metallic cations at every other body center. The disorder most likely involves the motion of the anions towards the empty body center sites at ½ ½ ½.

Several neutron diffraction studies on powdered samples of different fluorites have been performed but the various authors disagree as to the location of the interstitial ion. THOMAS [3.28] performed diffraction studies on CaF_2, BaF_2, and SrF_2, and DICKENS et al. [3.29] studied PbF_2 and $SrCl_2$. Both authors used exclusively a least squares refinement of the powder data consisting of 15 or so reflections. They found that the MSD of anions starts to increase anomalously at the temperature where the specific heat anomaly begins. These results, however,

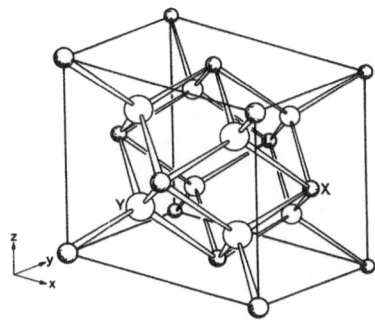

<u>Fig.3.4.</u> The fluorite structure of XY_2 compounds

were obtained from a model where the fluorines were fixed in their low temperature equilibrium sites. THOMAS [3.28] did allow the occupancy to vary for SrF_2 and found a 10% decrease of the population at the ¼ ¼ ¼ site but could not locate the interstitial fluorine. DICKENS [3.29] attempted to place an atom at the ½ ½ ½ site but found little improvement in the fit to their data.

AXE et al. [3.30] attempted a Fourier analysis of their powder data of BaF_2 and PbF_2 in an effort to avoid the assumptions of a particular model and the large number of parameters inherent in a least squares approach. The disadvantage of this method is that termination errors are severe and a cutoff function is necessary which further reduces the limited resolution. They found that near the melting temperature in both samples about 20% of the F^- ions leave the normal site and accumulate near ½ ½ ½. The precise position could not be determined because of the termination errors. A recent analysis of their data on PbF_2 and BaF_2 using a proper combination of Fourier analysis and least squares refinement was undertaken by REIDINGER et al. [3.31]. Possible interstitial sites were found by $\Delta\rho(\underline{r})$ maps (3.9) and their reality was established by a least squares analysis. If the quality of the fit improved significantly, occupation of the new site was considered a true effect.

This analysis yielded interesting results. Firstly, as shown in Fig.3.5a, the occupation of the normal fluorine site decreases with temperature. At the upper limit of the specific heat anomaly, $T \sim 625$ K, about 30% of the fluorines have left the ¼ ¼ ¼ site corresponding to a depletion rate of 0.3%/K. This rate decreases to 0.05%/K above 625 K. Near the melting temperature the depletion is about 60%. The MSD (Fig.3.5b) now shows a linear behavior with temperature in contrast to the other studies [3.28,29,32], with a change in slope near T_c.

REIDINGER et al. [3.31] were able to determine the location of the interstitial fluorine ions. The results of their analysis are given in Table 3.3 and in the one-dimensional plot of the fluorine density at $T = 775$ K shown in Fig.3.6. The dominant interstitial site is at 1/3 1/3 1/3 which is in the center of the face of a Pb tetrahedron. As the temperature is increased additional fluorine density begins to appear at the site (0.43, 0.43, 0.43) which is close to the expected octahedral site.

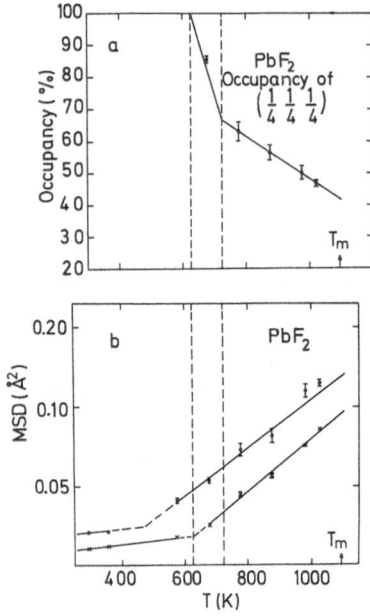

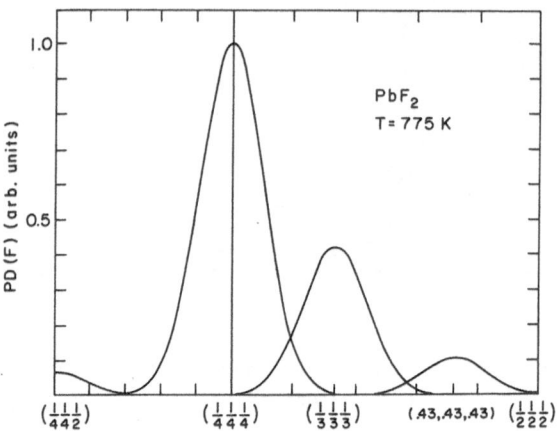

Fig.3.6. One-dimensional plot showing the fluorine density at the different interstitial positions for PbF_2 at 775 K [3.31]. This PD(F) is summed over all equivalent sites

Fig.3.5. (a) Occupancy of the ¼ ¼ ¼ fluorine site for PbF_2. The dashed vertical lines correspond to the temperature range where a specific heat anomaly is observed. (b) Temperature dependence of the mean square displacements of the fluorine atoms (●) and the lead atoms (x) for PbF_2 [3.31]

Occupation of this site was shown to occur in single crystals of CaF_2 containing excess fluorines [3.33]. A third site at ¼ ¼ ½, which is the center of an edge of Pb tetrahedron was found to be occupied only at 775 K and 871 K. The occupancies of these sites do not exhibit a simple temperature dependence when compared with the depletion of the main site. Also, the MSD of these sites were assumed to be the same as the main site since the limited number of reflections did not warrant the inclusion of extra Debye-Waller factors. Similar results were found in BaF_2 at 1327 K [3.33]. It will be interesting to have these measurements repeated for single crystals.

It is important to note that in PbF_2, the disorder is interpreted as occupation of new equilibrium sites. This implies that the pair potential changes as a function of temperature. The consequences are important for the molecular dynamics calculations [3.34] where the potential generally used is that of the ordered phase.

The results in the fluorites contrast with those of AgI in which the disorder can be explained by occupation of a single type of site, with the conducting ion undergoing large anisotropic vibrations.

Table 3.3. Temperature dependence of fluorine occupational parameters in PbF_2 [3.31]. Number in parentheses are errors in the least significant digits

Fluorine positions	Temperature[K]							
	296 [a]	344	573	673	775	871	977	1023
¼ ¼ ¼	1.008(5)	1.0	1.0	0.86	0.63	0.57	0.50	0.47
xxx x=0.43(1)	0.0	0.0	0.0	0.14(3)	0.261(2)	0.23(2)	0.34	0.38(1)
xxx x=0.43(1)	0.0	0.0	0.0	0	0.065(14)	0.12(1)	0.161(4)	0.151(2)
¼ ¼ ½	0.0	0.0	0.0	0.0	0.042(10)	0.078(8)	0.0	0.0

[a]Single crystal data

3.2.3 β-Alumina

Of all the superionic conductors, those of the beta alumina class are most important since they have already found technological uses in a variety of devices such as the soidum sulfur battery and the sodium heat engine [3.35]. There are significant differences between the beta alumina conductors and the Ag based and fluorite conductors discussed above. These are:

(I) All phases of the β-aluminas are nonstoichiometric.

(II) The anomalous mobility of the monovalent cation is constrained to two dimensions by the characteristic layered structure.

(III) The mobile ions constitute only about 4% of the total number of atoms.

(IV) Interaction between the mobile ions and the rest of the structure is weak and different mono- and divalent ions can serve as the conducting species.

(V) No well-defined phase transition occurs in these systems.

The most widely studied is sodium beta alumina which has the nominal chemical composition $Na_2O \cdot 11\ Al_2O_3$. The general features of this structure were solved by BRAGG et al. and BEEVERS and ROSS [3.36]. The structure is shown schematically in Fig.3.7a. The compound has hexagonal symmetry ($P6_3/mmc$) and is composed of two $Al_{11}O_{16}$ spinel blocks aligned along the c axis and separated by a sparsely populated mirror plane containing Na and O atoms in which the Na^+ ion is free to move. There are three sites in which the cation can be positioned and these are called the BEEVERS-ROSS (BR) site, the anti-BEEVERS-ROSS (aBR) site, and the mid-oxygen (mO) site, all shown in Fig.3.7b.

A careful study of the phase diagram [3.37] showed that β-alumina typically contains an excess of 20-35% of Na and that the phase of nominal composition does not exist. A good part of the X-ray and neutron diffraction studies have centered upon determining the type of compensating defects required to conserve electrical neu-

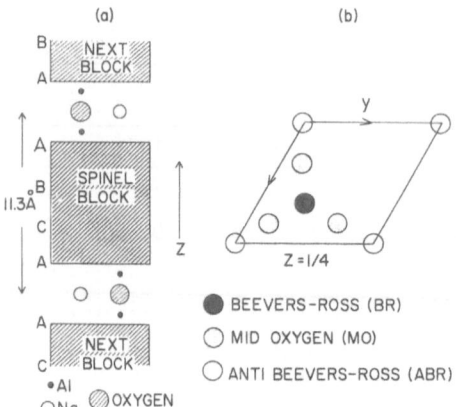

Fig.3.7a and b. Schematic structure
of the β-alumina system in the (a)
x-z plane and (b) the x-y conducting
plane

trality. The defect was thought to be either an Al^{3+} vacancy [3.38] or an interstitial oxygen [3.39]. Refinement of single crystal neutron data on Na β-alumina
[3.40] showed the existence of a complex defect in the spinel block. The defect has
the formula

$$Al_V - Al_I - O_I - Al_I - Al_V$$
$$\xrightarrow{\text{c-axis}}$$

where $Al_V - Al_I$ is a vacancy interstitial pair, i.e., a Frenkel defect. This defect
plays an important role in determining the Na disorder and the ionic conduction within
the conducting plane. Because the interstitial Al neutralizes the Al vacancy, excess
Na is compensated by the interstitial oxygen within the mirror plane. The model for
the disorder within the conduction plane consists of the interstitial oxygen and
four Na atoms near mO sites which are arranged in close pxoximity to the oxygen
[3.40]. From quantitative analysis of the diffuse X-ray scattering in K β-alumina
[3.41], however, it was concluded that the cell with two Na ions is more likely to
be a second nearest neighbor to the oxygen interstitial.

The disorder discussed above refers to the nonconducting state in which the Na
ions are distributed in the BR and mO sites with the aBR sites empty. If other
cations replace the sodium, the distribution among the three sites varies. Figure
3.8 shows the cation densities calculated from X-ray studies of Ag^+ [3.39], Na^+
[3.38], and K^+ [3.42] β-alumina. As the temperature increases in Na β-alumina, diffraction studies provide evidence that the mobile Na ions begin to spend some time
at the aBR site. Figure 3.9 shows the one-dimensional Fourier synthesis of the
sodium density at several temperatures. The transport of ionic charge proceeds when
an activated Na displaces a neighboring sodium from its equilibrium position into
an aBR site. This collective mechanism for diffusion was first proposed from the

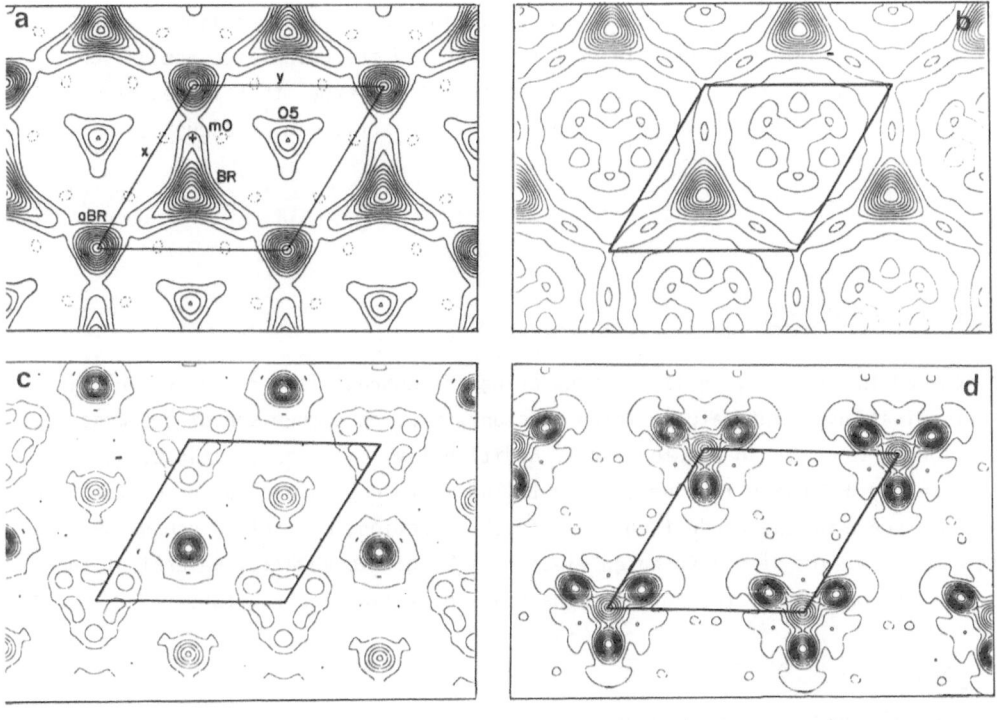

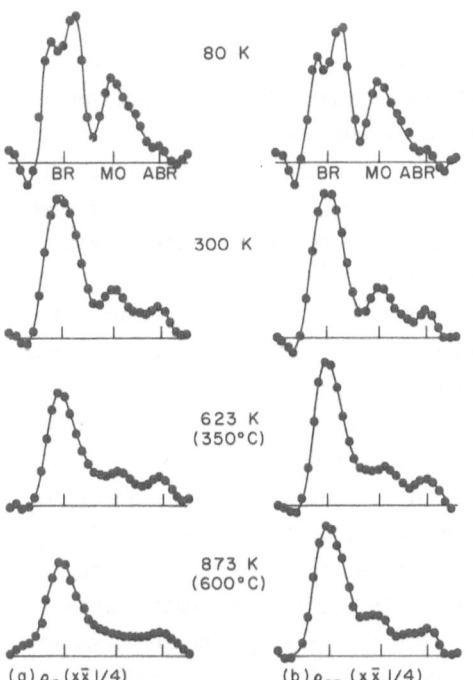

Fig.3.8a-d. Nuclear densities in the mirror plane for different β-aluminas obtained from Fourier synthesis of X-ray data. (a) Ag β-alumina showing major densities at the BR and aBR sites. The oxygen density at 0.5 is also shown [3.39]. (b) Na β-alumina showing densities at the BR and mO sites [3.38]. (c) K β-alumina showing mainly occupancy at the BR and O5 site [3.42]. (d) The same as (c) but showing only the smaller density about the aBR site

Fig.3.9. One-dimensional Fourier synthesis of the sodium density at different temperatures. $\rho_p(x\ \bar{x}\ 0\ \tfrac{1}{4})$ is a 1-D section through BR, MO and aBR sites. $\rho_{pp}(x\ \bar{x}\ 0\ \tfrac{1}{4})$ is a partial projection to clarify the density near the BR site at low temperature and the aBR site at elevated temperature

Haven ratio of 0.61 observed in Na and Ag β-alumina [3.43]. Calculations by WANG et al. [3.44] strongly support this mechanism because the calculated activation energy for single particle diffusion is a factor of 10 larger than the experimental value. Their calculation, however, proposes that the activated Na occupies the mO site.

3.3 Inelastic Studies

As discussed in Sect.3.1, inelastic scattering of neutrons is used extensively in the determination of interatomic force constants through comparison of the measured dispersion curves with an assumed model. Additionally, the quasielastic scattering of neutrons by diffusing ions provides detailed microscopic information on the time and length scale of the conducting ions. In this section, we shall discuss in-elastic measurements of the Ag-based electrolytes AgI and $RbAg_4I_5$, several of the fluorites and members of the beta-alumina class of solid electrolytes.

3.3.1 AgI

The phonon dispersion curves measured by BÜHRER et al. [3.19,45] on single crystals of the low-temperature phase of AgI are shown in Fig.3.10. The most striking feature of this dispersion curve is the very low lying dispersionless excitation near 2.0 meV (1 meV = 0.24 THz = 8.07 cm^{-1}) which is almost decoupled from the other vibrations. Strong anharmonicity and large Debye-Waller factors were re-ported. Additionally, no precursive effects related to the β-α phase transition have been observed. These curves were fitted to different models. The rigid ion model gave a very poor fit to the data (dotted line), but the valence shell model using four short range parameters, a variable ionic charge and mechanical and elec-trical polarizabilities for both ions gave a good fit to the data (solid line). The parameters obtained from the fit are given in Table 3.4.

Since the phase transition to the highly conducting cubic phase is abrupt and destructive for large crystals, no measurements have been performed in the α phase. However, time-of-flight studies have been performed for polycrystalline samples of AgI in both the β and α phase. In this type of experiment, one measures an average over momentum space and the spectra are proportional to the phonon density of states. The results of the measurements of FUNKE et al. [3.46] are shown in Figure 3.11. The β phase spectra, Fig.3.11a shows a pronounced peak at $\hbar\omega \sim 2.5$ meV, presumably due to a peak in the phonon density of states from the low frequency excitation observed in the single crystal studies (Fig.3.10). Upon heating into the conducting α phase, this feature has disappeared (Fig.3.11b). The spectrum now exhibits a broadened quasi-elastic central peak centered around $\hbar\omega = 0$ and a large

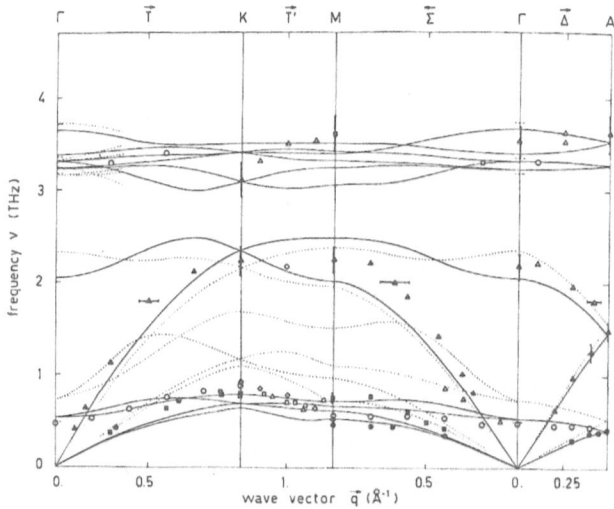

Fig.3.10. Dispersion curve of β-AgI. The dotted line is the calculated dispersion curve using a rigid ion model. The solid line is the calculated curve using a velence shell model with the parameters given in Table 3.4 [3.45]

Table 3.4. Valence shell model parameters of β-AgI [3.45]

	Stretch: Ag-I	$\lambda = 3.7 \times 10^4$ dyn/cm
	Stretch: I $-$ I	$N = 4.5 \times 10^3$ dyn/cm
Short range	Bend Ag/bend: I	$K_{\theta_1} = K_{\theta_2} = -8.5 \times 10^2$ dyn/cm
	Stretch-bend: Ag	$K_{r_{\theta_1}} = 0$
	I	$K_{r_{\theta_2}} = -7 \times 10^2$ dyn/cm
	Ionic charge:	$Z = 0.56e$
Long range	polarizabilities: Ag	$\alpha_1 = 1.0 \text{ Å}^3$
		$d_1 = 0.015e$
	I	$\alpha_2 = 3.0 \text{ Å}^3$
		$d_2 = 0.050e$

tail extending out to 6.0 meV. This latter feature is consistent with the optical studies where a broad band of low frequency modes have been observed (see Chap.5). The narrower quasi-elastic scattering is believed to arise from the translational diffusion and has been studied in greater detail using higher resolution in order to obtain detailed information about the diffusion mechanisms [3.47-49]. In these studies, at small scattering angles (small momentum transfers) only a narrow quasi-elastic line is observed which is slightly broader than their resolution (Fig.3.12a). At larger angles, a second, broader component becomes observable in

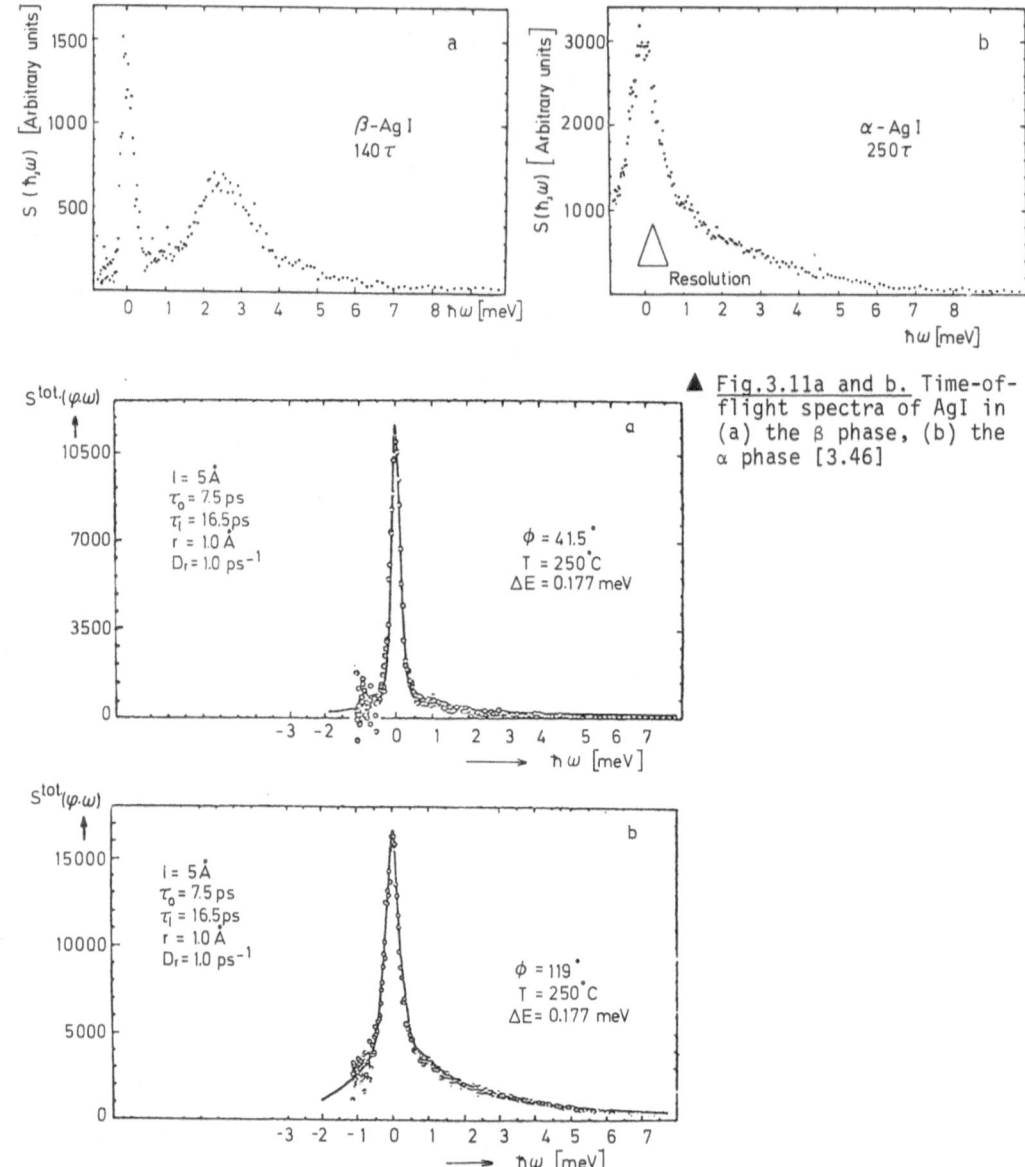

Fig.3.11a and b. Time-of-flight spectra of AgI in (a) the β phase, (b) the α phase [3.46]

Fig.3.12a and b. High resolution time-of-flight spectra of α-AgI taken at T = 250°C and two different scattering angles (a) φ = 41.5° (b) φ = 119°. The parameters obtained from a fit to a translational-rotational diffusion model are also given [3.47]

addition to the broadening of the quasi-elastic part. FUNKE et al. [3.46,48] have interpreted these results in terms of a local-plus-translational diffusion of the cations. The narrow peak is associated with the long time (τ_1) translational motion and the broader peak with the shorter time (τ_0) local motion. On this basis the cation motion in α-AgI was described as follows. Each silver ion is confined to a cage determined by the iodine sublattice. Within the cage the cation is performing a local motion for a time τ_0 while neighboring silver ions are blocking the exits of this cage. If one of the blocking silver ions moves away, the cation can exit the cage and undergo diffusive motion for a time τ_1 in a channel connecting one cage to another cage. The parameters obtained from this model are shown in Fig.3.12. The distance ℓ is the jump distance from cage to cage and the value is very close to one lattice parameter. The radius of the local diffusion region r is the size of the cage centered at an octahedral site within which the local motion is performed.

It should be noted that unique values of these parameters could not be obtained from the neutron scattering experiment alone. The final choice depended upon the assumption that the observed maximum in the microwave spectrum of α-AgI corresponds to the jump time τ_1 [3.47]. If this is fixed, then the other parameters in Fig.3.12 are obtained. Difficulties arise in this model in light of the diffraction studies discussed in Sect.3.2.1. They show that the jumps between near neighbor tetrahedral sites separated by 1.79 Å is the basic step in the diffusion process whereas the present model gives the jump distance between second neighbor octahedral sites as the basic step.

A major feature in the single-crystal work in the β phase is the low-lying branch near 2.0 meV which disappears in the α phase (Fig.3.11). This was also noted in the Raman scattering experiments (see Chap.5). BÜHRER and BRÜESCH [3.19] suggest that this mode favors the promotion of the cation to an interstitial site and is essential for the occurence of the cation order-disorder transition. There has, however, been little discussion of its disappearance in the α phase. Interpretation of the high temperature spectra of AgI in terms of different models is given in Chaps.4 and 8.

3.3.2 RbAg$_4$I$_5$

RbAg$_4$I$_5$ is a unique ionic conductor in that it has the highest conductivity at room temperature. The 16 conducting silver ions in a unit cell distribute themselves over 56 allowed sites. At T_2 = 208 K, there is a phase transition from one type of disorder of the Ag$^+$ ions to another with only a slight change in conductivity. At T_1 = 122 K, there is a discontinuous reduction of $\sim$100 in the ionic conductivity corresponding to a more ordered state of the Ag$^+$ cations. Early ultrasonic studies showed a singularity in C_{44} at T_2 [3.50]. Neutron scattering studies [3.51] showed no change in the elastic constant which implies that very long wavelength strains play an important role in the transition.

A search for a soft optic mode or some dynamical effect related to the phase transition was made [3.51]. At room temperature, a broad feature centered around $\hbar\omega$ = 0 with a 6.0 meV full width at half maximum was observed at every point in Q space studied (Fig.3.13). The intensity appeared to increase with momentum transfer. As the temperature is lowered a propagating mode becomes visible with $\hbar\omega \sim 2.8$ meV. Figure 3.13 shows spectra taken at $Q = (0, 5.7, 5.7)$, a position in Q space where the intensity of the acoustic mode is weak. No critical change in linewidth or intensity occurs at either T_1 or T_2. Instead, the behavior depends more upon the absolute temperature than on the temperature differences. Because this feature is observable at several unrelated Q values, we interpret this as an Einstein-like excitation related to the Ag^+ ions moving in an uncorrelated fashion. GEISEL [3.52], (see also Chap.8), has shown that this temperature-dependent behavior is predicted from a simple model of Brownian motion of the diffusing ion moving in a sinusoidal potential well. Figure 8.12 shows calculated spectra similar to those observed in Fig.3.13.

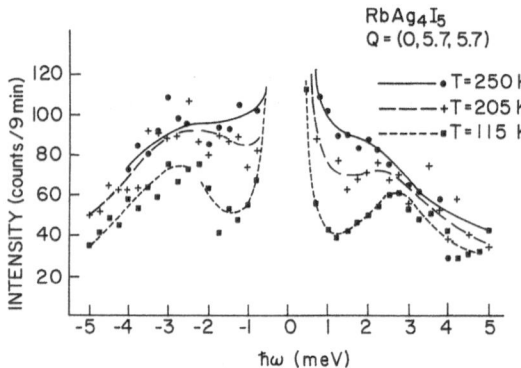

Fig.3.13. Inelastic spectra of RbAg$_4$I$_5$ at several temperatures [3.51]

3.3.3 Fluorites

As discussed in Sect.3.2.2, the fluorite system may very well be the ideal system to study ionic disorder and the dynamics of the ionic motion since it is structurally very simple and large single crystals, with generally favorable neutron properties, are readily available. The systems most extensively studied are BaF_2, PbF_2, and $SrCl_2$. PbF_2 has the largest conductivity of the fluorite group and the disordering of the anion sublattice occurs at the lowest temperature. Consequently, we shall discuss in more detail the experiments performed on this material.

The complete dispersion curves of BaF_2 and $SrCl_2$ were previously measured by inelastic neutron scattering and are discussed in detail by HAYES and STONEHAM [3.53]. A rigid-ion model of the short range interatomic forces gives a good fit to the acoustic modes; for the optic modes it was necessary to include the polarizability of the ions by means of a shell model or a polarizable-ion model. These measurements were performed prior to the recent emphasis on the large ionic conductivity in these materals. Recently, DICKENS and HUTCHINGS [3.54] measured the complete dispersion curves of PbF_2 at 10 K and their results are shown in Fig.3.14a.

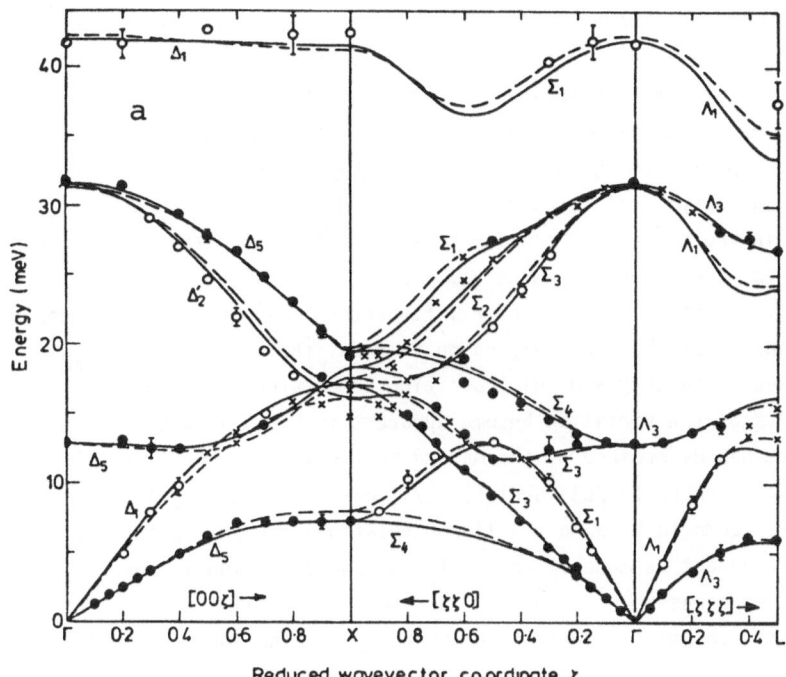

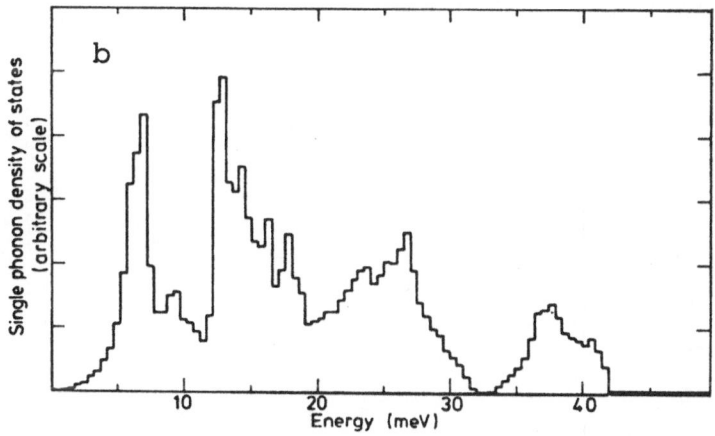

Fig.3.14. (a) Dispersion curve of PbF_2 measured at $T = 10$ K. Closed (open) circles are predominantly transverse (longitudinal) modes. The dotted curve is the calculated dispersion curve using a shell model with 8 parameters and the solid curve uses 13 parameters [3.54]. (b) The calculated phonon density of states for PbF_2 at $T = 10$ K using the parameters obtained from a fit to the dispersion curve [3.54]

68

This low temperature was chosen because at room temperature there is significant broadening of the optic modes due to anharmonic effects. The q = 0 modes are in good agreement with the more precise optical measurements (see Chap.5). The curves were fit to a shell model containing 13 adjustable parameters and the results of two slightly different models are shown in Fig.3.14a. From the determined force constants, the phonon density of states can be calculated and this is shown in Fig.3.14b. The peaks in the calculated density of states correspond to flat portions of the dispersion curve, i.e., regions where $\nabla_q \omega = 0$.

In comparing the dispersion curves of several fluorites, the striking difference is the variation in energy of the lowest-lying optic mode, the TO mode. In PbF_2 this mode has the lowest frequency and the transition temperature is likewise lowest. The systematics of the variation become apparent in Fig.3.15 where the transition temperature, as determined by the maxima in the specific heat anomaly, is plotted against ω_{TO} for PbF_2 [3.55], $SrCl_2$ [3.27], BaF_2 [3.56], SrF_2 [3.56], and CaF_2 [3.27]. There seems to be almost a linear relation between ω_{TO} and T_c. Also plotted is the Raman active mode ω_R versus T_c and the variation is less systematic. The Raman mode has been extensively studied, especially in PbF_2 [3.57]. This mode was considered the most important in understanding the ionic conductivity since it corresponds to the anion sublattice vibrating against the stationary cations. However, on the basis of the linear relation of ω_{TO} versus T_c, it appears that the TO mode could be more significant and suggests that the cations also play an important role in governing the anion disorder.

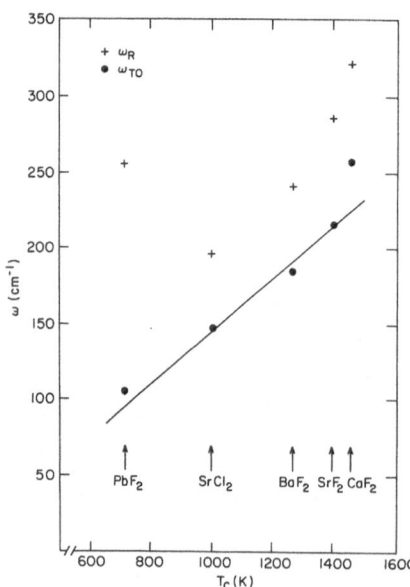

Fig.3.15. Variation of the frequencies of the Raman active mode (ω_R) and the TO mode (ω_{TO}) with the transition temperature T_c for several ionic conductors. T_c is taken as the maximum in the heat capacity

The major change that occurs in the phonons on heating the crystal is a large
increase in the phonon linewidth. The difficulty in interpreting the spectra is
the separation of the various contributions to the linewidths. They are:
(I) anharmonic effects; (II) dynamic defect hopping; and (III) time-averaged
static disorder. ELLIOTT et al. [3.57] attempted this separation for the Raman
studies and showed that near T_c, defect-induced scattering plays a major role.
DICKENS et al. [3.58] employed a similar process in the study of the acoustic
modes of PbF_2 at temperatures from 10 K to 900 K. Figure 3.16 shows the temperature
dependence of the transverse acoustic modes propagating along the [0 0 1] direction
with polarization along [1 1 0]. This mode, corresponding to the elastic constant
C_{44}, exhibits a decrease in frequency and a broadening of the line as the temperature increases. The linewidth was also studied as a function of q and it was found
that at each temperature the linewidth, Γ, was proportional to q^2 which suggests
that the phonons are being measured within the hydrodynamic regime. The slope,
$d\Gamma/dq^2$, is plotted as a function of temperature in Fig.3.17. This curve gives an
average variation of the linewidths of the acoustic phonons as a function of
temperature. For temperatures less than 600 K, the curve is linear implying that
anharmonic effects dominate the damping of the acoustic phonons; while at higher
temperatures, defect processes begin to contribute significantly. This is the same
temperature region where the specific-heat anomaly occurs. Similar conclusions are
obtained from the temperature variation of the elastic constants [3.58-60].
DICKENS et al. [3.58] performed a quantitative analysis of their data based on
the method of ELLIOTT et al. [3.57] and found that the relaxation of the phonon
due to the hopping ions [process (II)] contributes little to the broadening. The
major contribution to the linewidth above 600 K is the static disorder of the anion
sublattice, which, as we have seen, becomes large at these elevated temperatures.

3.3.4 β-Alumina

As is evident from the structural studies, the β-alumina compounds are some of the
more complicated superionic conductors. This complicates any microscopic understanding of the lattice dynamics of these systems. The zone-center lattice dynamics have
been extensively studied by optical techniques and they are discussed in Chap.5.
The major result was the identification of a low frequency peak in the ir conductivity which was dependent upon the kind of conducting cation [3.61]. It was also
shown that the optical spectra were sensitive to the method of crystal growth [3.62].

Inelastic neutron scattering experiments were performed on large single crystals
of Na^+ and Ag^+ β-alumina which were grown from the melt [3.63]. The low-frequency
part of the dispersion curve for modes propagating within the conduction plane is
shown in Fig.3.18 for Na β-alumina. Apparent are the acoustic modes and some optic
modes with strong dispersion. A dominant feature is the flat, almost dispersionless
branch with energy $\hbar\omega \sim 6.0$ meV. This energy corresponds closely to the local fre-

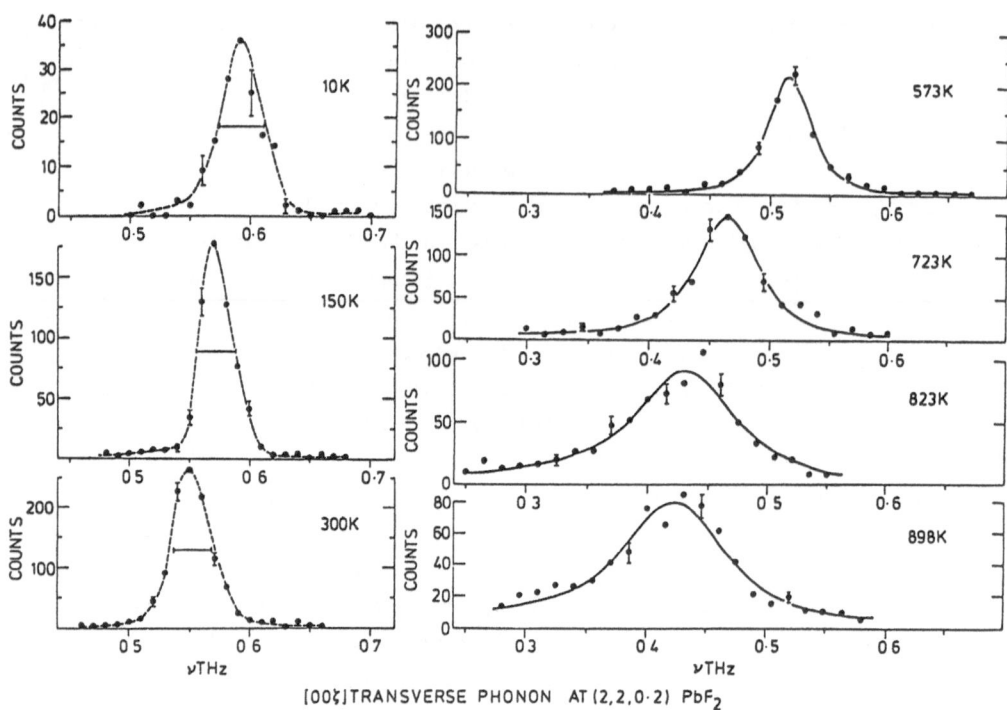

<u>Fig.3.16.</u> Neutron groups for typical constant-Q scans for PbF$_2$ of the [001]-TA mode measured at Q = (2, 2, 0.2). The horizontal bars for the lowest three temperatures denote the calculated instrumental limited width. [3.58]

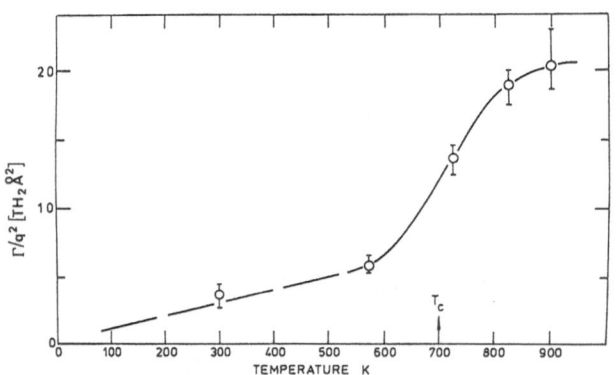

<u>Fig.3.17.</u>
Variation of the phonon linewidth divided by q^2 for PbF$_2$ [3.58]

quency of the Na$^+$ ion as observed in the infrared studies [3.61]. By comparing the intensity of this mode with the acoustic modes, it is apparent that only the Na ions are contributing to the intensity, which accounts for only ~4% of the total atoms. The dispersionless character of the branch is convincing evidence that this mode is a localized excitation with little correlation among the different

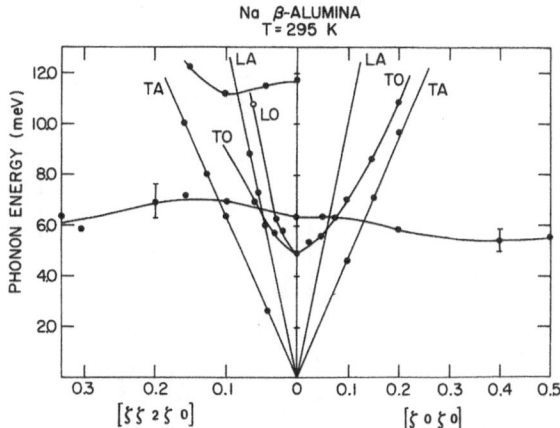

Fig.3.18. Low-frequency dispersion
curve of Na β-alumina taken at
room temperature [3.63]

atoms. This is consistent with the nature of a localized vibration of the cation in
a potential well set up by the rest of the lattice.

From the optical studies [3.61] it is known that the local frequency of Ag
β-alumina is 2.0 meV, less than that of Na β-alumina. The results of the inelastic
scattering Ag β-alumina are shown in Fig.3.19. There is no well-defined peak around
2.0 meV, but a broad overdamped excitation whose intensity decreases in temperature
following the Bose-Einstein thermal function. The optical studies also show that
the linewidth of the excitation in the Ag β-alumina is less than in Na β-alumina. In
the neutron measurements because of the overdamped nature of the mode, it appears
that the damping is still quite large. This could be due to the fact that the neutron
measurements are probing a different region in Q space than the optical probes.

The elastic scattering of neutrons by Ag β-alumina displays interesting behavior
on cooling. Figure 3.20 shows diffuse scattering measured between [11$\bar{2}$0] and
[22$\bar{4}$0] at several temperatures. The appearance of this diffuse scattering originates
from the disorder of the Ag ions and has been extensively studied by X-ray scattering
[3.42,64]. Near room temperature the spectrum is broad, but as the temperature is
lowered a sharp peak near $Q = (1/3, 1/3, 2\overline{/3}, 0)$ develops. This suggests an ordering
of the Ag^+ ions in a superstructure of $\sqrt{3}a \times \sqrt{3}a$ lattice. A major difference between
the neutron and the X-ray experiment is that a sharp diffraction spot is observable
in the X-ray studies at room temperature and none is observed in the neutron scattering. In the latter, only neutrons with energies less than the instrumental resolution
~1.0 meV are detected whereas by the nature of the X-ray experiment, an integration
is made over all energies. Thus the peak observed in the X-ray experiment at room
temperature must be dynamical in origin with energy greater than 1.0 meV. Only at
low temperatures does a truly elastic component develop.

The transverse acoustic modes propagating along the c direction, and polarized
perpendicular to the plane, have been measured on small single crystals of beta
alumina with Na^+, Ag^+, K^+, and Rb^+ as the conducting ions [3.65]. The results are

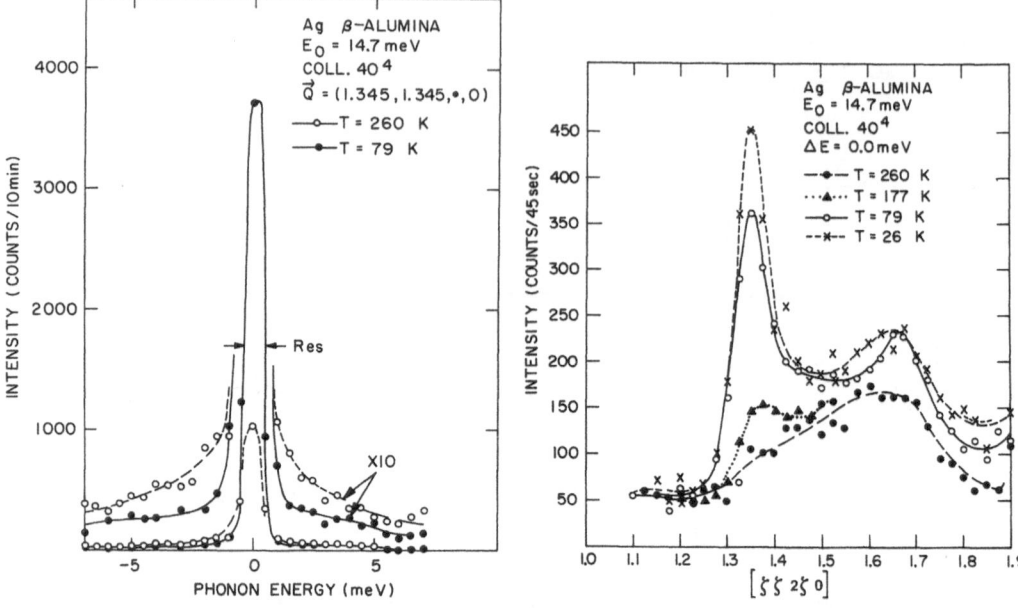

Fig.3.19. Inelastic spectra for Ag β-alumina taken at T = 79 K and T = 260 K [3.63]

Fig.3.20. Elastic ΔE = 0 scan in Ag β-alumina for several temperatures [3.63]

shown in Fig.3.21 along with the calculated elastic constant C_{44}. These results show a systematic increase in C_{44} with increasing size of the diffusing ion from Na → Ag → K → Rb. This is reasonable if we recall that this mode corresponds to a shear motion of the spinel blocks which is a "rolling" of the two spinel blocks over the mirror plane. The larger the cation the more difficult this "rolling" becomes and the larger the energy of the mode.

The most interesting aspect of the understanding the ionic transport is to probe the diffusive motion of the cations as was done for AgI by quasi-elastic scattering. The macroscopic measurements suggest that at 300°C, $D \sim 2 \times 10^{-5}$ cm^2/s which should be measurable in a quasi-elastic neutron scattering experiment. A recent experiment on Na β-alumina utilizing high energy resolution spectroscopy measured a linewidth considerably smaller than the macroscopically measured D would predict [3.66].

Diffusion constant measurements were performed on NH_4^+ β-alumina powders [3.66]. The idea behind this experiment was to use the large incoherent scattering of the hydrogen to make it possible to follow the NH_4^+ diffusion. A spectrum is shown in Fig.3.22, where two components are observed: a sharp resolution limited peak at ΔE = 0 sitting atop a broader Lorentzian quasi-elastic component. The latter corresponds to the jump reorientation of the NH_4^+ ion, from whose linewidth at room temperature a mean resident time $\tau_r = 6.6 \times 10^{-13}$ s was obtained. The narrow

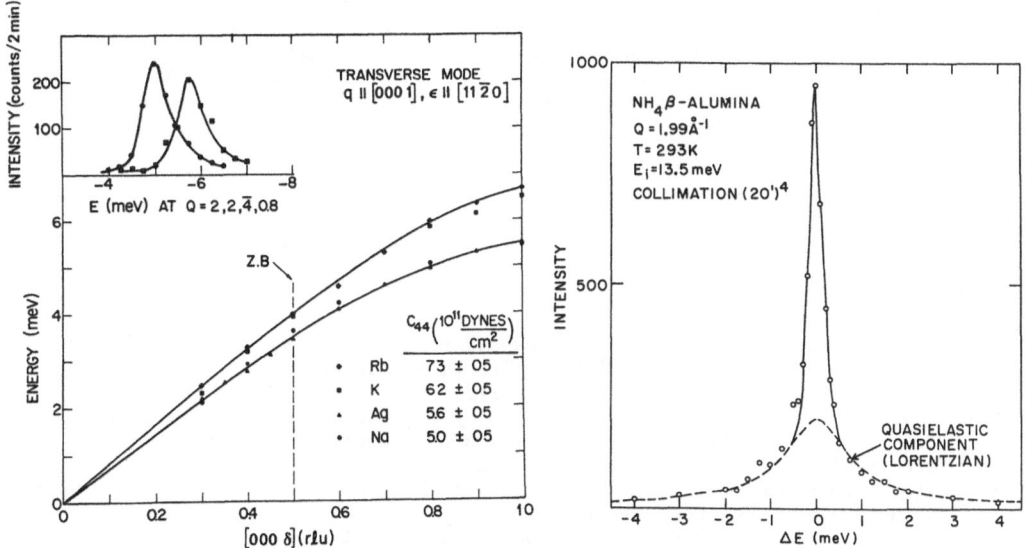

Fig.3.21. Variation of the transverse mode with $q_\parallel$ [0 0 0 1] and polarization $\hat{e}_\parallel$ [1 1 $\bar{2}$ 0] for different diffusing ions substituted in β-alumina. The inset shows typical energy loss spectra for Na (circles) and K (squares) at Q = (2,2,$\bar{4}$,0.8). [3.65]

Fig.3.22. Inelastic scan for NH_4 β-alumina showing the elastic and quasi-elastic components [3.67]

elastic component corresponds to a translational diffusion of the NH_4^+ which is the quantity of interest for the ionic conductivity. Unfortunately, the diffusion is too slow to be measured by neutron scattering and only an upper limit on the diffusion constant of $D \leq 4 \times 10^{-6}$ cm/s at 473 K can be deduced.

3.4 Conclusions

Neutron and X-ray diffraction techniques provide fundamental input in characterizing the disorder in SI conductors. They yield a time-averaged picture of the ionic densities which is basic knowledge needed to construct models explaining the ionic transport. The lattice dynamics of SI conductors yield information about the rigid framework as well as the important local vibrational modes of the conducting ions. The fundamental diffusive aspects of the ionic motion can be probed by quasi-elastic neutron scattering where the energy width is proportional to the diffusion constant. As more SI conductors are discovered, the diffraction and scattering techniques discussed above will be necessary to probe the microscopic nature of these most interesting and valuable materials.

Acknowledgements. We are indebted to our collaborators in the studies reviewed here: J.D. Axe, R.J. Cava, D.B. McWhan, J.J. Reilly, W.L. Roth, M.B. Salamon, D. Semmingsen, G. Shirane, N. Wakabayashi, and B.J. Wuensch. We also thank W. Bührer, M. Dickens, and R. Lechner for sending us preprints of their work. Research was supported by the Division of Basic Energy Sciences, Department of Energy, under Contract No. EY-76-C-02-0016.

References

3.1 T. Springer: *Quasielastic Neutron Scattering for the Investigation of Diffusive Motions in Solids and Liquids*, Springer Tracts in Modern Physics, Vol.64 (Springer, Berlin, Heidelberg, New York 1972)

3.2 S.W. Lovesey, T. Springer (eds.): *Dynamics of Liquids and Solids by Neutron Scattering*, Topics in Current Physics, Vol.3 (Springer, Berlin, Heidelberg, New York 1977)

3.3 L. Koster: *Neutron Physics*, Springer Tracts in Modern Physics, Vol.80 (Springer, Berlin, Heidelberg, New York 1977)

3.4 H. Dachs (ed.): *Neutron Diffraction*, Topics in Current Physics, Vol.6 (Springer, Berlin, Heidelberg, New York 1978)

3.5 P. Coppens: In [Ref.3.4, p.71]

3.6 S.W. Lovesey: In [Ref.3.2, p.127]

3.7 H. Dachs: In [Ref.3.4, p.1]

3.8 M.J. Cooper, K.D. Rouse, B.T.M. Willis: Acta Crystallogr. A *24*, 484 (1968)

3.9 C.K. Johnson: Acta Crystallogr. A *25*, 187 (1969)

3.10 C.K. Johnson, H.A. Levy: *International Tables for X-Ray Crystallography*, Vol.4 (Kynoch, Birmingham 1974)

3.11 C.K. Johnson: Acta Crystallogr. A *33*, 879 (1977)

3.12 W. Busing, H.A. Levy: ORNL Report 59-4-39 (Oak Ridge National Laboratory, Oak Ridge, TN 1959)

3.13 C. Tubandt, E. Lorenz: Z. Phys. Chem. (Leipzig) *87*, 513 (1914)

3.14 G. Aminoff: Z. Krist. *57*, 180 (1922)

3.15 N.H. Kolkmeijer, J.W.A. Van Heugle: Z. Krist. *88*, 317 (1934)

3.16 L. Helmholtz: J. Chem. Phys. *3*, 740 (1935)

3.17 G. Burley: J. Chem. Phys. *38*, 2807 (1963)

3.18 R.J. Cava, F. Reidinger, B.J. Wuensch: Solid State Commun. *24*, 411 (1977)

3.19 W. Bührer, P. Brüesch: Solid State Commun. *16*, 155 (1975)

3.20 R.J. Cava, B.J. Wuensch: In *Superionic Conductors*, ed. by G.D. Mahan, W.L. Roth (Plenum Press, New York 1976) p.217; Acata Crystallogr., submitted (1978)

3.21 R.J. Cava: Thesis (Department of Materials Science and Engineering, MIT, Cambridge 1978)

3.22 L.W. Strock: Z. Phys. Chem. B *25*, 411 (1934); B *31*, 132 (1936)

3.23 F. Reidinger, J.J. Reilly, R.W. Stoenner: In *Superionic Conductors*, ed. by G.D. Mahan, W.L. Roth (Plenum Press, New York 1976) p.427; to be published

3.24 R.J. Cava, F. Reidinger, B.J. Wuensch: J. Solid State Chem., submitted

3.25 A. Kvist, A.M. Josefson: Z. Naturforsch. *25*a, 625 (1968)

3.26 K.W. Kehr: Jülich Report No.1211, Jülich (1975)

3.27 A.S. Dworkin, M.A. Bredig: J. Phys. Chem. *72*, 1277 (1968)

3.28 M.W. Thomas: Chem. Phys. Lett. *40*, 111 (1976)

3.29 M.H. Dickens, W. Hayes, M.T. Hutchings: J. Phys. (Paris) *37*, C7-353 (1976)

3.30 J.D. Axe, S.M. Shapiro, N. Wakabayashi: Unpublished

3.31 F. Reidinger, J.D. Axe, S.M. Shapiro, N. Wakabayashi: (To be published)

3.32 A. Sadoc, Y. Allain: Solid State Commun. *25*, 739 (1978)

3.33 A.K. Cheatham, B.E.F. Fender, M.J. Cooper: J. Phys. C *4*, 3107 (1971)

3.34 A. Rahman: J. Chem. Phys. *65*, 4845 (1976)
     M. Dixon, M.J. Gillan: J. Phys. C *11*, L165 (1978)

3.35 N. Weber: In *Superionic Conductors*, ed. by G.D. Mahon, W.L. Roth (Plenum Press, New York 1976) p.37

3.36 W.L. Bragg, C. Gottfried, J. West: Z. Krist. *77*, 255 (1931)
C.A. Beevers, M.A.S. Ross: Z. Krist. *97*, 59 (1937)

3.37 J.T. Kummer: Prog. Solid State Chem. *7*, 141 (1974)

3.38 C.R. Peters, M. Bettman, J.W. Moore, M.D. Glick: Acta Crystallogr. B *27*, 1826 (1971)

3.39 W.L. Roth: J. Solid State Chem. *4*, 60 (1972)

3.40 F. Reidinger: Thesis (SUNY-Albany 1979)
W.L. Roth, F. Reidinger, S. La Place: In *Superionic Conductors*, ed. by G.D. Mahan, W.L. Roth (Plenum Press, New York 1976) p.223

3.41 D.B. McWhan, S.J. Allen, Jr., J.P. Remeika, P.D. Dernier: Phys. Rev. Lett. *35*, 953 (1975)
D.B. McWhan, P.D. Dernier, C. Vettier, A.S. Cooper, J.P. Remeika: Phys. Rev. B *17*, 4043 (1978)

3.42 G. Collin, J.P. Boilot, A. Kahn, J. Théry, R. Comès: J. Solid State Chem. *21*, 283 (1977)

3.43 M.S. Whittingham, R.A. Huggins: J. Chem. Phys. *54*, 414 (1971); J. Electrochem. Soc. *118*, 1 (1971)

3.44 J.C. Wang, M. Gaffari, S. Choi: J. Chem. Phys. *63*, 772 (1975)

3.45 W. Bührer, R.M. Nicklow, P. Brüesch: Phys. Rev. B *17*, 3362 (1978)

3.46 K. Funke, J. Kalus, R. Lechner: Solid State Commun. *14*, 1021 (1974)

3.47 G. Eckold, K. Funke, J. Kalus, R. Lechner: Phys. Lett. *55* A, 125 (1976); J. Phys. Chem. Solids *37*, 1097 (1976)

3.48 K. Funke, G. Eckold, R. Lechner: In *Proc. of Summer School on Microscopic Structure and Dynamics of Molecular Liquids*, ed. by J. Dupuy, A.J. Dianoux (To be published)

3.49 R.E. Lechner, G. Eckold, K. Funke: In [3.48]

3.50 C.J. Graham, R. Chang: J. Appl. Phys. *46*, 2 (1975)

3.51 S.M. Shapiro, D. Semmingsen, M. Salamon: In *Proc. of the Intern. Conf. on Lattice Dynamics*, ed. by M. Balkanski (Flammarion, Paris 1978) p.538

3.52 T. Geisel: In [Ref.3.51, p.549]

3.53 W. Hayes, A.M. Stoneham: In *Crystals with the Fluorite Structure*, ed. by W. Hayes (Oxford University Press, London 1974) Chap.2

3.54 M.H. Dickens, M.T. Hutchings: J. Phys. C *11*, 461 (1978)

3.55 C.E. Derrington, N. Navrosky, M. O'Keefe: J. Solid State Chem. *18*, 47 (1976)

3.56 U.R. Belosludov, R.I. Efremuza, E.U. Matizen: Sov. Phys. Solid State *16*, 847 (1974)

3.57 R.J. Elliott, W. Hayes, W.G. Kleppmann, A.J. Rushworth, J.F. Ryan: Proc. R. Soc. London A *360*, 317 (1978)

3.58 M.H. Dickens, M.T. Hutchings: Neutron Inelastic Scattering (IAEA Vienna,1978) Vol.2, p.285

3.59 R.T. Harley, W. Hayes, A.J. Rushworth, J.F. Ryan: J. Phys. C *8*, L530 (1975)

3.60 C.R.A. Catlow, J.D. Comins, F.A. Germano, R.T. Harley, W. Hayes: J. Phys. C (To be published)

3.61 S.J. Allen, Jr., J.P. Remeika: Phys. Rev. Lett. *33*, 1478 (1974)

3.62 S.J. Allen, Jr., A.S. Cooper, F. De Rosa, J.P. Remeika, S.K. Ulasi: Phys. Rev. B *17*, 4031 (1978)

3.63 S.M. Shapiro, D.B. McWhan, J.P. Remeika: To be published

3.64 Y. Le Cars, R. Comès, L. Deschamps, J. Théry: Acta Crystallogr. A *30*, 305 (1974)
J.P. Boilot, J. Théry, R. Collongues, R. Comès, A. Guinier: Acta Crystallogr. A *32*, 250 (1976)

3.65 D.B. McWhan, S.M. Shapiro, J.P. Remeika, G. Shirane: J. Phys. C *8*, L487 (1975)

3.66 J.D. Axe, R. Pynn, R. Lechner, D.B. McWhan, S.M. Shapiro: Private communication

3.67 J.D. Axe, L.M. Corliss, J.M. Hastings, W.L. Roth, O. Muller: J. Phys. Chem. Solids *39*, 155 (1978)

4. Statics and Dynamics of Lattice Gas Models

H. U. Beyeler, P. Brüesch, L. Pietronero, W. R. Schneider,
S. Strässler, and H. R. Zeller

With 8 Figures

Superionic conductors are characterized by an almost liquid-like mobility of one
of the constituent ion species. The simplest theoretical concept of particle mobility in such a system is based on Brownian motion of particles in a periodic
potential. Within this picture it is possible to identify two limiting cases:

(i) Barrier height large compared with k_BT and not too large damping of the
particle oscillations. For this case standard free rate theory applies and the
dc conductivity is thermally activated.

(ii) Barrier height small compared with k_BT. The ion moves as quasi-free particle and the potential wells provide scattering instead of localization. The dc
conductivity decreases with increasing temperature and exhibits a Drude-type
frequency dependence.

Depending on conditions (damping etc.) the maximum conductivity σ_{dc}^{max} is obtained for a barrier height $\simeq k_BT$; for reasonable values of the parameters one
finds $\sigma_{dc}^{max} \simeq 10 \ \Omega^{-1}cm^{-1}$. All experimental systems show thermally activated behavior and dc conductivities up to $\simeq 2 \ \Omega^{-1}cm^{-1}$. From this it is clear that a
hopping model is a correct starting point but that the best conductors are sufficiently close to the maximum conductivity to cause a breakdown of free rate
concepts.

In this paper we introduce a new method of statistical mechanics which fills
the gap between free-rate theory and quasi-free particle motion and enables us to
describe high mobility materials on a microscopic level.

To make the basic ideas clear we discuss them in this introduction in terms
of single particle jumps. To account for many-body effects, the concept of a reaction coordinate is used in the bulk of the paper.

We start from a lattice gas model. The crystal volume is divided up into cells
which completely fill the volume and contain either zero or one ion. We define
several characteristic times crucial for a correct microscopic description of the
system. τ_{res} is the average residence time of a particle in its potential well.
τ_{rel} is the time required for a system to relax after a particle jump. Standard
free-rate theory or its equivalent for an interacting many-body system, the master
equation, holds only in the limit $\tau_{rel}/\tau_{res} \to 0$. In general τ_{rel} has a complex
spectrum, but obviously it is the longest relevant τ_{rel} that determines the validity

of the master equation. For the best superionic conductors $\tau_{rel}/\tau_{res} \geq 1$ and the master equation does not apply.

In Sect.4.1 we introduce a generalized lattice gas model which has three essential novel features. First it explicitly includes the effects of the internal dynamics of the system on the transition reate. The conductivity can be written as a sum of four terms of which the first two represent the standard free-rate or master-equation result. The third term treats the influence of arbitrary ratios τ_{rel}/τ_{res}, i.e., the detailed lattice dynamics. The fourth term becomes important if the actual jump distance deviates from the elementary jump distance. This can occur in two ways: either a particle performs multiple jumps before becoming thermalized, or at high temperature, the scattering is due predominantly to thermal fluctuations and not to the static potential. In this latter case the jump distance is shorter than the elementary distance. It is the fastest relaxation process of the system which controls the thermalization or trapping time τ_{trap}, and we note that for all experimental systems $\tau_{trap} \ll \tau_{transit}$ where $\tau_{transit}$ is the transit time across a potential well. Hence we never reach the multiple jump or propagating-particle limit, but approach the Smoluchowsky or viscous limit as the temperature is raised or the barrier height decreased.

Jump diffusion models based on the master equation are restricted to small frequencies. The second novel feature in our model consists in the introduction of an effective medium Liouvillian. This allows a description of the full frequency range and provides a microscopic basis for the memory function approximation introduced in earlier phenomenological models. The third point consists in the introduction of a rope model. This makes use of the fact that local charge neutrality has to be conserved and reduces the number of allowed configurations to a manageable level.

Section 4.2 stresses the importance of the conductivity terms not included in the master equation in the dc to microwave frequency regime and discusses the possibility of soliton-type transport in cases where the diffusion reaction coordinate has a strong collective component. For topological reasons collective effects are expected to be important in one-dimensional systems.

Finally, in Sect.4.3 we apply the formalism to AgI and to the one-dimensional system $K_{2x}Mg_xTi_{8-x}O_{16}$ (hollandite).

Many of the ideas presented in this paper are based on concepts introduced by FLYNN who has pioneered the theoretical understanding of diffusion in solids.

4.1 General Theory of the Lattice Gas Model for Superionic Conductors

4.1.1 Definition of the Lattice Gas Model

We consider a classical system built up of N particles with charges e_n and masses m_n, n = 1,2, ..., N. An ideal crystal at zero temperature is characterized by the fact that the position vector $q_0 = (q_{n\alpha}^0)$ of all particles defines a periodic lattice. $q_{n\alpha}^0$ is the component of the coordinate vector of the particle n in the α direction. We assume that for a small increase of the temperature, the representative point of the system lies within a small volume γ_0 of configuration space around q_0. In such a case, the potential energy V of the system can be represented by the harmonic approximation,

$$V = V(q_0) + \frac{1}{2} \sum_{\substack{n,\alpha \\ m,\beta}} (q_{n\alpha} - q_{n\alpha}^0) m_n^{\frac{1}{2}} \phi_{\alpha\beta}^0(n,m) m_m^{\frac{1}{2}} (q_{m\beta} - q_{m\beta}^0) \quad . \tag{4.1}$$

All properties of the system are now characterized by the mass-renormalized force-constant matrix ϕ^0. In particular the free energy is given by

$$F_0 = V(q^0) + k_B T \sum_{q,j} \ln \frac{[\lambda^0(q,j)]^{\frac{1}{2}}}{k_B T} \quad , \tag{4.2}$$

in units with $\hbar = 1$, where $\lambda^0(q,j)$ are the eigenvalues of the matrix ϕ^0 indexed by the wavevector q and the branch index j.

In a nearly perfect crystal, atomic diffusion is always connected with the existence of lattice defects. A crystal defect can be described by a point q_a in configuration space which corresponds to a local minimum of the potential.

We can define a small volume γ_a in configuration space by the set of all points lying on force lines (-grad V) which end up at q_a. The equilibrium probability p_a^{eq} for the occurrence of the configuration a is then given by

$$p_a^{eq} = Z_a/Z \quad , \tag{4.3}$$

where

$$Z = \sum_a Z_a \quad , \tag{4.4}$$

and

$$Z_a = \int_{\gamma_a} d^{3N}q \, \exp(-V/k_B T) \quad . \tag{4.5}$$

$\hbar = h/2\pi$ (normalized Planck's constant)

In real space, q_a represents a particular distribution of interstitials and vacancies. A system initially within γ_a makes transitions from γ_a to γ_b at a thermal mean rate given exactly [4.2-3] by

$$\Gamma_{ab}^{eq} = <\Gamma_{ab}>_a \quad , \tag{4.6}$$

where

$$\Gamma_{ab} = \int_{S_{ab}} t_{ab}(q)(\underline{v}\cdot\underline{n}_{ab})\Theta(\underline{v}\cdot\underline{n}_{ab})ds \quad , \tag{4.7}$$

$$t_{ab}(q) = \delta^{3N}(q - q_s) \quad . \tag{4.8}$$

Here $\underline{v} = (v_{n\alpha})$ is the velocity vector of the system. The integration in (4.8) is over the common surface S_{ab} of the configuration volumes γ_a and γ_b, $\underline{n}_{ab}$ is the normal to this surface in the direction $\gamma_a \rightarrow \gamma_b$, and Θ is the Heaviside stepfunction. Finally, the average in (4.6) is defined by

$$<B>_a = \prod_n \left(\frac{2\pi k_B T}{m_n}\right)^{\frac{1}{2}} \cdot Z_a^{-1} \int d^{3N}v d^{3N}q \, \exp(-H_a/k_B T)B \quad , \tag{4.9}$$

where H_a is the original Hamiltonian of the system within configuration γ_a but with an infinite potential outside. For a superionic conductor the defect density is large, comparable to the density of ions.

The lattice gas model for superionic conductors is based on choosing an appropriate subset of configurations according to the following construction.

It is assumed that the volume Ω of the system can be divided up into cells, $\Omega_c, c = 1, \ldots, M$, such that to a good approximation a cell is either empty or occupied by one particle only. Clearly there must be more cells than particles, $M > N$.

We now introduce the "lattice gas representation" for the configuration,

$$a = (n_1, n_2, \ldots, n_c, \ldots, n_M) \quad . \tag{4.10}$$

Here n_c takes the values $0, 1, \ldots, t, \ldots$. The value $n_c = 0$ indicates that cell c is empty. $n_c = t$ indicates that cell c is occupied by a particle of type t.

The identification (4.10) represents the "lattice gas model" for superionic conductors. In general we cannot expect that all configurations defined by (4.10) are stable, e.g., there may not exist a point q_a in the subspace defined by a for which $\delta V = 0$, $\delta^2 V \geq 0$. The probability of the occurrence of a particular configuration is given by (4.3) as for the nearly perfect crystal. The thermal mean rate for the system to make a transition from a $\rightarrow$ b is again given by (4.6).

Two particularly illustrative examples of the realization of (4.10) are discussed in Sect.4.3 where we also introduce the concept of an effective lattice gas
Hamiltonian to simplify the calculation of p_a^{eq} as given by (4.3). Related problems
in connection with the static properties are discussed in [4.4-8].

4.1.2 Liouvillian Approach to Lattice Gas Dynamics

The calculations of the dynamic properties for the lattice gas model defined in
Sect.4.1.1 will be based on linear response theory. In particular, in the case of
the frequency-dependent conductivity $\sigma_{\alpha\beta}(\omega)$, we start with the Kubo formula

$$\sigma_{\alpha\beta}(\omega) = \frac{1}{\Omega}\,\mathrm{tr}\{QS_{\alpha\beta}(\omega)\}\quad. \tag{4.11}$$

Within our rigid ion model the oscillator strength matrix Q is given by

$$Q(n,m) = \frac{e_n}{m_n^{\frac{1}{2}}}\frac{e_m}{m_m^{\frac{1}{2}}}\quad. \tag{4.12}$$

The velocity correlation matrix $S_{\alpha\beta}$ is defined as

$$S_{\alpha\beta}(n,m) = \frac{1}{k_BT}\left\langle m_n^{\frac{1}{2}}v_{n\alpha}(-i\omega-L)^{-1}v_{m\beta}m_m^{\frac{1}{2}}\right\rangle\quad. \tag{4.13}$$

The Liouvillian L is given by

$$L = v\cdot\frac{\partial}{\partial q} - \sum_{n,\alpha} m_n^{-1}\frac{\partial V}{\partial q_{n,\alpha}}\frac{\partial}{\partial v_{n,\alpha}}\quad. \tag{4.14}$$

For an ideal crystal in the harmonic approximation V is given by (4.1), and (4.13)
can be evaluated directly

$$S = [-i\omega + \phi^0/(-i\omega)]^{-1}\quad. \tag{4.15}$$

Substituting (4.15) into (4.11) gives the standard expression for the optical conductivity,

$$\sigma_{\alpha\beta}(\omega) = \sum_j f_j[-i\omega + \lambda^0(0,j)/(-i\omega)]^{-1}\quad, \tag{4.16}$$

where f_j is the oscillator strength of the jth optical mode. In the case of superionic conductors we must deal with anharmonicity, disorder, and diffusion. We
shall now introduce a new concept which will make it possible to analyze these effects in a systematic way.

First consider the situation where infinite barriers prevent the system from
changing its configuration. In this case we have a Liouvillian L_a which describes

the system in every configuration. Anharmonicity and disorder are included in this description, but no diffusion occurs. In other words we have a Liouvillian L_{ab} for the complete system which is diagonal with respect to the configuration representation,

$$L_{ab} = \delta_{ab} \cdot L_a \quad , \tag{4.17}$$

where

$$L_a = v \cdot \frac{\partial}{\partial q} - \sum_{n,\alpha} m_n^{-1} \frac{\partial v^a}{\partial q_{n,\alpha}} \frac{\partial}{\partial v_{n,\alpha}} \quad . \tag{4.18}$$

It is shown in Appendix A that the Liouvillian for the original system including diffusion can be written in the configuration representation as

$$L_{ab} = \delta_{ab} \cdot L_a + L_{ab}^{tr} \quad , \tag{4.19}$$

where the transfer Liouvillian L_{ab}^{tr} is defined as

$$L_{ab}^{tr} = \Gamma_{ab} - \delta_{ab} \sum_c \Gamma_{ac} \quad , \tag{4.20}$$

and Γ_{ab} is given by (4.7).

This expression represents a very convenient starting point for the discussion of the general behavior of the frequency dependent conductivity of the lattice gas model.

Let us first illustrate the application of this formalism in the case of small and high frequencies. For sufficiently high frequencies, the diffusive part can be neglected. The Liouvillian is thus given by (4.17) and the velocity-velocity correlation matrix becomes

$$S_{\alpha\beta}(n,m) = \frac{1}{k_B T} \sum_a p_a^{eq} \left\langle m_n^{\frac{1}{2}} v_{n\alpha} (-i\omega - L_a)^{-1} v_{m\beta} m_m^{\frac{1}{2}} \right\rangle_a \quad . \tag{4.21}$$

The problem is now reduced to one of disorder and anharmonicity where standard methods can be applied. A further discussion of (4.21) is presented in Sect.4.1.4.

For sufficiently low frequencies, the oscillatory part can be neglected. In this case we can make use of (B.2) when evaluating (4.13). The expression for the conductivity is then given by

$$\sigma_{\alpha\beta}(\omega) = \frac{1}{\Omega} \, tr\{Q[\Lambda_{\alpha\beta}^{(0)}(\omega) + \Lambda_{\alpha\beta}'(\omega)]\} \quad , \tag{4.22}$$

where the matrix $\Lambda_{\alpha\beta}^{(0)}$ is

$$\Lambda_{\alpha\beta}^{(0)}(n,m) = \frac{m_n^{\frac{1}{2}}m_m^{\frac{1}{2}}}{k_BT}\left[\frac{1}{2}\sum_{ab} p_a^{eq}(\ell_{n\alpha}^b - \ell_{n\alpha}^a)\Gamma_{ab}^{eq}(\ell_{m\beta}^b - \ell_{m\beta}^a)\right] \quad . \tag{4.23}$$

Here p_a^{eq} and Γ_{ab}^{eq} are defined in (4.3) and (4.6), respectively; and

$$\ell_n^a = <q_{n\alpha}>_a \tag{4.24}$$

represents the average position of the particle n in configuration a.

The matrix $\Lambda_{\alpha\beta}'(\omega)$ is given by

$$\Lambda_{\alpha\beta}'(n,m) = -\frac{m_n^{\frac{1}{2}}m_m^{\frac{1}{2}}}{k_BT}\sum_{ab} p_a \left\langle v_{n\alpha}^a(-i\omega-L)_{ab}^{-1}v_{m\beta}^b\right\rangle_a \quad , \tag{4.25}$$

where

$$v_{n\alpha}^a = \sum_b \Gamma_{ab}(\ell_{n\alpha}^b - \ell_{n\alpha}^a) \quad . \tag{4.26}$$

To proceed in the analysis one is now forced to make various simplifying assumptions about the Γ_{ab} defined in (4.7), which we discuss in the next section. A method for evaluating the conductivity for all frequencies will be introduced in Sect.4.1.5.

4.1.3 Master-Equation Approximation

In this section we want to discuss a particular approximate evaluation of the low frequency conductivity as given by (4.22). In the simplest approximation Γ_{ab} in (4.26) is replaced by its average value Γ_{ab}^{eq}. This approximation is valid if a jump from one configuration to another has no influence on the following jumps. To see more clearly what this implies it is useful to introduce reaction coordinates (see [4.2], pp.327-332) by projecting the coordinate and velocity vectors on the direction $q_{ba} - q_a$, where q_{ba} refers to the saddle-point position on the surface S_{ab}.

Using this concept of the reaction coordinate Γ_{ab} is broken up into 3 terms

$$\Gamma_{ab} = \sum_{i=1}^{3} \Gamma_{ab}^{(i)} \quad , \tag{4.27}$$

where

$$\Gamma_{ab}^{(1)} = \Gamma_{ab}^{eq} \quad , \tag{4.28}$$

$$\Gamma_{ab}^{(2)} = <\Gamma_{ab} - \Gamma_{ab}^{eq}>_{ab} \quad , \tag{4.29}$$

and

$$\Gamma_{ab}^{(3)} = \Gamma_{ab} - \langle\Gamma_{ab}\rangle_{ab} \qquad . \tag{4.30}$$

The average in (4.29) is defined by

$$\langle B\rangle_{ab} = \int dq_r dv_r B \exp(-H_a/k_B T)/ \int dq_r dv_r \exp(-H_a/k_B T) \qquad . \tag{4.31}$$

Introducing (4.29) into (4.25), we obtain

$$\sigma_{\alpha\beta}(\omega) = \sum_{i=0}^{3} \frac{1}{\Omega} \text{tr}\{Q\Lambda_{\alpha\beta}^{(i)}(\omega)\} = \sum_{i=0}^{3} \sigma_{\alpha\beta}^{(i)}(\omega) \qquad , \tag{4.32}$$

where $\Lambda^{(0)}$ is defined by (4.23),

$$\Lambda_{\alpha\beta}^{(1)}(n,m) = -\frac{m_n^{\frac{1}{2}} m_m^{\frac{1}{2}}}{k_B T} \sum_{ab} p_a^{eq} \left\langle v_{n\alpha}^{a,(1)} (-i\omega - L^{eq})_{ab}^{-1} v_{m\beta}^{b,(1)} \right\rangle_a \qquad , \tag{4.33}$$

with

$$v_{n\alpha}^{a,(i)} = \sum_b \Gamma_{ab}^{(i)} (\ell_{n\alpha}^b - \ell_{n\alpha}^a) \qquad , \tag{4.34}$$

and

$$L_{ab}^{eq} = \Gamma_{ab}^{eq} - \delta_{ab} \sum_c \Gamma_{ac}^{eq} \qquad . \tag{4.35}$$

For $i = 2,3$ we have

$$\Lambda_{\alpha\beta}^{(i)}(n,m) = -\frac{m_n^{\frac{1}{2}} m_m^{\frac{1}{2}}}{k_B T} \sum_{ab} p_a^{eq} \left\langle v_{n\alpha}^{a,(i)} (-i\omega - L)_{ab}^{-1} v_{m\beta}^{b,(i)} \right\rangle_a \qquad . \tag{4.36}$$

Equation (4.32) represents the conductivity in the region where only the diffusive contribution to the conductivity is important. The first term is frequency independent and is the only term which survives for large frequencies. In particular it is shown in Appendix C that the real part of $\sigma^{(1)}$ always increases as a function of frequency. It is now of interest to note that we could have obtained the expression for the conductivity as given by the terms $i = 0$ and 1 directly from a master equation for the nonequilibrium probability distribution $p_a(t)$

$$\frac{dp_a(t)}{dt} = \sum_b p_b L_{ba}^{eq} [1 + \phi_b(t)/k_B T] \qquad . \tag{4.37}$$

Here $\phi_b(t)$ represents the potential energy of the system in the presence of an electric field. This result tells us once more that within the approximation

$$\sigma_{\alpha\beta}(\omega) = \sigma_{\alpha\beta}^{(0)}(\omega) + \sigma_{\alpha\beta}^{(1)}(\omega) \quad , \tag{4.38}$$

all dynamic effects except for those defined by the thermal transition rates Γ_{ab}^{eq} are neglected.

Simplified versions of the result (4.38) have been evaluated in various approximations in [4.9-13]. The other terms in (4.32) become important if the reaction coordinate does not thermalize before the next jump and if there exist modes which have a significant influence on Γ_{ab} and whose relaxation times are longer than an average residence time.

A further discussion of these contributions is presented in Sect.4.2.

4.1.4 High-Frequency Limit

For high enough frequencies we neglect the diffusive part in the Liouvillian. The motion of the system in a particular region γ_a of the configuration space is then described by the Liouvillian L_a given by (4.18).

It is convenient to represent the system by a Liouvillian $\bar{L}$ of an effective medium defined by

$$\bar{L} = v \cdot \frac{\partial}{\partial x} - \sum_{\substack{n,\alpha \\ m,\beta}} m_n^{-\frac{1}{2}} x_n \bar{\phi}_{\alpha\beta}(n,m;\omega) \frac{\partial}{\partial v_{m\beta}} m_m^{-\frac{1}{2}} \quad . \tag{4.39}$$

Here $x = (x_{n\alpha}) = q - q^a$ is the position vector for the particles of the effective medium, and $\bar{\phi}_{\alpha\beta}$ is the frequency-dependent force-constant matrix of the effective medium. The equation for $\bar{\phi}$ is obtained by requiring that the velocity correlation function of the effective medium is identical to the exact one,

$$\sum_a p_a^{eq} \left\langle m_n^{\frac{1}{2}} v_{n\alpha} (-i\omega - L_a)^{-1} v_{m\beta} m_m^{\frac{1}{2}} \right\rangle_a \equiv k_B T \bar{S}_{\alpha\beta}(n,m) \quad . \tag{4.40}$$

The matrix $\bar{S}$ is defined by

$$\bar{S}^{-1} = [-i\omega + \bar{\phi}/(-i\omega)] \quad . \tag{4.41}$$

The corresponding frequency-dependent conductivity is given by

$$\sigma_{\alpha\beta}(\omega) = \frac{1}{\Omega} \text{tr}\{Q\bar{S}_{\alpha\beta}\} \quad . \tag{4.42}$$

We now approximate the system in each configuration by the harmonic approximations given by (4.1) with the substitution $q_0 \to q_a$. Multiplying (4.40) by $\bar{S}^{-1}$ and performing the velocity averages, we obtain

$$\sum_a p_a^{eq} M_{\alpha\beta}^a(n,m) = \delta_{\alpha\beta} \cdot \delta_{nm} \quad . \tag{4.43}$$

Here the matrix M^a is given by

$$M^a = \left[\underline{1} + (\phi^a - \bar{\phi}) \frac{\bar{S}}{-i\omega} \right]^{-1} \quad , \tag{4.44}$$

where $\underline{1}$ is the units matrix, and ϕ^a is the force-constant matrix of configuration a. The virtual crystal approximation for $\bar{\phi}$ is obtained by keeping only the first term in the expansion of (4.44) in powers of $\phi^a - \bar{\phi}$,

$$\phi^V = \sum_a p_a^{eq} \phi^a \quad . \tag{4.45}$$

In order to evaluate an expression as given by (4.45) for a system with long-range forces, it is convenient to introduce a particular representation of (4.10) which explicitly excludes large density fluctuations. A particular group of cells Ω_c will form a primitive cell. On average such a primitive cell contains z_t particles of each type t. Let $\underline{r}(\ell)$ be the center of such a cell. We now assign such an average set of particles to each primitive cell and attach them to the center by a rope of length r. The indices for these particles are now denoted by $n \to (\ell, \kappa)$ where $\kappa = 1, 2, \ldots, \sum_{t=1,2,\ldots} z_t$. The configurations of the primitive cells are numbered as $a_\ell = 1, 2, \ldots$. These configurations are given by all possible arrangements of the attached particles in the cells Ω_c which can be reached for a given length r. The length must be choosen so as to avoid large density fluctuations, but large enough to have a reasonable amount of disorder. In this "rope" representation, (4.45) becomes

$$\phi_{\alpha\beta}^V \binom{\ell, \ell'}{\kappa, \kappa'} = \sum_{a_\ell; a_{\ell'}} p(a_\ell; a_{\ell'}) \phi_{\alpha\beta}^{a_\ell; a_{\ell'}} \binom{\ell, \ell'}{\kappa, \kappa'} \quad \text{for } \binom{\ell'}{\kappa'} \neq \binom{\ell}{\kappa}$$

$$= - \sum_{\binom{\ell''}{\kappa''} \neq \binom{\ell}{\kappa}} \phi_{\alpha\beta}^V \binom{\ell, \ell''}{\kappa, \kappa''} \quad \text{for } \binom{\ell'}{\kappa'} = \binom{\ell}{\kappa} \quad , \tag{4.46}$$

where $p(a_\ell; a_{\ell'})$ is the pair correlation probability for the cells ℓ and ℓ'. The summation is over nonequivalent configurations only. This result is applied to AgI in Sect.4.3.

Note that ϕ^V is only a function of $\underline{r}(\ell) - \underline{r}(\ell')$ and therefore can be diagonalized by a transformation to wavevectors and branches, $(\ell, \kappa) \to (\underline{q}, j)$.

Anharmonic effects can be included by expanding the left-hand side of (4.40) in powers of $L_a - \bar{L}$.

In the "rope" representation, (4.43) becomes

$$\sum_a p_a^{eq} M^a = \underline{1} \quad . \tag{4.47}$$

We shall make now an approximation which for mass defects is equivalent to the coherent potential approximation (CPA) [4.14], i.e., we set in (4.47)

$$\phi_{\alpha\beta}^{a}\binom{\ell,\ell'}{\kappa,\kappa'} - \bar{\phi}_{\alpha\beta}\binom{\ell,\ell'}{\kappa,\kappa'} = 0, \quad S_{\alpha\beta}\binom{\ell,\ell'}{\kappa,\kappa'} = 0 \quad \text{for } (\ell,\ell') \neq (0,0) \quad , \quad (4.48)$$

and in (4.41),

$$\bar{\phi}_{\alpha\beta}\binom{\ell,\ell'}{\kappa,\kappa'} \equiv \phi_{\alpha\beta}^{V}\binom{\ell,\ell'}{\kappa,\kappa'} \quad \text{for } (\ell,\ell') \neq (\ell,\ell) \quad . \quad (4.49)$$

To improve the theory one can go from the primitive cell to a larger unit cell and perform a standard perturbation expansion in $\bar{\phi} - \phi^{V}$ for $\ell \neq \ell'$. Other possible extensions for extended defects are discussed in [4.14]. A simplified form of (4.47) is evaluated to discuss some properties of AgI in Sect.4.3.

4.1.5 Extension to All Frequencies

The extension to all frequencies can be made by introducing the effective medium Liouvillian,

$$\bar{L}' = \bar{L} - \sum_{\substack{n,\alpha \\ m,\beta}} m_n^{-\frac{1}{2}} x_{n\alpha} \lambda_{\alpha\beta}(n,m;\omega) \frac{\partial}{\partial x_{m\beta}} m_m^{-\frac{1}{2}} \quad . \quad (4.50)$$

Here $\bar{L}$ is given by (4.39) and

$$\lambda = \Lambda\bar{\phi} \quad , \quad (4.51)$$

where

$$\Lambda = \sum_{i=0}^{3} \Lambda^{(i)} \quad , \quad (4.52)$$

with the $\Lambda^{(i)}$'s given by (4.35,36). The equation for $\bar{\phi}$ is [in analogy to (4.40)]

$$\sum_{ab} P_a \left\langle m_n^{\frac{1}{2}} v_{n\alpha} (-i\omega - L)_{ab}^{-1} v_{m\beta} m_m^{\frac{1}{2}} \right\rangle_a = k_B T S'_{\alpha\beta}(n,m) \quad , \quad (4.53)$$

where

$$S'^{-1} = -i\omega + 1/(-i\omega\bar{\phi}^{-1} + \Lambda) \quad .$$

All contributions from L_{ab}^{tr} are contained in the matrix Λ, therefore $\bar{\phi}$ in various approximations can be determined from (4.40-49).

The conductivity is now obtained from

$$\sigma \quad (\omega) = \frac{1}{-} \text{tr}[\sigma' \quad] \quad (4.55)$$

We have used the frequency-dependent conductivity to illustrate the development of a general theory for superionic conductors within a lattice gas concept. Other quantities can be obtained within the same context [4.15]. In particular the density of states is given by

$$N(\omega) = \frac{1}{\Omega\pi} \sum_{\alpha} \text{Re}\{\text{tr}(S'_{\alpha\alpha})\} \quad . \tag{4.56}$$

We complete the general discussion of the theory of the lattice gas model for super-ionic conductors by noting that a tracer conductivity [4.2] can be defined within our formalism by an oscillator strength matrix

$$Q(n,m) = (e^2/m) \cdot \delta_{n,m}\delta_{n,1} \quad . \tag{4.57}$$

4.2 Extended Dynamical Theory

The frequency-dependent conductivity in the diffusive regime (low frequencies) is of special interest because it involves the study of the diffusion dynamics for a many-body system. This interest is also motivated by the fact that peculiar structures are observed in the microwave conductivity of AgI type superionic conductors [4.16].

In this section we discuss the validity of the fast trapping condition $(\tau_{trap}/\tau_{res} \ll 1)$ showing also what kind of contributions to $\sigma(\omega)$ arise when this condition breaks down.

4.2.1 τ_{trap} and Its Relation to a Soliton Model

If the fast-trapping condition is not satisfied the terms $\sigma^{(2)}$ and $\sigma^{(3)}$ in (4.32) become important. We show here how this situation can give rise to free-particle behavior of the reaction coordinate even for a many-body system [4.17].

We illustrate this possibility with a one-dimensional model. In particular we consider the case of a rigid host lattice that defines channels within which ions can diffuse. The host lattice provides a potential with periodicity a and barrier height V_0. The diffusing ions have a density ρ (determined by the stoichiometry of the material) with respect to the period of the potential and interact with each other with a Coulomb potential. We shall see in Sect.4.3 how this model actually relates to specific materials.

The average distance between two nearest diffusing ions is b = a/ρ. Expanding the ion-ion interaction potential V(r) up to quadratic terms around the equilibrium distance r = b and defining the harmonic force constant

$$A = \frac{\partial^2 V(r)}{\partial r^2}\bigg|_{r=b} \quad , \qquad (4.58)$$

we can describe our system with the Hamiltonian

$$H = \sum_{j=1}^{N} \frac{1}{2} m\dot{x}_j^2 + \sum_{j=1}^{N-1} A(x_{j+1} - x_j - b)^2 + \sum_{j=1}^{N} \frac{V_0}{2} (1 - \cos \frac{2\pi}{a} x_j) \quad . \qquad (4.59)$$

The sinusoidal form of the potential is assumed for convenience. In general $\rho < 1$ so there is a number of empty sites in the potential. In the limit of a few vacancies ($\rho \le 1$) we can treat them as independent. In order to study the transport properties of such a system we have to determine the potential barrier for the motion of a vacancy. Actually, because of the ion-ion interaction, a distortion of the neighboring ions is always associated with a vacancy. The collective co-ordinate describing the system: vacancy + distortion is an example of the reaction coordinate defined in Sect.4.1.

The computation of the potential barrier, which we shall call pinning barrier, V_p, is performed numerically for the case of eight wells with seven ions keeping the two boundary ions fixed (similar to the calculation in [4.18]), and is shown in Fig.4.1. V_p is the energy difference between the configuration in which $\delta x_4 = a/2$ and the other ions can relax, and the ground state in which all the ions can relax. The limiting values are $V_p = V_0$ for $A = 0$ and $V_p \sim 0$ for strong coupling ($Ab^2 \gg V_0$). Since the displacements of the ions close to a vacancy are experimentally observable quantities, we have plotted V_p as a function of the average fractional displacement of the ions 4 and 5. The parameters used in the calculations and shown in Fig.4.1 are those in the range of superionic conductors. An observed displacement is $\delta \simeq 0.25$ [4.19]. This corresponds to values of V_p of the order of, or smaller than, thermal fluctuations at room temperature. Such small barriers suggest actually the possibility that the system: vacancy + distortion can propagate freely in the channel.

In order to investigate this possibility, it is convenient to use a continuum approximation, defining $\phi(j) = (2\pi/a)\delta x_j$. The system: vacancy + distortion can now be identified with a soliton described by the field [4.20]

$$\phi(x,t) = 4 \tan^{-1}\left[\exp\left(\frac{x-ut}{\ell_0\sqrt{1-u^2/c_0^2}}\right)\right] \quad , \qquad (4.60)$$

where u is the velocity ($0 \le |u| \le c_0$), and

$$c_0^2 = \frac{Ab^2}{m} \quad ,$$

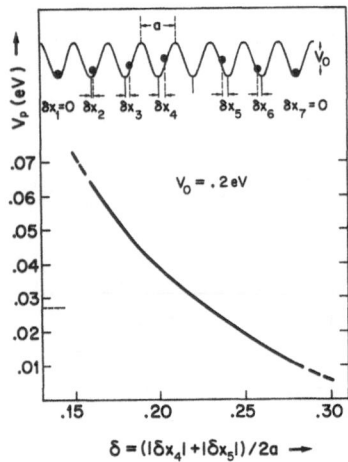

Fig.4.1. Pinning potential barrier V_p for the discrete model shown in the figure. V_p is plotted as a function of the average relative displacement of the ions close to the vacancy in the ground state. This is an observable quantity and is indirectly related to the ion-ion interaction

$$\ell_0 = \frac{c_0}{\omega_0} \, , \qquad \omega_0^2 = (\frac{2\pi}{a})^2 \frac{V_0}{2m} \quad . \qquad (4.61)$$

The displacement field given by (4.60) can be treated as an effective single particle [4.21] of mass

$$M_s = \frac{2m}{\pi^2} \frac{a}{\ell_0} \quad . \qquad (4.62)$$

To discuss the dynamics of these effective particles we have to include the effects of thermal motion of the ions. We simply assume that this gives rise to a phenomenological damping force $-\eta v$ for each particle in the equation of motion corresponding to (4.60). This results in an effective damping force $-\Gamma_s u$ for the motion of the soliton. A relationship between η and Γ_s can be obtained as follows. The energy dissipated by a soliton propagating with a constant velocity u over a distance L is equal to the energy dissipated by the displacement by a distance a of all the ions present in L. This gives [4.18]

$$\Gamma_s = \eta \left[\frac{1}{2} (\frac{a}{2\pi})^2 \int_{-\infty}^{+\infty} d\xi \, (\frac{d\phi}{d\xi})^2 \right] = \eta \left[\frac{2}{\pi^2} \frac{1}{\sqrt{1-u^2/c_0^2}} \frac{a^2}{b\ell_0} \right] \quad , \qquad (4.62)$$

where $\xi = x-ut$.

We can now write an effective equation of motion for the soliton including pinning potential and damping. In the nonrelativistic limit [u << c_0, the velocity of sound given by (4.61)], this is

$$M_s \ddot{X}_s + \Gamma_s \dot{X}_s + \nabla V_p(X_s) = f(t) \qquad (4.63)$$

where X_s is the soliton coordinate, $V_p(X_s)$ is the pinning potential, and $f(t)$ is the stochastic force corresponding to Γ_s. We have therefore reduced the problem of the dynamics of the collective excitations to that of the dynamics of an effective single particle. The frequency dependent mobility $\mu(\omega)$ for this problem has been recently studied in some detail and one can directly use at this point the results of [4.22,23].

We have now all the ingredients to discuss the various possibilities for the dynamics of the collective reaction coordinate as a function of the parameters τ_{trap}, V_p, and the temperature. τ_{trap} is related to Γ_s^{-1} and defines a mean free path λ for the reaction coordinate given by $\lambda \sim <v>\tau_{trap}$ where $<v>$ is the average thermal velocity. We can define τ_{trap} short or long corresponding respectively to λ shorter or longer than the distance between two minima of the potenial in configuration space. In the simple case of Fig.4.1 this distance is a. The following cases can then occur:

(I) τ_{trap} small ($\lambda \ll a$), $k_B T \ll V_p$. This is the case of pure hopping.

(II) τ_{trap} small ($\lambda \ll a$), $k_B T \geq V_p$. The potential is not able to localize the collective coordinate that moves with a random walk. (Smolukowski limit).

(III) τ_{trap} large ($\lambda > a$), $k_B T \ll V_p$. Two-fluid case. The potential mostly localizes the collective coordinate, but when a thermal fluctuation causes it to overcome the barrier, it can travel over several barriers.

(IV) τ_{trap} large ($\lambda > a$), $k_B T > V_p$. The collective coordinate is not localized by the potential and can travel freely because of the large τ_{trap}. For the case of a single particle this implies free propagation of the particle itself. For a many-body system like the one we discuss here this situation corresponds to the propagation of solitons.

The perturbation theory described in Sect.4.1 starts from the point of view of τ_{trap} small and therefore it is appropriate to describe only the cases (I) and (II). The first terms in the expansion correspond in fact to $\tau_{trap} = 0$ and the effects due to the fact that τ_{trap} is different from zero are treated as perturbations. In the next section we discuss an example of these effects.

4.2.2 Low-Frequency Conductivity

The first two terms of the total expression for the frequency dependent conductivity (4.32) correspond to the limit of infinitely fast relaxation. They are equivalent therefore to a master-equation description. In this section we discuss the general behavior of $\sigma(\omega)$ as given by these two terms and estimate the effect of the other terms.

In Appendix C it is shown that $Re\{\sigma^{(0)} + \sigma^{(1)}\}$ increases monotonically with frequency. A phenomenological form for $Re\{\sigma^{(0)} + \sigma^{(1)}\}$ which is compatible with this condition is

$$\text{Re}\{\sigma^{(0)} + \sigma^{(1)}\} \simeq \sigma_0 \left(1 - R \frac{\Gamma^2}{\omega^2 + \Gamma^2}\right) \tag{4.64}$$

The parameter Γ represents an effective eigenvalue of the matrix L_{ab}^{eq} (4.35), and $R(0 \leq R \leq 1)$ depends on the specific problem. In particular one has $R = 0$ (flat conductivity) for a system with equivalent sites and hard-core interaction between particles. Explicit evaluation of σ_0 for various models can be found in [4.7,11-12].

In general the fast-relaxation condition is not satisfied in superionic conductors, and it is not correct to replace the transition rates Γ_{ab} by their equilibrium values Γ_{ab}^{eq}. The modulations of these transition rates are contained in $\Gamma_{ab}^{(2)}$ and reflect the internal local dynamics of the system. Once $\Gamma_{ab}^{(2)}$ is included, the low-frequency $\sigma(\omega)$ no longer increases monotonically. As a specific example, consider a fluctuation of the reaction coordinate described by a harmonic mode with frequency ω_s. Equation (4.32) gives in this case

$$\text{Re}\{\sigma^{(2)}(\omega)\} \propto \frac{\Gamma}{2} \left[\frac{1}{(\omega + \omega_s)^2 + \Gamma^2} + \frac{1}{(\omega - \omega_s)^2 + \Gamma^2}\right]$$

$$= \int_0^\infty \exp(-i\omega t) \exp(-\Gamma t) \cos\omega_s t \quad . \tag{4.65}$$

Equation (4.65) results from the first-order term of an expansion of $\Gamma_{ab}^{(2)}$ in powers of the oscillator coordinate. Its structure is of special interest because: (I) it can increase the dc conductivity with respect to σ_0; and (II) for $\omega_s > \Gamma/\sqrt{3}$, it gives rise to structure at finite frequency, while for $\omega_s < \Gamma/\sqrt{3}$ there is a Drude-like behavior.

This is probably the mechanisms responsible for the structure observed in the microwave conductivity of AgI [4.16].

4.3 Applications to Silver Iodide and Hollandite

4.3.1 Silver Iodide: Structural Properties, Lattice Gas Representation

The high-temperature (α) phase of silver iodide is the archetype of a superionic conductor. Over the last 50 years a wealth of experimental data on the static and dynamic properties of this model system has been accumulated. In this section we present an analysis of some principal features in terms of the generalized lattice gas model.

As first determined by STROCK [4.25] the α phase of silver iodide consists of a bcc iodide lattice. On the basis of powder-diffraction data, no definite sites could be assigned to the silver ions. This had been taken as evidence for a liquid dis-

order among the silver ions. Recent neutron diffraction (see Chap.3) and EXAFS
(see Chap.2) results [4.26,27], however, clearly show that the silver ions reside
in the neighborhood of the tetrahedral sites. A detailed reinspection [4.29] of
X-ray data shows that in fact all of them are compatible with the silver ions occupying basically the tetrahedral 12d sites (see Fig.4.2) whereby the silver-silver
interaction shifts the ions considerably off the crystallographic positions, analogous to the situation found in other systems.

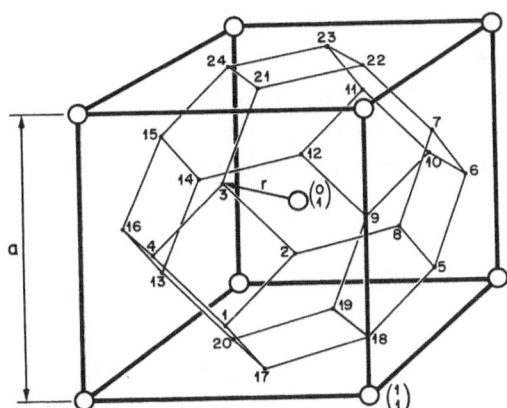

Fig.4.2. Cubic unit cell of α-AgI.
($\begin{smallmatrix}0\\1\end{smallmatrix}$) is the central iodine ion in
unit cell 0. 1,2,...,24 indicate
the 24 tetrahedral sites of the
silver ions, a is the lattice constant and r = ($\sqrt{5}/4$)a is the rope
length

We therefore base our lattice gas description of α-AgI on the configurations
formed by distributing N silver ions over 6N tetrahedral sites.

4.3.2 The Disorder Entropy of AgI

At 147°C, silver iodide transforms from the fully ordered hexagonal (β) phase to
the disordered α phase. The transition temperature is related to the enthalpy
change ΔH and the total entropy change ΔS by

$$T_c = \frac{\Delta H}{\Delta S} \quad . \tag{4.66}$$

Because the phonon frequencies do not change much from the β to the α phase, and,
furthermore, the disorder is negligibly small in the β phase, we conclude that ΔS
is mainly given by the disorder entropy of the α phase.

From the experimental value of ΔH [4.29] follows ΔS = 1.8 per silver ion. This
value is directly related to the total number of possible configurations.

To proceed further, we choose an appropriate form for the interaction and then
compute the corresponding disorder entropy. We consider a set of hard-core interactions between the silver ions defined by

$$V_n(r) = \begin{cases} \infty & r < d_n \\ 0 & r \geq d_n \end{cases} \qquad , \tag{4.67}$$

where r is the distance between the ions and d_n is the distance between n^{th} neighbors. $V_n(r)$ thus excludes simultaneous occupation of sites closer than the nth neighbor sites. Equation (4.67) includes only the short-range part of the potential. Qualitatively we can account for the long-range Coulomb interaction by using the rope representation introduced in Sect.4.2, which considers only configurations in which each silver ion is attached by a rope to a particular iodine ion; it can thus only occupy the sites within reach of the given rope length. The rope therefore excludes large charge-density fluctuations. Let us now choose the shortest possible rope so that each silver ion is bound to occupy one of the 24 nearest neighbor sites of its associated iodine ion. The entropy (in units of k_B) can be computed from

$$\Delta S = \lim_{N \to \infty} \left(\frac{1}{N} \log W_N \right) \qquad , \tag{4.68}$$

where W_N is the number of configurations that are possible in a crystal with 6N sites. As (4.68) can be computed for finite crystals only we perform the computations for a finite crystal directly (giving a upper bound for ΔS) and for a periodically continued crystal (giving a lower bound). In Table 4.1 the results of this computation for N = 4 are given for various values of the hard sphere parameter n.

Table 4.1. Upper and lower bound for the disorder entropy for the hard core model defined by (4.67)

n	1	2	3	
ΔS upper	2.81	2.22	1.59	finite crystal
ΔS lower	2.09	1.87	1.24	periodic crystal

The closest approximation to the experimental value of $\Delta S = 1.8$ is given by n = 2 which means that nearest-neighbor silver sites are not occupied simultaneously. This is in reasonable agreement with steric considerations: d_1 is equal to 1.79 Å which is far less than twice the ionic radius of Ag^+.

The remaining discrepancy between the experimental value of 1.8 and the values for n = 2 can be accounted for by the fact that simultaneously occupied second-nearest-neighbor sites will still suffer an appreciable repulsive interaction of the order of $k_B T$ increasing the equilibrium distance from 2.5 to about 2.8 Å and lowering the entropy.

4.3.3 Dynamic Properties of α-AgI

In this section we study the dynamic properties of α-AgI in the oscillatory regime. The vibrations of the ions in this solid electrolyte are complicated by the high degree of disorder of the silver ions and by anharmonic effects. Anharmonicity will be of prime importance in the diffusion-controlled regime but less important in the oscillatory regime of interest. The aim is to discuss the far infrared conductivity and the recent inelastic neutron scattering experiments [4.30] which both demonstrate the existence of very-low-energy lattice excitations between ~0.5 and 5 meV.

The lattice dynamics of α-AgI have been studied before by ALBEN and BURNS [4.31], who performed lattice dynamical calculations in the harmonic approximation. In order to account for the disorder of the silver ions, the problem was attacked with the equation-of-motion method for large clusters containing about 250 unit cells of different configurations for the silver ions. Using a short-range model with two parameters it was possible to account qualitatively for the observed broad far-infrared and Raman spectra [4.16,32,33].

Our approach is quite different. We attempt here to account for the effects of disorder which are observed in experiments by considering the harmonic motion of the ions in a *small unit cell only*. It is shown that this approach yields new physical insight into the ion dynamics which is not available from the "computer approach" of ALBEN and BURNS. We shall discuss two models: the first emphasizes the dynamical aspects, while the second emphasizes disorder.

In the first model, we consider the unit cube which contains two iodine and two silver ions. Disorder is taken into account only within this cube, that is, the dynamical problem as represented by the Liouvillian (4.18) in the harmonic approximation is replaced by one with only four different configurations a. They correspond to the 4 possible configurations of the two silver ions in the cubic unit cell which are periodically repeated with the periodicity of the cube. We are therefore dealing with the lattice dynamics of four perfectly ordered structures. Within this model, the observed broad optical spectra and the large halfwidths observed by inelastic neutron scattering result from a superposition of the phonon frequencies $\omega(\underline{q})$ of the four configurations. The large number of possible configurations is here drastically reduced to only four configurations. The advantage of this model is, however, that the small unit cell allows the 12 × 12 dynamical matrices to be diagonalized, which yields not only the eigenvalues but also the eigenvectors. The latter give us some insight into the ion motion of the low-frequency modes which provides a basis for the construction of reaction coordinates for jumps of silver ions.

The second method illustrates the CPA outlined in Sect.4.1. In a first step we construct a dynamical matrix for the primitive unit cell (containing one iodine and one silver ion) of a virtual crystal which has the full symmetry of the bcc lattice.

Although the phonon density of states of this virtual crystal has pronounced structures in the region of the residual ray absorption, there is no appreciable mode
density in the low-frequency region of interest. A better approximation is obtained
if the fluctuations of the force constants with respect to the virtual crystal are
introduced. As compared to the first model, this approach is able to account well
for disorder but in the actual calculations discussed below the dynamics are drastically simplified.

We start with the first model. The four different configurations of the silver
ions in the cubic unit cell are shown in Fig.4.3. We have excluded all configurations
with two silver ions on nearest-neighbor sites (see Sect.4.3.2). Configuration
1 is important because its statistical weight is four, whereas the statistical
weights of configurations 2,3, and 4 are one only. The force field used for all
four configurations was determined from the observed phonon dispersion of β-AgI
[4.34]. It includes both short-range and long-range (Coulomb) interactions. The
justification for using force constants obtained for β-AgI for α-AgI comes from the
fact that the local structure and the interatomic distances as well as the observed
frequencies of the TO modes are approximately the same in both phases. Thus, there
are no new parameters introduced in the α phase.

Based on this force model phonon dispersion curves were calculated for q parallel
to [100], [010], [001], [110], and [111] for all four configurations. It turns out
that the frequencies of the TA phonons in the [110] direction are particularly low
for all four configurations ranging between 14 and 30 cm^{-1} at the zone boundary.
The [110] TA branch is very flat (Fig.4.4) implying a high density of states around
20 cm^{-1}. The shape of this branch is in good agreement with the mean TA branch in
the [1$\bar{1}$0] direction as observed by recent inelastic neutron scattering experiments
which also show a very large spread in frequencies ranging between ~5 and 40 cm^{-1}
at the zone boundary. On the other hand the [111] direction is a "hard direction":
the calculated frequencies of the [111] TA phonons are about twice as high as those
for the [110] TA phonons. No neutron scattering experiments have yet been performed
in the [111] direction.

The eigenvectors of the low frequency zone boundary [110] TA modes are depicted
in Fig.4.3. Inspection shows that the displacements of the ions are such that little
or no changes occur in the nearest Ag-I and I-I distances minimizing the mutual repulsion between nearest-neighbor ions. These modes are undoubtedly responsible for
the extreme softness of α-AgI. In reality, disorder will break the q selection
rules, but locally, low-frequency modes such as those shown in Fig.4.3 will still
exist although they will now couple to other local modes resulting in a large spread
of vibrational frequencies. Due to the breakdown of the q selection rules as well
as of the symmetry of the unit cell, many of these local modes will gain oscillator
strength which explains the low-frequency structures observed around 20 cm^{-1} in the
far-infrared spectrum.

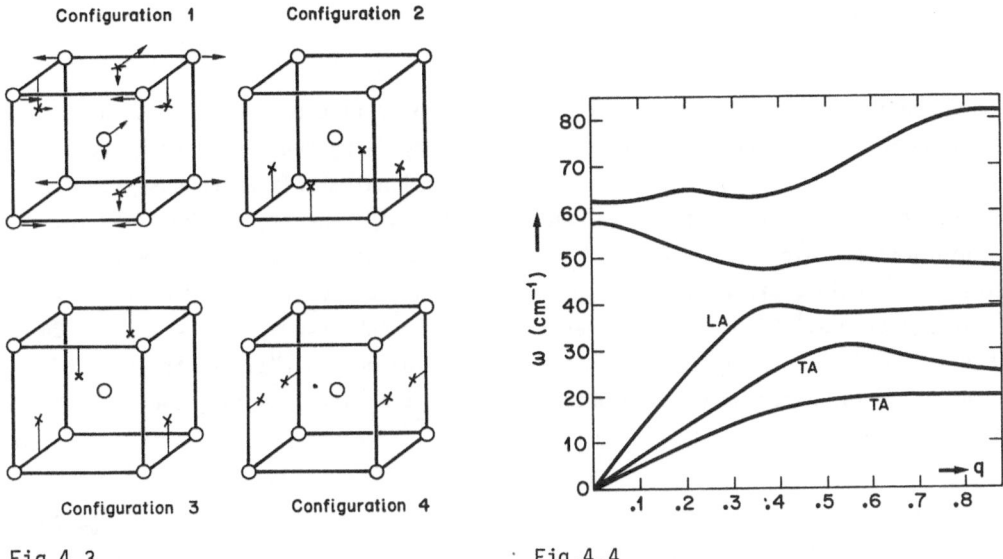

Configuration 1   Configuration 2

Configuration 3   Configuration 4

Fig.4.3   · Fig.4.4

Fig.4.3. The four different configurations of the silver ions in the cubic unit cell of α-AgI. All configurations with two silver ions on nearest neighbor sites are excluded. The statistical weight of configuration 1 is four, that of configurations 2,3, and 4 is one only. O iodine ions, × occupied silver sites. The arrows in configuration 1 indicate the eigenvector of the [110] TA zone boundary mode with a calculated frequency of 20 cm^{-1}

Fig.4.4. Calculated phonon dispersion of the acoustic and low-lying optical branches of α-AgI for q parallel to [110] of configuration 1. The low lying TA branch agrees well with the recent inelastic neutron scattering data [4.34]

It should be emphasized here that the potential provided by a rigid iodine sublattice would never allow for low-frequency modes of the silver ion. Only if both the silver *and* the iodine ions are moving in the way illustrated in Fig.4.3 are low frequency modes possible. Such local modes or simple combinations of them are easily shown to result in "reaction coordinates" for the jump of silver ions into nearest neighbor empty sites. It is clear, however, that for such large-amplitude motions, anharmonicity will be of prime importance and will further decrease the frequencies of these local lattice excitations. Such highly anharmonic lattice excitations might well be of importance in the diffusive regime. In fact, fluctuations of the reaction coordinates are expected to give rise to the observed structure in the microwave conductivity [4.16].

We now discuss our second model. The density of states is given by (4.56) and (4.54). In a first approximation we replace $\bar{\phi}$ in (4.54) by the mass-renormalized force-constant matrix ϕ^V of the virtual crystal defined by (4.46); in addition we neglect the effects of anharmonicity ($\Gamma = 0$) and of diffusion ($\Lambda = 0$). This approximation yields the density of states $N^V(\omega)$ and the conductivity $\sigma^V(\omega)$ of the virtual crystal. The dynamical matrix of the virtual crystal is defined by

$$D_{\alpha\beta}^V\left(\genfrac{}{}{0pt}{}{q}{\kappa\kappa'}\right) = \sum_{\ell'} \phi_{\alpha\beta}^V\left(\genfrac{}{}{0pt}{}{0\ell'}{\kappa\kappa'}\right) \exp[i\underline{q}\cdot\underline{r}(\ell')] \quad . \tag{4.69}$$

We have constructed the elements $\phi_{\alpha\beta}^V\left(\genfrac{}{}{0pt}{}{0\ell'}{\kappa\kappa'}\right)$ for the primitive unit cell containing one iodine ion ($\kappa = 1$) and one silver ion ($\kappa = 2$) by using the rope representation discussed in Sect.4.2. This results in a six-dimensional matrix which possesses the full symmetry of the bcc lattice. We consider here only nearest-neighbor I-Ag and I-I interactions; other short-range forces such as Ag-Ag interactions as well as Coulomb interactions are neglected. In order to construct $\phi_{\alpha\beta}^V\left(\genfrac{}{}{0pt}{}{00}{12}\right)$, for example, we have to sum over all 24 possible silver sites of the cube (numbered $i \doteq 1, \ldots, 24$ in Fig.4.2) which can be reached from the central iodine ion $\left(\genfrac{}{}{0pt}{}{0}{1}\right)$ with the rope of length $r = \sqrt{5}a/4$. Assuming equal probabilities $p(a_\ell; a_{\ell'}) = 1/24$ in (4.46), we obtain

$$\phi_{\alpha\beta}^V\left(\genfrac{}{}{0pt}{}{00}{12}\right) = \frac{1}{24}\sum_{i=1}^{24} \phi_{\alpha\beta}^{(i)}\left(\genfrac{}{}{0pt}{}{00}{12}\right) \quad , \tag{4.70}$$

where $\phi_{\alpha\beta}^{(i)}\left(\genfrac{}{}{0pt}{}{00}{12}\right)$ is the usual mass-renormalized force constant which describes the coupling of the central iodine ion $\left(\genfrac{}{}{0pt}{}{0}{1}\right)$ with the nearest silver ion at site i. The elements $\phi_{\alpha\beta}^V\left(\genfrac{}{}{0pt}{}{0\ell'}{12}\right)$ with $\ell' \neq 0$ are given similarly by

$$\phi_{\alpha\beta}^V\left(\genfrac{}{}{0pt}{}{0\ell'}{12}\right) = \frac{1}{24}\sum_{i=1}^{24} \phi_{\alpha\beta}^{(i)}\left(\genfrac{}{}{0pt}{}{0\ell'}{12}\right) \quad . \tag{4.71}$$

For the construction of $\phi_{\alpha\beta}^V\left(\genfrac{}{}{0pt}{}{01}{12}\right)$ for example, the rope of length r is now attached to the iodine ion at $\ell' = 1$ (Fig.4.2) and the summation formally extends over all the 24 nearest silver sites surrounding the iodine ion at $\ell' = 1$. Within the nearest-neighbor approximation, however, the sum extends only over the common nearest neighbors of cell $\ell = 0$ and cell $\ell' = 1$, in the example the sum runs over the six sites i = 1,2,5,8,17,18 (Fig.4.2). Since there is no disorder in the iodine sublattice, $\phi_{\alpha\beta}^V\left(\genfrac{}{}{0pt}{}{0\ \ell'\neq 0}{1\ \ \ 1}\right) = \phi_{\alpha\beta}\left(\genfrac{}{}{0pt}{}{0\ \ell'\neq 0}{1\ \ \ 1}\right)$. According to (4.46), the "diagonal elements" are given by the relations

$$\phi_{\alpha\beta}^V\left(\genfrac{}{}{0pt}{}{00}{\kappa\kappa}\right) = -\sum_{\substack{\ell' \\ \left(\genfrac{}{}{0pt}{}{\ell'}{\kappa'}\right)\neq\left(\genfrac{}{}{0pt}{}{0}{\kappa}\right)}} \phi_{\alpha\beta}^V\left(\genfrac{}{}{0pt}{}{0\ell'}{\kappa\kappa'}\right) \quad . \tag{4.72}$$

Neglecting Ag-Ag interactions we obtain, for example,

$$\phi_{\alpha\beta}^V\left(\genfrac{}{}{0pt}{}{00}{22}\right) = -\sum_{\ell'} \phi_{\alpha\beta}^V\left(\genfrac{}{}{0pt}{}{0-\ell'}{1\ \ 2}\right) \quad . \tag{4.73}$$

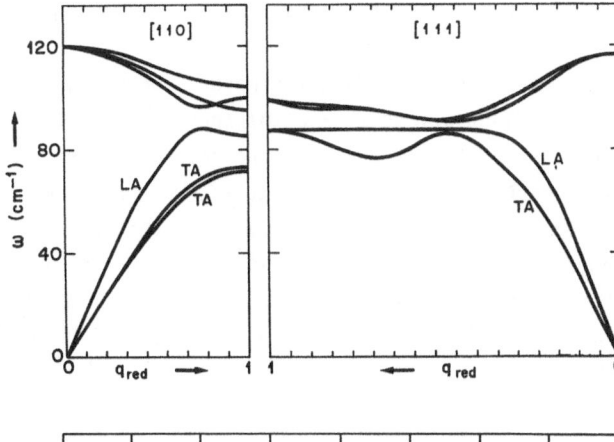

Fig.4.5. Phonon dispersion of the virtual crystal of α-AgI for q parallel to [110] and [111]. The TA branch in the [110] direction is much too hard compared with experiment

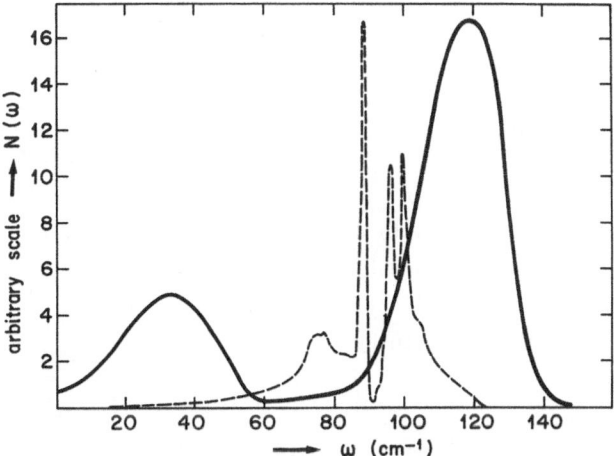

Fig.4.6. Phonon density of states of the virtual crystal of α-AgI (-----) and density of states calculated with CPA (———). The parameters for the CPA calculations are: Δ_1 = 0.388, Δ_2 = -056, p_1 = 0.6, p_2 = 0.4, Λ = 0.1, Γ = 0. Different normalizations are used for the two densities of states

In the actual calculations we have expressed the force constants $\phi_{\alpha\beta}^{(i)}\binom{0\ell\,'}{\kappa\kappa}$ in terms of the Ag-I stretching force constant F_1 = 0.370 mdyn/Å and the nearest-neighbor I-I stretching force constant F_2 = 0.045 mdyn/Å.

Figure 4.5 shows the phonon dispersion of the virtual crystal for q ∥ [110] and q ∥ [111]. The model predicts frequencies of the q = 0 TO modes of about 120 cm^{-1}, in agreement with experiments. However, the sound velocity of the [110] TA phonons is high and the TA zone-boundary frequencies reach values as high as 88 cm^{-1}. This is in sharp disagreement with the neutron scattering experiments which show a very flat dispersion of the TA branch in the [1$\bar{1}$0] direction and a mean zone boundary frequency of only 16 cm^{-1}. Figure 4.6 illustrates the phonon density of states $N^V(\omega)$ of the virtual crystal. As expected from the dispersion curves, $N^V(\omega)$ shows some pronounced peaks in the region of the TO branches but the density of states in the physically interesting region is very small in contrast to experiments.

We therefore conclude that the virtual crystal is much too hard. A better approximation is obtained if we allow for fluctuations $\phi^a - \phi^v$ of the force constants of a given configuration a with respect to the force constants of the virtual crystal. The problem is treated using the CPA method, that is, we replace $\bar{\phi}$ in (4.54) by ϕ_{CPA} defined by (4.47-49) which yields the density of states $N_{CPA}(\omega)$ and the conductivity $\sigma_{CPA}(\omega)$. We have performed such a calculation for the primitive unit cell but we realized that this cell is not large enough to contain those low-frequency local modes which are relevant for the real system. For simplicity we make the ansatz

$$\bar{\phi}_{\alpha\beta}\binom{\ell\ell'}{\kappa\kappa'} = \phi^v_{\alpha\beta}\binom{\ell\ell'}{\kappa\kappa'} + \delta_{\ell\ell'}\delta_{\kappa\kappa'}\delta_{\alpha\beta}\phi(\omega) \quad , \tag{4.74}$$

and replace (4.47) by

$$\text{tr}\{\sum_a p_a^{eq} Q_{exp} M^a\} = \text{tr}\{Q_{exp}\} \quad , \tag{4.75}$$

where Q_{exp} is the unit matrix in case of the density of states, and the oscillator strength matrix in case of frequency-dependent conductivity. This implies that all local modes are coupled in the same way to the effective medium. The equation for $\phi(\omega)$ for the case of the density of states is given by

$$\sum_{a,i} p_a \{1 + [\Delta_{ai} - \phi(\omega)]D^{-1}\}^{-1} = 1 \quad , \tag{4.76}$$

with

$$D = \int d\omega' N^v(\omega')[-\omega^2 + \omega'^2 + \phi(\omega)]^{-1} \quad , \tag{4.77}$$

and where Δ_{ai} denotes the eigenvalues of $\phi^a - \phi^v$.

The density of states is then given by (4.56) and a corresponding expression holds for $\phi(\omega)$. As shown in Sect.4.1 the influence of diffusion and anharmonicity can be included by choosing $\Lambda \neq 0$ and by writing $\bar{\phi} = \phi_{CPA} - i\omega\Gamma$.

To demonstrate what CPA can produce, we drastically simplify the eigenvalue spectrum Δ_{ai} of $\phi^a - \phi^v$ and use only two eigenvalues Δ_1 and Δ_2 corresponding to a high-frequency Ag-I stretching mode and to a low-frequency "bending mode". Based on the information from our first model, we can estimate rough values for the eigenvalues Δ_1, Δ_2 the probabilities p_1, p_2 and the oscillator strengths f_1 and f_2 of these two modes. The results of the calculations are illustrated in Figs.4.6 and 4.7 which show $N_{CPA}(\omega)$ and $\sigma_{CPA}(\omega)$. The model predicts the main experimental features: a peak at high frequencies corresponding to the residual ray absorption

and a structure at low frequencies which originates from the low-frequency modes.
A high density of states at low frequencies is to be expected from far-infrared
[4.16,4,33], Raman [4.33] (see also Chap.5), and inelastic neutron scattering experiments [4.34]. In the far-infrared spectra, a very strong disorder-induced absorption at low frequencies has been observed, while in the Raman experiments a
large increase in intensity is observed for $\omega < 70$ cm^{-1} when heating through T_c.
The very flat and broad $1\bar{1}0$ TA branch observed by inelastic neutron scattering
is also qualitatively consistent with the calculated $N_{CPA}(\omega)$ shown in Fig.4.6. It
should be emphasized that our oversimplified model on which $N_{CPA}(\omega)$ and $\sigma_{CPA}(\omega)$ of
Figs.4.6 and 4.7 are based does not allow qualitative comparison with the experi-
ments. Better agreement of $\sigma_{CPA}(\omega)$ with the experiment (Fig.4.8) would be obtained
if more than two eigenvalues and anharmonicity were taken into account ($\Gamma \neq 0$). We
believe that a realistic calculation should be done with a cubic unit cell which
corresponds to a 12-dimensional problem for each configuration a.

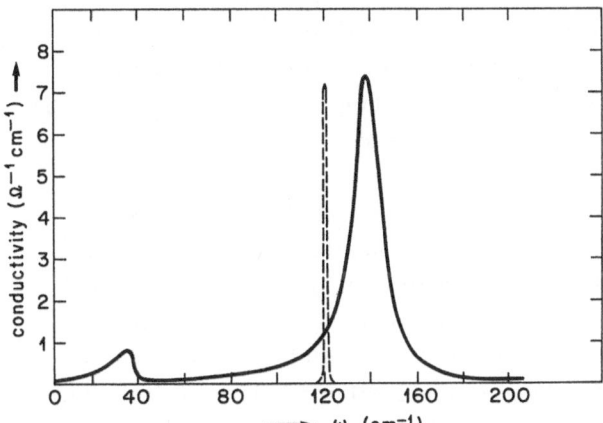

Fig.4.7. Frequency-dependent
conductivity $\sigma(\omega)$ of α-AgI.
(-----) $\sigma(\omega)$ of the virtual
crystal. (———) $\sigma_{CPA}(\omega)$ from
CPA calculations with para-
meters $\Delta_1 = 0.388$, $\Delta_2 = -0.56$,
$p_1 = 0.6$, $p_2 = 0.4$, $f_1 = 1.0$,
$f_2 = 0.3$, $\Lambda = 0.1$, $\Gamma = 0$. All
energies are measured in units
of the residual ray energy

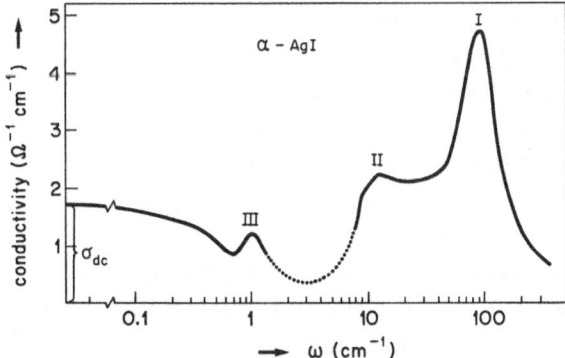

Fig.4.8. Experimentally observed
conductivity of α-AgI at T=453 K.
The high frequency ($\omega \geq 7$ cm^{-1}) results are our own data, the re-
sults below $\omega \sim 1.5$ cm^{-1} are from
[4.35]

4.3.4 Collective Excitations in One-Dimensional Systems: Hollandite

There are ionic conductors in which the ionic motion is confined to noncommunicating, fractionally occupied channels, for example, hollandite [4.36] or β-eucryptite [4.37]. Because of their low dimensionality, these systems are excellent models for a detailed study of some particular aspects of superionic conductors. In particular we study the possibility of collective excitations due to the ion-ion interaction.

In what follows, we assume such a system to consist of a channel containing N equivalent sites with a mutual spacing a and a total of ρN ions (ρ < 1). Assuming a uniformly charged background we may then ask about the equilibrium distribution of the ions in the presence of a two-body electrostatic potential

$$ V(r) = \frac{q^2}{\varepsilon(r) \cdot r} \quad , \tag{4.78} $$

where q is the charge, ε the dielectric constant, and r the distance between the two ions. Instead of dealing here with the $\binom{N}{\rho N}$ configurations of the complete system as defined in Sect.4.2 it is advantageous to break them up into arrays, each array A_n (with associated probability P_n) consisting of a vacancy and (n - 1) occupied sites up to the next vacancy (excluded). The computation of the probability P_n is described in detail in [4.8,19].

Applying this model of the static state of order in a one-dimensional system to the case of hollandite ($K_{1.54}Mg_{0.77}Ti_{7.23}O_{16}$) one has to account for the exceptionally high dielectric constant of the framework (originating from the high polarizability of the TiO_6 octahedral) which is effective beyond the first Ti shell around a channel. Within the channel the effective dielectric constant is very much lower, and is determined by the electronic polarizability of the potassium and oxygen ions. We approximate this situation by a dielectric cylinder of radius R with an external dielectric constant ε_a = 100 and in internal value ε_i = 2.

Experimental information on order in the hollandite channels was derived from an analysis of the diffuse X-ray scattering experiments which yielded approximate values for the probabilities P_n for channels with an average occupancy ρ = 0.77.

Within the independent-array approximation, the best fit to the experimental P_n values was found with R = 4.3 Å. This completely determines the effective ion-ion potential V(r). The fitted V(r) corresponds to a function ε(r) increasing rapidly from 2 at r = 0 to 100 for r ≥ 5a [4.19].

We have thus far assumed the ions to reside exactly on the crystallographic channel sites. One of the first experimental results was, in fact, the conclusion that some of the ions are as far as a/4 off the regular sites [4.38]. Within the lattice gas concept introduced in Sect.4.2 (defined as a parceling in configuration space rather than as a set of discrete sites in real space), this effect can be dealt with by adding to the effective Hamiltonian a sinusoidal term of period a and amplitude V_0 representing the interaction with the framework lattice. A fit of this extended model to the experimental data yields $V_0 \sim 0.2$ eV, which is a typical value for a barrier height in a superionic conductor.

We have now all the ingredients to study the collective excitations of this system by using the analysis described in Sect.4.2.1.

The period of the background periodic potential is a = 2.9 Å. The density ρ = 0.77 corresponds to an average distance between nearest potassium ions b = a/ρ = 3.77 Å. From the effective ion-ion potential V(r) we obtain the elastic constant [4.19]

$$A = \left. \frac{\partial^2 V}{\partial r^2} \right|_{r=b} = \frac{1.72 \text{ eV}}{b^2} \quad . \tag{4.79}$$

These values correspond to the following parameters for the collective excitation consisting of a vacancy plus the associate chain distortion [see (4.56-61)]

$$\ell_0 = 0.67 \text{ a} \quad ,$$

$$M_s = 0.31 \text{ m}_k \quad , \tag{4.80}$$

where m_k is the mass of potassium ion. The pinning barrier for the propagation of this excitation can be deduced from Fig.4.1. In hollandite we have δ = 0.25, which corresponds to

$$V_p \simeq 0.018 \text{ eV} < k_B T \tag{4.81}$$

at room temperature. The conversion factor from the single particle damping η to the damping of the collective excitation Γ_s is [see (4.62)]

$$\Gamma_s = 0.92\eta \quad . \tag{4.82}$$

The fact that this conversion factor is in general close to unity for realistic interactions is a very interesting result. It shows in fact that in order to have real solitonlike propagation, small single-particle damping is needed. This condition is satisfied, for example, in the case of dislocations [4.37], but in general is not satisfied for superionic conductors.

For hollandite, far-infrared measurements indicate than η is quite large: $\eta/m_k \gtrsim 50$ cm^{-1} [4.40]. This together with the other parameters gives an extremely damped ($\Gamma_s/M_s \gtrsim 150$ cm^{-1}) Drude-like behavior for the dynamics of collective excitations. The corresponding mean free path at room temperature is in fact

$$\lambda_s \simeq \sqrt{\frac{k_B T}{M_s} \frac{M_s}{\Gamma_s}} \lesssim 1 \, \text{Å} \quad .$$ (4.83)

A last important point to be mentioned concerns the number of such excitations. We try to make this clear with an example. Let us consider the case $\rho = 1$ ($a = b$); then solitons can be created only in pairs with a creation energy $2E_s$, and one of the two solitons is actually a moving vacancy with the associated lattice distortion. If ρ is slightly less than one, we can directly identify the solitons with the *already existing* vancancies for which the creation energy is obviously zero. For $\frac{1}{2} < \rho < 1$ the situation is in general complex, but certainly for $\rho = \frac{1}{2}$ a finite creation energy is needed to have soliton-type excitations. In this case there is no relation between the holes in the ground state and the number of soliton-type excitations. In the case $\rho = 0.77$ we are probably too far from the dilute limit to conclude that the creation energy for these excitations is small.

4.4 Conclusions

We have shown that for highly conducting solid electrolytes, free-rate theory or its equivalent for an interacting system, for example, the master equation is not adequate. The model introduced in this paper begins to fill the gap between the low-temperature limit and the quasi-free particle limit.

The terms beyond the master equation become important whenever the transition rate becomes comparable to characteristic relaxation rates of the system. If the residence time is shorter than the relaxation time of a particular lattice or local mode, then the transition rate effectively couples to this mode. This in turn leads to a retarded effective particle-particle interaction mediated by vibronic fluctuations of the system. A retardation in the velocity-velocity correlation function is equivalent to an oscillation in the frequency-dependent conductivity. This is included in the third term of (4.32) and forms the basis for explaining the microwave conductivity of AgI type conductors.

For strongly interacting systems the interaction may drastically reduce the activation energy for conduction from its single particle value, and introduce a dominant collective component to the reaction coordinate. This can reach the point where the single-particle character of a transition is completely lost and the activation energy becomes smaller than $k_B T$. If the collective entity, defect plus associated distortion field, can propagate in a quasi-free fashion then it is called a soliton. In other systems (Bloch walls, charge density waves, etc.) the interaction which leads to the existence of the soliton also effectively suppresses scattering processes and enables a propagation of the soliton. In superionic conductors the scattering mechanism is the particle-phonon interaction and is operative in very much the same way for single particles and solitons. The soliton in superionic conductors is thus not a propagating particle but performs a highly viscous Brownian motion. Soliton transport is a case where the fourth conductivity term of (4.32) is important.

Whereas the third term of (4.32) describes a situation in which the system does not reach equilibrium within a mean residence time, the fourth term describes a situation in which the particle (or reaction coordinate) does not thermalize after a transition. If this latter situation prevails, then a particle (or reaction coordinate) may undergo multiple jumps before thermalizing. In the low-damping limit this leads to quasi-free propagation, and in the strong-damping limit, to viscous Brownian motion with an effective jump distance much shorter than an elementary jump distance. If the crystal potential can be treated as a perturbation, then it is easy to calculate the forth term, and we find that for solitons it leads to the strong-damping limit. The parameters of all experimental systems are such that upon raising the temperature they would reach the strong damping limit in the absence of melting. A genuine free-particle limit can never be obtained.

A calculation of the conductivity terms beyond the master equation requires information on the lattice dynamics of the system. The lattice dynamics are complicated by disorder and anharmonicity which are inherent properties of superionic conductors. We have shown that already in relatively small unit cells, disorder generates oscillator strength at small frequencies. Disorder thus not only provides empty lattice sites but also low-lying modes conditional for fast diffusion.

The principal results of this paper with regard to experimental systems can be illustrated in Fig.4.8, which shows the experimentally observed conductivity of AgI. The harmonic lattice leads to a pronounced reststrahl resonance around 100 cm^{-1}(I). Disorder transfers oscillator strength to low frequencies [shoulder around 15 cm^{-1}(II)] an effect which is further enhanced by anharmonicity. The resulting low-frequency fluctuations couple to the diffusive motion and lead to correlated particle jumps responsible for the structure at $\sim 1 \text{ cm}^{-1}$(III). It is important to note that a large part of the observed dc conductivity is due to terms beyond the master equation. This is directly evident when comparing Fig.4.8 with the master equation result $d\sigma/d\omega \geq 0$ in the diffusive regime.

In conclusion we have for the first time introduced a model which deals with the dynamics of classical diffusing particles on a microscopic level. The corrections to the standard master equation results are large for good conductors and may be overwhelming in the case of collective transport (solitons).

Many of the results presented here are sketchy and clearly require substantial further work. Nevertheless we feel that we have moved a major step in the direction of understanding highly conducting systems.

Acknowledgements. It is a pleasure to thank J. Bernasconi, P. Fulde, and H.J. Wiesmann for numerous discussions, and W. Bührer for making available unpublished experimental results as well as for his computer calculations of the density of states of the virtual crystal.

Appendix A

Let (Γ, ϕ_t, μ) be a classical dynamical system with phase space Γ, flow ϕ_t, and invariant measure μ. If $(x_a)_{a \in I}$ is a partition of unity, i.e., x_a is measurable, with range $x_a = \{0,1\}$ and (almost everywhere)

$$x_a(\gamma)x_b(\gamma) = 0 \, , \quad a \neq b \, , \quad \sum_{a \in I} x_a(\gamma) = 1 \quad , \tag{A.1}$$

we define (this is the crucial point)

$$[U_{ab}(t)f](\gamma) = x_a(\gamma)x_b(\phi_t\gamma)f(\phi_t\gamma) \tag{A.2}$$

for any (measurable) function f on Γ. A simple calculation yields

$$\sum_{b \in I} U_{ab}(t)U_{bc}(s) = U_{ac}(t + s) \quad . \tag{A.3}$$

To $f : \Gamma \to \mathbb{C}$ we associate $F : \Gamma \to \mathbb{C}^{|I|}$ by

$$F_a(\gamma) = f(\gamma)x_a(\gamma) \quad , \tag{A.4}$$

and we define $U(t)F$ by

$$[U(t)F]_a = \sum_{b \in I} U_{ab}(t)F_b \quad , \tag{A.5}$$

and a scalar product (F,G) by

$$(F,G) = \sum_{a \in I} \int_{\Gamma_a} d\mu(\gamma)\overline{F_a(\gamma)}G_a(\gamma) \quad , \tag{A.6}$$

where $\Gamma_a = \{\gamma | x_a(\gamma) = 1\}$.
   Another simple calculation yields

$$(U(t)F, U(t)G) = (F,G) \tag{A.7}$$

Hence, by (A.3) $[U(t)]_{t \in \mathbb{R}}$ is a one-parameter unitary group in the Hilbert space of square-integrable $\mathbb{C}^{|I|}$-valued functions with scalar product given by (A.6). Its generator L is given formally by

$$(LF)_a = \sum_{b \in I} L_{ab} F_b \quad , \tag{A.8}$$

with

$$L_{ab} = L^0_{ab} + L^1_{ab} \quad . \tag{A.9}$$

The first term of (A.9) is given by

$$(L^0_{ab} f)(\gamma) = \lim_{t \downarrow 0} \chi_a(\gamma)\chi_b(\phi_t\gamma)t^{-1}[f(\phi_t\gamma) - f(\gamma)], \text{ i.e.,}$$

$$L^0_{ab} = \chi_a \chi_b L^0 \quad , \tag{A.10}$$

where L^0 is the generator of the unitary group $[U^0(t)]_{t \in \mathbb{R}}$ definded by $[U^0(t)f](\gamma) = f(\phi_t\gamma)$ on the Hilbert space of square-integrable functions. It vanishes unless $a = b$ where it is essentially given by L^0 apart from boundary conditions. The second part of (A.9) is given by

$$(L^1_{ab} f)(\gamma) = \lim_{t \downarrow 0} \chi_a(\gamma)t^{-1}[\chi_b(\phi_t\gamma) - \chi_b(\gamma)]f(\gamma) \quad . \tag{A.11}$$

It may be called transfer part of the generator as it describes transfer from Γ_a to Γ_b.

For a Hamiltonian system with configuration space $\Gamma = \mathbb{R}^n \times \mathbb{R}^n$ and Hamilton function

$$H = \sum_{k=1}^{n} \frac{p_k^2}{2m_k} + V(q_1, \ldots, q_n) \quad , \tag{A.12}$$

the generator L^0 ("Liouvillian") is given by

$$L^0 = \sum_{k=1}^{n} \frac{p_k}{m_k} \frac{\partial}{\partial q_k} - \frac{\partial V}{\partial q_k} \frac{\partial}{\partial p_k} \quad . \tag{A.13}$$

We assume that the partition of unity $(\chi_a)_{a \in I}$ is generated by open disjoint sets $D_a, a \in I$, in configuration space such that the boundaries ∂D_a, $a \in I$, satisfy either:

(I) the intersection $\partial D_a \cap \partial D_b$ is empty,

or

(II) $\partial D_a \cap \partial D_b$ is a (smooth) k-dimensional manifold, $k \le n - 1$; D_a and D_b are called neighbors if $\partial D_a \cap \partial D_b = S_{ab}$ is a $(n - 1)$-dimensional manifold.

Under these assumptions the explicit form of the transfer Liouvillian given by (4.20) in Sect.4.2 may be obtained from (A.11).

Appendix B

Let L be a skew-adjoint operator in a Hilbert space H with domain D(L). Then, in $|\text{Re}\{z\}| \geq \varepsilon|z|$, $0 < \varepsilon < 1$,

$$\lim_{z \to 0} (z - L)^{-1}Lf = Pf - f, \quad f \in D(L) \quad , \tag{B.1}$$

where $P = E(\{0\})$ if E denotes the spectral measure associated with L. If 0 is isolated from the rest of the spectrum of L, the limit is in the norm, otherwise strong.

If $H = L^2(\Gamma,\mu)$ for some probability measure space (Γ,μ) and if 0 is a simple eigenvalue with eigenfunction 1 of a skew-adjoint operator L in H, then, by (B.1),

$$\lim_{z \to 0} (z - L)^{-1}Lf = <f> - f \tag{B.2}$$

for f in D(L), where $<f>$ is the expectation of f with respect to μ. This result has to be used in deriving (4.22).

Appendix C

If A is a positive self-adjoint operator in a Hilbert space H then the following inequalities hold for positive ω and all f in the domain of A:

$$\text{Re}\{(f,(i\omega + A)^{-1}f)\} \geq 0 \quad , \tag{C.1}$$

$$\frac{d}{d\omega} \text{Re}\{(f,(i\omega + A)^{-1}f\} \leq 0 \quad . \tag{C.2}$$

Using the spectral resolution of A

$$A = \int_0^\infty \lambda \, dE_\lambda \tag{C.3}$$

leads to

$$\text{Re}\{(f,(i\omega + A)^{-1}f)\} = \int_0^\infty \frac{\lambda}{\lambda^2 + \omega^2} (f,dE_\lambda f) \quad , \tag{C.4}$$

and

$$\frac{d}{d\omega} \text{Re}\{(f,(i\omega + A)^{-1}f)\} = -2\omega \int_0^\infty \frac{\lambda}{(\lambda^2 + \omega^2)^2} (f,dE_\lambda f) \quad . \tag{C.5}$$

From (C.4) it follows that there exists a nonnegative γ with

$$\text{Re}\{(f,(i\omega + A)^{-1}f)\} = \gamma(\gamma^2 + \omega^2)^{-1} \quad . \tag{C.6}$$

Let now H be the Hilbert space of sequences $f = (f_a)_{a\in I}$ with a finite or countable index set I. The scalar product is given by

$$(f,g) = \sum_{a\in I} p_a \bar{f}_a g_a \quad , \tag{C.7}$$

where $0 < p_a \leq 1$, and

$$\sum_{a\in I} p_a = 1 \quad . \tag{C.8}$$

Let A be given by

$$(Af)_a = \sum_{b\in I} \Gamma_{ab}(f_a - f_b) \tag{C.9}$$

where $\Gamma_{ab} \geq 0$,

$$\sum_{b\in I} \Gamma_{ab} < \infty \quad , \tag{C.10}$$

and

$$p_a \Gamma_{ab} = p_b \Gamma_{ba} \tag{C.11}$$

for $a,b \in I$, $a \neq b$. It follows that

$$(f,Af) = \frac{1}{2} \sum_{a,b} p_a \Gamma_{ab} |(f_a - f_b)|^2 \quad , \tag{C.12}$$

i.e., A is positive. As $\exp(-tA)$, $t \geq 0$ is a Markov transition-matrix function [4.39], A is also self-adjoint. Hence, (C.1) and (C.2) are applicable.

References

4.1 L. Landau, E. Lifshitz: *Statistical Physics* (Oxford University Press, New York 1938) p.184
4.2 C.P. Flynn: *Point Defects and Diffusion* (Clarendon Press, Oxford 1972) pp.306-327
4.3 R. Vargas, M.B. Salamon, C.P. Flynn: Phys. Rev. Lett. *37*, 1550 (1976)
4.4 M.J. Rice, S. Strässler, G.A. Toombs: Phys. Rev. Lett. *32*, 546 (1974)
4.5 W.J. Pardee, G.D. Mahan: J. Solid State Chem. *15*, 310 (1975)

110

4.6　H. Sato: In *Solid Electrolytes*, Topics in Appl. Phys. Vol.21, ed. by S. Geller. (Springer, Berlin, Heidelberg, New York 1977) p.3
4.7　H. Sato, R. Kikuchi: J. Chem. Phys. *55*, 677 (1971)
4.8　L. Pietronero, S. Strässler: Z. Phys. B *32*, 339 (1979)
4.9　W. Dieterich, I. Peschel, W.R. Schneider: Commun. Phys. *2*, 175 (1977) Eq. (13)
4.10 D.L. Huber: Phys. Rev. B*15*, 533 (1977)
4.11 P.M. Richards: Phys. Rev. B*16*, 1393 (1977)
4.12 J.C. Kimball: Bull. Amer. Phys. Soc. *22*, 369 (1977)
4.13 R. Kikuchi: *Fast Ion Transport in Solids, Solid State Batteries and Devices*, ed. by W. van Gool. (North-Holland, Amsterdam 1973) p.555
4.14 R.J. Elliot, J.A. Krummhansl, P.L. Leath: Rev. Mod. Phys. *46*, 465 (1974)
4.15 S. Strässler, W.R. Schneider, G.A. Toombs: To be published
4.16 K. Funke: Prog. Solid State Chem. *11*, 345 (1976)
4.17 L. Pietronero, S. Strässler: Solid State Commun. *27*, 1041 (1978)
4.18 J.C. Wang, D.F. Pickett, Jr.: J. Chem. Phys. *65*, 5278 (1977)
4.19 H.U. Beyeler, L. Pietronero, S. Strässler, H.J. Wiesmann: Phys. Rev. Lett. *38*, 1532 (1977)
4.20 A. Barone, F. Esposito, C.J. Magee, A.C. Scott: Riv. Nuovo Cimento *1*, 227 (1971)
4.21 M.B. Fogel, S.E. Trullinger, A.R. Bishop, J.A. Krummhansl: Phys. Rev. B*15*, 1587 (1977)
4.22 P. Fulde, L. Pietronero, W.R. Schneider, S. Strässler: Phys. Rev. Lett.*26* , 1776 (1975)
4.23 W. Dieterich, I. Peschel, W.R. Schneider: Z. Phys. B*27*, 177 (1977)
4.24 L.W. Strock: Z. Physik. Chem. B*25*, 441 (1934)
4.25 R.J. Cava, F. Reidinger, B.J. Wuensch: Solid State Commun. *24*, 411 (1977)
4.26 J.C. Boyce, T.M. Hayes, W. Stutius, C. Mikkelsen: To be published
4.27 W. Bührer, W. Hälg: Helv. Phys. Acta *47*, 27 (1974)
4.28 H.U. Beyeler: Unpublished
4.29 J. Nölting: Ber. Bunsenges. Phys. Chem. *67*, 172 (1963)
4.30 W. Bührer, P. Brüesch: To be published
4.31 R. Alben, G. Burns: Phys. Rev. B*16*, 3746 (1977)
4.32 P. Brüesch, S. Strässler, H.R. Zeller: Phys. Stat. Sol. (a) *31*, 217 (1975)
4.33 G. Burns, F.H. Dacol, M.W. Shafer: Phys. Rev. B*16*, 1416 (1977)
4.34 W. Bührer, P. Brüesch: Phys. Rev. B, to be published
4.35 K. Funke, A. Jost: Ber. Bunsenges. Phys. Chem. *75*, 436 (1971)
4.36 A. Byström, A.M. Byström: Acta Cristallogr. *3*, 146 (1950)
4.37 H.G.F. Winkler: Acta Cristallogr. *1*, 27 (1948)
4.38 H.U. Beyeler: Phys. Rev. Lett. *37*, 1557 (1976)
4.39 J.L. Doob: *Stochastic Processes* (Wiley and Sons, New York, London, Sidney 1967)
4.40 H.U. Beyeler: Unpublished

5. Light Scattering in Superionic Conductors

M. J. Delaney and S. Ushioda

With 10 Figures

In the past dozen or so years, inelastic scattering of light has become one of the most powerful means of investigating elementary excitations in condensed matter, along with such other methods as inelastic neutron scattering and photon absorption. Light scattering has been used in numerous ways to study both acoustic and optical phonons, and electronic and magnetic excitations of solids [5.1,2]. In this chapter we review the information on superionic conductors obtained by light scattering spectroscopy and attempt to correlate these results with the data gained by different techniques described in other chapters of this book. Before we go into a discussion of specific results for individual superionic materials, we shall outline the mechanisms involved in the light scattering process and the kind of information one gains by light scattering experiments.

Traditionally, Raman scattering refers to inelastic light scattering from optical phonons and Brillouin scattering, to scattering from acoustic phonons. However, in the modern context the distinction is made more on the basis of the magnitude of the accompanying frequency shift of the light and the experimental techniques used in measuring the frequency shift in the scattering process. In a Raman scattering experiment, one usually measures a frequency shift of 1 cm^{-1} (1 meV = 8 cm^{-1}) or greater by using a grating spectrometer, while in Brillouin scattering the typical range of frequency shift is 10^{-3} cm^{-1} (30 MHz) to 1 cm^{-1} (30 GHz) and a Fabry-Perot interferometer is employed. Quasi-elastic light scattering with a frequency shift less than a few tens of MHz is referred to as Rayleigh scattering, and here the standard method of spectrum analysis is a photon beating technique.

From light scattering experiments one determines the energy, momentum, lifetime, and symmetry of an elementary excitation. The energy ($\hbar\omega$) and momentum ($\hbar k$) are obtained from the difference ($\omega = \omega_i - \omega_s$) between the frequency of the incident light (ω_i) and the scattered light (ω_s) and the difference in the wave vector ($\underline{k} = \underline{k}_i - \underline{k}_s$). The lifetime τ is the inverse of the spectral linewidth Γ. The symmetry of an elementary excitation is found from the polarization selection rules

---

$\hbar$ = h/2π (normalized Planck's constant)

of the scattering process; that is, the scattering intensity is finite only for certain combinations of the polarization of the incident light $(\hat{e}_i)$ and the scattered light $(\hat{e}_s)$. The coupling constant of the elementary excitation to the incident photons, which determines the scattering intensity, is an important parameter in light scattering experiments. In a typical Raman scattering experiment the incident laser beam interacts very weakly with the sample, with only one in 10^{14} photons being scattered and detected.

The scattering of light in a crystal is caused by local fluctuations of the dielectric constant of the medium. With a typical wavelength of visible light of $\lambda \simeq 5000$ Å, light scattering experiments detect dielectric fluctuations having an extent of the same order as λ. Thus the magnitude of the wave vector $|k|$ of the fluctuations measured by light scattering is on the order of 10^5 cm^{-1}, which is three orders of magnitude smaller than the typical size of the first Brillouin zone in a crystal. In terms of the local fluctuations of the dielectric constant, $\delta\varepsilon(\underline{r},t)$, the light scattering intensity at frequency shift ω and wave-vector change $\underline{k}$ is given by [5.3]

$$I(\underline{k},\omega) = A \int_{-\infty}^{\infty} dt\, e^{-i\omega t} \int_V\!\!\int d^3r\, d^3r'\, e^{i\underline{k}\cdot(\underline{r}-\underline{r}')} \left\langle \delta\varepsilon(\underline{r}',0)\delta\varepsilon(\underline{r},t)\right\rangle , \qquad (5.1)$$

where A is a proportionality constant, and < > signifies the time-correlation function. Sometimes it is more convenient to deal with the polarizability $\alpha(\underline{r},t)$, and a similar expression for $I(\underline{k},\omega)$ can be written in terms of $\delta\alpha(\underline{r},t)$. Thus the scattered light at the wave vector and frequency shift $(\underline{k},\omega)$ arises from the dielectric constant fluctuations of the same wave vector and frequency shift.

When one considers light scattering from atomic vibrations in solids, where the harmonic approximation is appropriate [5.4], one expands $\delta\varepsilon(\underline{r},t)$ in terms of the phonon normal coordinates and decomposes the scattering intensity $I(\underline{k},\omega)$ into first-order scattering, second-order scattering, etc., according to the number of phonons involved in the scattering process [5.3]

$$I(\underline{k},\omega) = I_1(\underline{k},\omega) + I_2(\underline{k}' + \underline{k}'',\omega' + \omega'') + \text{higher order terms} . \qquad (5.2)$$

If the crystal lattice is perfect and lattice translations are good symmetry operations, the crystal momentum is conserved and we have $\underline{k} = \underline{k}_i - \underline{k}_s$ for Stokes scattering, where $\underline{k}$ is the wave vector of the phonon responsible for the first-order scattering. Since $|\underline{k}_i| = |\underline{k}_s| \simeq 10^5$ cm^{-1} and $|k| \approx 10^5$ cm^{-1}, a first-order process involves only phonons at the Brillouin zone center. In a second order scattering process, one has the relation $\underline{k}' + \underline{k}'' = \underline{k}_i - \underline{k}_s$ for the wave vectors of the two phonons $\underline{k}'$ and $\underline{k}''$. Thus $\underline{k}'$ and $\underline{k}''$ can have a large magnitude with opposite directions, and a pair of phonons with opposite wave vectors at any point in the Brillouin zone can scatter light. The peaks in the second-order scattering intensity I_2 gen-

erally correspond to the peaks in the two-phonon density of states, which usually
occur at high symmetry points at the zone boundary. These higher-order processes
are, in general, considerably weaker in intensity and broader in spectral width
than the one-phonon processes.

In liquids, in contrast to solids, the constituent molecules or atoms are in
motion and not fixed in a harmonic well; further, there is no microscopic translational symmetry. Thus the harmonic approximation which enables one to decompose
atomic motions in solids into phonon normal modes is not appropriate for the translational motions in liquids. Atomic vibrations within individual molecules (intramolecular vibrations) are well described by the harmonic approximation, and in
light scattering experiments, one observes sharp peaks due to first-order scattering by the quanta of intramolecular vibrations. The fluctuations of the dielectric
constant in a liquid that scatter light are produced by several mechanisms including intermolecular vibrations and rotations. In monoatomic liquids molecular vibrations and rotations are absent, and scattering is caused by the dielectric-constant
variations due to density fluctuations [5.6] and atomic collisions [5.7]. Of course,
these mechanisms are also present in molecular liquids. The dielectric constant is
proportional to the liquid density in a first approximation. Thus dielectric-constant fluctuations are proportional to density fluctuations. Atomic collisions in
a liquid cause a distortion of the electron distribution about the colliding atoms,
and the altered electron distribution in turn produces dielectric-constant fluctuations. Experimentally observed light scattering spectra of liquids are super-
positions of scattering caused by these mechanisms and they are not as well understood as the solid spectra.

The dynamics of ionic motion in superionic materials have similarities to those
of both liquids and solids. As is now generally recognized, the highly mobile
metal ion can reside at one of several available sites created by the rigid lattice
of the nonconducting species. In this lattice framework the mobile species of ions do
not stay localized in a potential well, but move through the lattice of stationary
ions, although they may stay localized long enough to execute many oscillatory
cycles in a given potential well. An important parameter, which can be easily estimated, is the ratio of the time spent at a given site t_d to the time spent between
sites t_f. If $t_d > t_f$, a hopping or jump diffusion model is appropriate; on the
other hand if $t_d < t_f$, we have a continuous diffusion or fluid model. Elementary
considerations from diffusion theory [5.8] give $t_d \simeq a_0^2/6D$ where a_0 is the hopping
distance and D is the diffusion constant. Time-of-flight estimates from thermal considerations give $t_f \simeq (ma_0^2/k_BT)^{1/2}$ where m is the mass of the diffusing ion and T is
the temperature. In most superionic conductors, with $a_0 \cong 2$ Å and $D \cong 10^{-5}$ cm^2/s,
$t_d(\simeq 3 \text{ ps}) \geq t_f(\simeq 0.5 \text{ ps})$, and a jump diffusion model can be considered. Within
this approximation estimates indicate that the jump and vibrational frequencies of
the mobile species are comparable to those of the normal lattice vibrations and

hence within the realm of light scattering experiments. Since the potential wells
that the mobile species see are very shallow ($E \simeq 0.1$ eV), the oscillation amplitude is very large and the motions are quite anharmonic. Thus one may not expect
the harmonic approximation to give a good picture of the dynamics of ions in these
materials. Another important feature of solids that facilitates simplification of
theories is the translational symmetry. This feature is lost to varying degrees in
superionic materials because of the disorder caused by the large number of defects
which characterizes the highly conducting state. Thus discussions of light scattering spectra in terms of phonons with a definite wave vector $\underline{k}$ and frequency ω
seem to be quite inadequate, although one tends to think in terms of the familiar
concept of phonons in the absence of better concepts. In the following sections we
discuss the Raman and Brillouin spectra of several superionic materials in terms
of "phonons" for convenience, but one should keep in mind that "phonons" in this
context are very much an approximation to reality. Unfortunately, on the other hand,
microscopic theories of the dynamics of atoms and molecules in liquids do not seem
to be well enough developed to replace the phonon concept. One method which goes
beyond the harmonic approximation is the method of molecular dynamics by computer
simulation [5.9] but this method has the difficulty of lacking simple picture of the
physics involved. Several innovative approaches have recently been taken to arrive
at an appropriate theory of light scattering from the highly anharmonic and disordered superionic materials (see Chap.8). None of these theories is sophisticated
enough to describe the dynamics in detail nor to predict new correlations among
different experimental results.

Superionic materials, interesting to study in their own right, in addition represent an interesting example of a highly anharmonic and disordered structure
bridging the gap between solid and liquids. In terms of understanding the dynamics
of the mobile ions, studies of superionic materials have just begun, and we review what has been learned so far through light scattering studies.

5.1 Raman Scattering

Raman scattering experiments, which have been performed on the three classes of
superionic conductors —the metal halides, the β-aluminas, and the fluorites, show
broad, structureless spectra with only weak temperature dependences. Each class,
however, has properties that sets it apart from the others. There are many similarities among the metal halide compounds and comparisons of the spectra aid in
the determination of the origin of the features. In the layered ceramic β-aluminas,
the segregation of the spectral features due to mobile cations at low frequencies
simplifies the interpretation of the data. In the fluorites, maintenance of the

symmetry of the structure over the entire temperature range simplifies the identification of the contribution of the disordered anion sublattice to the spectrum. The similarities and differences in the properties of the three classes of superionic materials collectively enhances the understanding of the Raman spectrum of the individual materials.

5.1.1 Silver Iodide

Silver iodide, long studied for its unique mobile cation effects, was a logical choice for one of the first in a long sequence of optical studies of superionic conductors. The relatively simple structure of this crystal, with a sharp first-order transition into the highly conducting α-phase, was seen as an invitation for light scattering studies. As we shall see, however, the Raman spectrum of α-AgI is by no means simple or completely understood. We shall attempt to bring together the various interpretations [5.10-17] of the spectrum into a unified picture.

The Raman spectrum of α-AgI reflects the disordered nature of the superionic phase. The spectrum is very broad, extending out to approximately 240 cm^{-1} without exhibiting the sharp structure normally observed in the spectra of crystals. As seen in Fig.5.1a the Raman intensity decreases rapidly, almost exponentially, away from zero frequency at the laser line to a broad shoulder at approximately 110 cm^{-1} and then decreases monotonically to the background level at 240 cm^{-1}. A weak shoulder is observed at approximately 30 cm^{-1}. We shall see that this spectral shape occurs in many other superionic conductors and in the melts of metal halides.

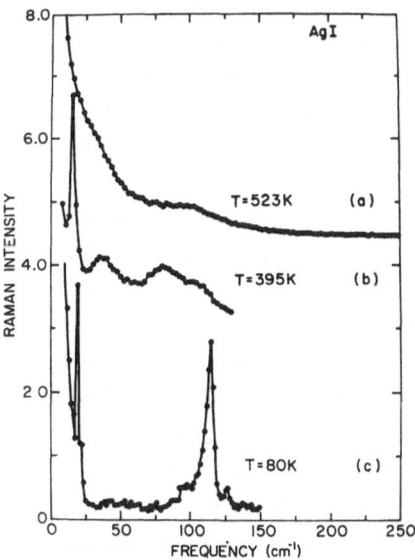

Fig.5.1. Raman spectra of AgI: (a) α-AgI, (b) and (c) β-AgI. A backscattering configuration was used for these unpolarized spectra. [5.17]

To obtain an initial understanding of the spectrum of α-AgI it is useful to compare the changes in the spectrum at the β-α transition in light of the structural changes. In the "normal" β phase, AgI has the hexagonal wurtzite structure (space group C_{6v}^4) with a well-defined phonon spectrum comprised of $A_1 + E_1 + 2E_2$ Raman active modes as seen in Fig.5.1b,c [5.17,18]. The intense, sharp peak at low frequency (17 cm^{-1}) is an E_2 phonon. The modes at high frequency are separated by only 6 cm^{-1} and hence difficult to resolve above liquid nitrogen temperature (A_1TO = E_1TO = 106 cm^{-1}, E_2 = 112 cm^{-1}). The intermediate structure, in Fig.5.1b, is due to second-order processes.

The structure of α-AgI has been extensively studied [5.19] and can be described as having a rigid body-centered iodine lattice which creates a large number of excess silver sites. In Chap.2 Boyce and Hayes discuss in detail the positions of the cations which they conclude are near the centers of iodine tetrahedra, a conclusion supported by neutron diffraction (see Chap.3).

The most striking change in the Raman spectrum upon traversing the β-α transition is the large increase in scattering intensity at lower frequencies. It is this sudden increase which brought about speculation that the mobile cations give rise to scattering through hopping or diffusion mechanisms. We also note the extension of scattering intensity to higher frequencies in the α phase. Although the α and β phases have very different symmetries, the nearest-neighbor Ag-I environment is actually very similar [5.20,21]. One can then speculate that the optical modes will have similar frequencies. Therefore, we may tentatively identify the region of the shoulder at 110 cm^{-1} in the Raman spectrum and, as we shall see in Sect.5.3, the peak in the frequency-dependent ionic conductivity with the optical modes similar to those of β-AgI. A comparison to the other silver and copper halide-based compounds with tetrahedral halide coordination about the cation, RbAg$_4$I$_5$ in particular, suggest that the high-frequency shoulder is due to the symmetric breathing mode of the cage (the iodine tetrahedra). The selection rule observed for this frequency region — loss of intensity when polarization of incident and scattered light is analyzed in crossed configuration — is additional evidence for a symmetric mode [5.10]. The lack of sharp structure is due, in part, to heavy damping which is expected in these materials. Aside from the highly mobile cation, X-ray experiments [5.19] have found large Debye-Waller factors which indicate that the iodine ions have rather large rms deviations from their lattice sites so that damping and anharmonic contributions will be important.

Let us now consider the more direct mobile cation effects. As noted above, we can estimate, from diffusion and thermal considerations, the diffusion and flight times of the cations. For α-AgI at 523 K, the diffusion constant D = 2.14×10^{-5} cm^2/s [5.22], and the distance between 'd' sites (center of the iodine tetrahedra) a_0 = 1.78 Å, we obtain the diffusion (or hopping) time $t_d \cong a_0^2/6D$ = 2.5 ps, and the time of flight between sites $t_f \cong (m_{Ag}a_0^2/k_BT)^{\frac{1}{2}}$ = 0.88 ps. Since the time spent at the site is greater than the time between sites, α-AgI is in the hopping regime. We therefore expect a response more closely resembling a solid than a liquid although we may actually find contributions from both.

The cations, when hopping from site to site, would modulate the polarizibility of the surrounding lattice. Hence we expect a response at the hopping frequency $\omega_d \simeq 2$ cm^{-1}. This is too low a frequency for Raman scattering and, in fact, the entire view is too simplistic as we shall see. The attempt frequency — the frequency of oscillation of the cation in its potential well prior to jumping — can be obtained via the relation $\omega_A = \omega_d \exp(E/k_B T) = 17$ cm^{-1} where E (= 0.094 eV) is the activation energy [5.22] and ω_d is the diffusing frequency. It must be noted that these values are estimates and that we have assumed equal hopping distances and ignored correlation effects. This type of attempt mode may be heavily damped in which case the spectral shape would be extremely distorted as seen in scattering experiments involving soft modes [5.23].

The postulation of an attempt frequency has, in the past, been used to identify peaks in the frequency-dependent conductivity and Raman spectra. This idea was more readily adapted to the β-aluminas, as we shall see in Sect.5.1.3, where the spectrum shows a definite separation between cation motion and that of the host lattice. In α-AgI the attempt frequency has been identified with the peak in the frequency-dependent conductivity at 110 cm^{-1} [5.26]. However, we now believe this to be due to the stretching mode of the iodine cage because of the symmetry selection rules and comparison of the high-frequency stretching modes of β-AgI and RbAg$_4$I$_5$. The frequency is also a factor of six larger than expected from our elementary calculation. The attempt mode is the cause of the strong low-frequency Raman scattering. This identification is substantiated by comparison to both Raman and neutron studies of β-AgI and RbAg$_e$I$_5$.

In β-AgI the low frequency E_2 mode is, as noted earlier, at 17 cm^{-1}. Neutron data has shown that this mode is in fact dispersionless and it has been identified as the mode responsible for the β-α transition [5.20,24]. The mode is due to bond bending rather than bond stretching; hence the low frequency. Similar dispersionless, low-frequency modes are observed ($\omega \simeq 20$ cm^{-1}) in RbAg$_4$I$_5$ in recent neutron experiments [5.25]. The lack of dispersion suggests that this mode is due to individual Einstein oscillators (the cations) which have little coupling to the lattice. Such a low energy mode is directly related to the diffusive motion of the cations. In addition a peak in $\sigma(\omega)$, α-AgI, in this low-frequency region at approximately 20 cm^{-1} [5.26] has been observed (see Fig.5.10a).

The scattering mechanisms we have discussed so far do not in themselves completely account for the Raman spectrum of α-AgI. Other mechanisms that contribute to the spectrum include second-order (two-phonon) and higher-order Raman processes, disorder-induced one-phonon scattering, and anharmonic effects on lattice vibrations.

It has been proposed by the authors [5.14] that a large portion of the Raman spectrum of α-AgI is due to multiphonon processes. The evidence that has led to this conclusion includes the spectral extend compared with other phases of AgI and AgI-based compounds, the dominant second-order scattering in β-AgI at high temperatures and comparison to the spectra of the high-temperature solid and melts of AgBr and AgCl via mass-scaling of the phonon frequencies. We have already noted that the high-frequency stretching mode of the iodine lattice in α-AgI should be close in frequency to that of the β phase (ω = 110 cm^{-1}) because of the tetrahedral coordination about the cation sites and the similarity in Ag-I distances. The spectrum of α-AgI, however, extends out to 240 cm^{-1}, which indicates either very different bond strengths or second-order processes. The first-order spectrum of the high-pressure trigonal phase of AgI does not extend beyond 140 cm^{-1}, whereas the spectrum of the high-pressure NaCl face-centered-cubic phase, which is completely due to second-order scattering, extends to 250 cm^{-1} [5.10]. In β-AgI, as noted earlier, the spectrum is well defined at low temperatures and completely first order, Fig.5.1c. As the temperature is raised, however, second-order features appear at 37 and 80 cm^{-1} (below the TO frequency) which are attributed to combinations of acoustic phonons and the low-frequency E_2 mode. It has been found by neutron scattering that these modes have a high density of states at the zone edge [5.20,24]. An interesting point is that just below the β-α transition, the second-order scattering is very strong indicating anharmonicity, and abscures the high-frequency first-order peaks as seen in Fig.5.1b. As is the case in the two-phonon spectra of all silver and copper halides [5.27,28] the combinations of optical phonons contribute little to the Raman spectrum of α-AgI at high temperatures and only give a weak high-frequency tail. The sharp structure in the second-order spectrum, normally expected from special points in the Brillouin zone, is not seen in either α-AgI or AgBr and AgCl at high temperatures due to the anharmonicity. We therefore believe that the intermediate and high-frequency region the Raman spectrum of α-AgI is due to second-order processes.

The comparison of the second-order spectra of AgBr and AgCl to that of their melts, whose spectra are also due primarily to second-order processes [5.14], suggests that a high degree of damping and disorder already exists at high temperatures. Figure 5.2 compares the Raman spectrum of α-AgI to that of the melts of AgBr and AgCl. The similarity in the spectral shape, which also occurs in the copper halides, suggests that the same scattering processes are involved in both the solid and melts. In addition, scaling of the frequency axis by the appropriate mass factor—inverse square root of the halide mass ratio for combinations of optical phonons—for zone boundary phonons involved in two-phonon processes substantiates this conjecture [5.14]. This agrees with a description of the β-α transition as a cation sublattice melting. This relation between second-order processes across the melting transition

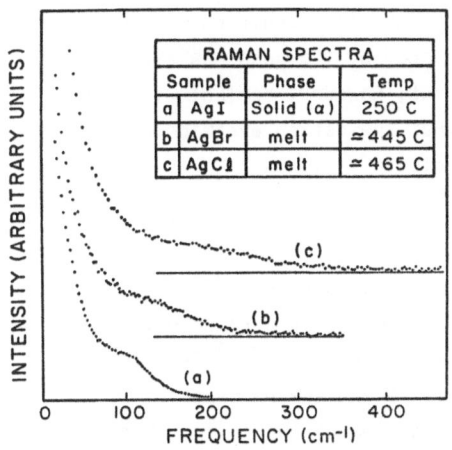

RAMAN SPECTRA		
Sample	Phase	Temp
a	AgI	Solid (α)	250 C
b	AgBr	melt	≈ 445 C
c	AgCl	melt	≈ 465 C

Fig.5.2. Raman spectra of (a) α-AgI at 523 K, (b) AgBr melt at 717 K and (c) AgCl melt at 738 K. All spectra were taken in a right-angle scattering geometry [5.17]

has also been observed in the alkali halides and rare gases [5.29-30] and shows the importance of short-range as compared with long-range order in superionic materials.

A lattice dynamical calculation has recently been performed by ALBEN and BURNS [5.16] in which the density of states, Raman spectral density, and frequency-dependent conductivity were calculated for silver iodide. The calculation used the "equation-of-motion" method with an interaction potential in the harmonic approximation. Only nearest-neighbor forces are considered and the central and noncentral forces are determined by fits to β-AgI. The polarizibility tensor is defined in terms of linear displacements, the coefficients of which are separated into three components corresponding to polarized and depolarized scattering due to bond stretching and depolarized scattering due to bond bending. The model assumes a static lattice with the cations distributed over the d sites. The results for α-AgI indicate a large density of states at low frequency, peaked at 20 cm^{-1}, which corresponds to TA modes arising from the bond bending. This agrees well with our above discussion. The calculation also indicates a broad peak centered a 110 cm^{-1} without sharp optical-mode structure. The disorder of the cations removes the translational symmetry and lifts the momentum selection rule. Because of this effect the Raman spectrum should closely reflect the density of states. The calculated Raman spectrum has the general shape of the experimental results below 150 cm^{-1}. Differences between calculation and experiment can be explained by the absence of second-order Raman processes and anharmonic effects. Including defects, such as shifts of the cation from the center of the tetrahedra, as suggested in several experiments [5.20,21], does not substantially change the density of states.

We have shown that the very broad and structureless Raman spectrum of α-AgI is due to the contributions of several effects; the first-order oscillatory (attempt) modes of the cations at low frequency, and the symmetric stretching (cage) modes at high frequency, both of which are broadened and reflect the density of states due to the cation disorder. We have also seen that there is a contribution to the spectrum from second-order processes which are enhanced by the anharmonicity and lack of long-range order. The Raman spectrum does not, however, reflect directly the cation motion and we shall see in Sect.5.2.2 that the cation dynamics are observed in the low-frequency (less than 10 cm^{-1}) region.

5.1.2 $M^+Ag_4I_5$ ($M^+ = Rb^+$, K^+, NH_4^+)

The subclass of superionic conductors $M^+Ag_4I_5$ is most extraordinary. Members of this group have a high ionic conductivity over a wide temperature range but centered at very low temperature. From the light-scattering point of view, these crystals can be very complicated due to their structure and the presence of phase transitions, as well as the cation disorder. In view of the similarities of the properties, we shall restrict our discussion to $RbAg_4I_5$ and refer the reader to Table 5.1 and references [5.12,17,31-36] for detailed information on the K and NH_4 isomorphs.

Unlike pure AgI, $RbAg_4I_5$ has a complicated cubic structure at room temperature with four formula units per unit cell. The sixteen silver ions are distributed over fifty-six sites of three nonequivalent types. As in AgI the silver sites are in iodine tetrahedra, although slighthly distorted. Each Rb ion is surrounded by six iodine ions forming a distorted trigonal prism, the faces of which connect to the silver sites. From elementary considerations we expect 3N-3 = 117 (N = 40) optical modes in this structure. Group theoretical calculations have not been completed, both because the exact space group is not known and because the cations do not have fixed sites. We would, however, expect a large number of Raman active modes. It is possible, by considering the point-group symmetry and simple molecular models, to identify most of the features of the spectrum as having a particular symmetry character.

Several authors have reported Raman experiments on $RbAg_4I_5$, the most extensive having been completed by GALLAGHER et al. [5.34-36]. At toom temperature and above the spectrum is very broad, lacking sharp structure, as seen from Fig.5.3a. The extent and shape of the spectrum, although narrower, is very similar to α-AgI. The spectrum does not narrow appreciably with decreasing temperature; however, a greater separation in the high- and low-frequency structure is seen near the β-α transition at 208 K (Fig.5.3b). The allowed Raman modes for a crystal with O symmetry have A_1, E, and T_2 character and can be observed separately by proper choice of sample orientation (Fig.5.3a). The broad peak at 107 cm^{-1} has A_1 character and is associated with the fully symmetric breathing mode of the iodine tetrahedra. With the large amount of disorder one can consider a model in terms of AgI_4 molecules which have T_2 symmetry. In such tetrahedral molecules the breathing mode has A_1 symmetry. This and other simplifying models have been considered by GALLAGHER [5.36]. The lower-frequency structure seen in the E and T_2 symmetries are assigned

Table 5.1. Properties of metal halides

Compound	Phase	Range [K]	Structure	Space group
AgI		0-420	wurtzite	C_{6v}^4
	α [a]	420-829	body-centered cubic	O_h^9
	melt	829-		
AgBr	-	0-707	face-centered cubic	O_h^5
	melt	707-		
AgCl	-	0-728	face-centered cubic	O_h^5
	melt	728-		
RbAg$_4$I$_5$	γ	0-122	trigonal	D_3^2
	β	122-208	rhombohedral	D_3^7
	α [a]	208-501	cubic	O^6
	melt	501-		
KAg$_4$I$_5$	γ	0-139	trigonal	D_3^2
	β	139-194	rhombohedral	D_3^7
	α [a]	194-526	cubic	O^6
	melt	526-		
NH$_4$Ag$_4$I$_5$	γ	0-135	trigonal	D_3^2
	β	135-199	rhombohedral	D_3^7
	α [a]	199-505	cubic	O^6
	melt	505-		
CuI	γ	0-642	zinc blende	T_d^2
	β	642-680	wurtzite	C_{6v}^4
	α [a]	680-878	face-centered cubic	O_h^5
	melt	878-		
CuBr	γ	0-664	zinc-blende	T_d^2
	β [a]	664-744	wurtzite	C_{6v}^4
	α [a]	744-765	body-centered cubic	O_h^9
	melt	765-		
CuCl	γ	0-681	zinc blende	T_d^2
	β	681-703	wurtzite	C_{6v}^4
	melt	703-		

[a]Superionic phase

by Gallagher to the cation oscillatory motion. Several "attempt frequency" modes should exist because of the differences in the available sites —both hopping distances and activation energies. In fact several distinct peaks are observed at the lowest temperatures in the α phase. This conclusion has also been reached in neutron scattering experiments [5.25] where a broad dispersionless peak was observed at low frequencies. The very low frequency spectrum will be discussed in Sect.5.3.

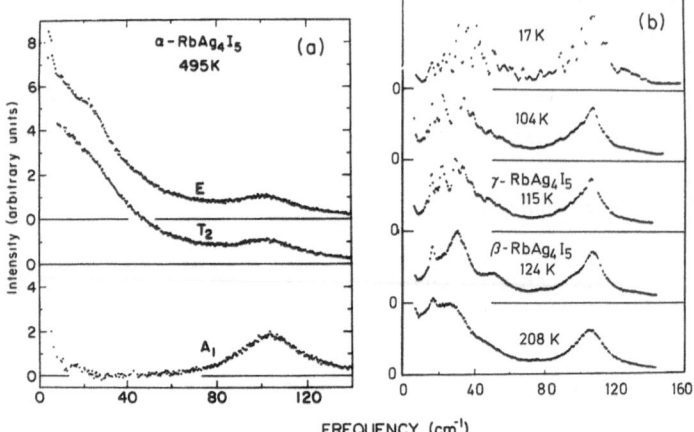

<u>Fig.5.3a and b.</u> Raman spectra of RbAg$_4$I$_5$: (a) polarized spectra of α-RbAg$_4$I$_5$, (b) unpolarized spectra of β and γ phases [5.36]

Little change is noted in the spectral shape when the β-α transition, the nature of which is discussed in Chap.7, is traversed. It was initially thought that the change in site occupancy and activation energy would be reflected by either mode frequency or damping. This was not observed; however, the broad structure could easily conceal small shifts in frequency or width. At still lower temperatures, the spectra narrows with several distinct peaks becoming visible just above the γ-β transition (Fig.5.3b). The spectrum is then formed by many closely spaced, heavily damped modes which at higher temperatures cannot be resolved. Unfortunately in the β-phase, birefringent domains are formed, restricting the use of polarization selection rules to determine the symmetry character of the resolved structure.

Below the first-order transition at 112 K there is a discontinuous change in several of the low-frequency structural features with new modes appearing. This is expected due to the change in crystal structure at this first-order transition and the resulting increase in the size of the unit cell [5.32]. At 17 K the low-frequency region is resolved into fourteen peaks and the high-frequency region into eleven peaks. An interesting point is that although the conductivity changes by two orders of magnitude at the γ-β transition, the mode damping does not change appreciably. This further implies that cation diffusion is not directly related to the Raman spectrum.

Recently the spectrum of nonsuperionic Rb$_2$Ag I$_3$ has been measured [5.37]. This is the only other stable form of the RbI-AgI system [5.38]. In the Raman spectrum there is considerable structure below 80 cm^{-1} with ten peaks and several shoulders resolved. The structure is orthorhombic (D_{2h}^{16}) with 36 Raman modes allowed. The relevance of this work is that there is little intensity in the vicinity of 110 cm^{-1}.

This is understandable when the structure is considered, in that there is no tetra-hedral coordination of iodine around either Rb or Ag ions. This work reinforces our understanding of the high-frequency region of $RbAg_4I_5$.

5.1.3 Copper Halides

The copper halides form another interesting group of superionic conductors of the metal halide type. CuI and CuBr have superionic phases while CuCl does not, although it is useful to discuss its Raman spectrum for comparison. All three of the copper halides have the zinc-blende structure at low temperatures and a narrow wurtzite phase at higher temperatures. CuI and CuBr then have cubic phases (super-ionic) at still higher temperatures; CuI has a face-centered-cubic iodine lattice, whereas CuBr has a body-centered-cubic bromine lattice. Of particular interest is the fact that the hexagonal β phase of CuBr is also superionic, with cation conduc-tivity comparable to α-AgI; however, this does not change the spectrum relative to nonsuperionic β-CuCl. Table 5.1 gives the relevant data on the extent of the phases of the copper halides and comparison to the silver halides. The cations, in all the phases, are at or near the center of halide tetrahedra as in the silver halides. This accounts, as we shall see, for the similarity of the Raman spectra of the different phases.

There have been many Raman studies of the various phases of the copper halides [5.28,39,40] including a recent thorough investigation by NAMENICH [5.39] of all the phases including the melts. In the zinc-blende γ phase there is one allowable Raman mode which is split into transverse and longitudinal components. At room temperature these peaks are very broad due to anharmonic effects [5.28] and nearly obscured by strong second-order scattering, as shown in Fig.5.4. Just below the γ-β transition the spectra become smooth and nearly structureless. In the β phase, the phonon modes are as in β-AgI ($A_1 + E_1 + 2E_2$). The characteristic low-frequency E_2 mode is observed in CuI but is just a shoulder in both CuBr and CuCl. The same is true for the high-frequency region of the spectrum, the stronger damping in CuBr and CuCl being due to the larger anharmonicity of the lattice. As in β-AgI second-order Raman scattering also contributes to the spectrum and further broadens the structure. The α-CuBr spectrum is identical to that of α-AgI in shape; recall that the crystal structures are the same. The spectrum of α-CuI is similar in shape but has sharper features than in the lower-temperature phase. As in the silver hal-ides the high-temperature spectra of these solids are nearly identical to those of the melts, indicating the same lack of long-range order occurs in the solid.

The interpretation of these spectra is the same as for silver iodide. The broad, structureless spectral shape in the α phase is due to strong anharmonic broadening of the one-phonon cage and cation vibrational features and second-order scattering. There is also the one phonon density of states contribution due to lifting of the momentum conservation selection rule by the disorder. The similarity of these spec-

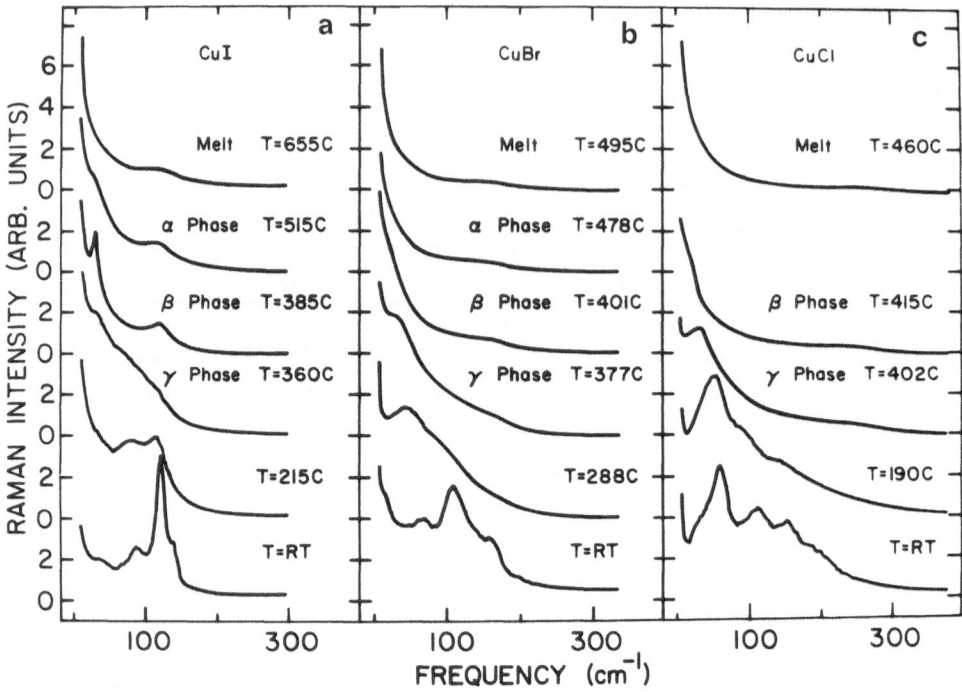

Fig.5.4a-c. Raman spectra of Copper Halides: (a) CuI, (b) CuBr and (c) CuCl [5.39]

tra to the AgI spectra enhance our understanding of the scattering mechanisms through comparison to the different phases of the copper halides.

As in α-AgI the scattering intensity increases rapidly below 20 cm^{-1}. Experiments are currently being conducted to determine if similar relaxation modes exist in the copper halides [5.39] as are discussed in Sect.5.2.2 for α-AgI.

5.1.4 β-Aluminas

The two-dimensional cation conductors of the form M_2O: 11 X_2O_3 (M = K, Ag, Na, Li, Rb; and X = Al, Ga, Fe) are of considerable interest due to their high ionic conductivities and unique structure. The monovalent cations have high mobility in a two-dimensional hexagonal network formed by oxygen bridges between spinel blocks. The structure of the aluminas and gallates has been discussed in great detail elsewhere (see Chap.3) and we shall limit our discussion of structural aspects to that necessary for description of the light scattering spectrum [5.41].

In the stoichimometric structure all the cations are at Beavers-Ross (BR) sites on three corners of a hexagon, the anti-Beavers-Ross (aBR) sites at the other three corners are vacant, and an oxygen ion is in the center. Cation diffusion is along the edge of the hexagon between the anions, passing through the aBR sites. As grown, the samples have between 15-40% excess cations, the amount depending on the method and temperature of preparation. Determination of where the excess cations reside has been of great interest lately. Lattice potential calculations [5.42] and X-ray experiments [5.43-44] suggest that cation pairs are centered at BR sites with the cations near the mid-oxygen (between BR and aBR) sites. The excess cations are related to an excess of oxygen in the planes and with Frenkel defects in the spinel blocks. These deviations from stoichiometry only serve to complicate the analysis of the Raman spectrum.

Very complete Raman spectra of the β-aluminas of Na, Ag, K, Rb, and Li have been obtained by HAO et al. [5.45-47]. The spectral density due to the alumina alone is easily identifiable and is almost completely above 200 cm^{-1}. This is plainly seen in the different isomorphs and we shall not discuss this region. The scattering due to the cations is at low frequency, from 28 to 82 cm^{-1} depending on the cation involved. If the spinel blocks and the in-plane oxygen are ignored and only the BR sites are considered, the crystal has D_{6h}^4 symmetry. The unit cell contains two nonequivalent planes with one cation in each. There is then one Raman active E_{2g} mode corresponding to the out-of-plane vibrations of the cations in neighboring planes. The in-plane motion of the cations in neighboring planes corresponds to the infrared-active E_{1u} mode.

In Fig.5.5 we show the spectrum of Ag β-alumina at several temperatures; a single broad line at high temperatures is observed, which narrows and splits with decreasing temperature. The peak has the expected E_{2g} symmetry and the different isomorphs scale with the inverse square root of the mass. The close proximity of the peak in the ionic conductivity (28 cm^{-1}) as determined from infrared absorption [5.48] compared with the Raman mode indicates that the planes are weakly coupled, breaking the selection rule. These in-plane vibrations have been identified as single-particle attempt frequencies [5.48] in the same light as discussed in Sect.5.1.1. As in the silver halides, a dispersionless phonon mode is observed in neutron scattering experiments [5.49]; however, it lies below both the Raman and infrared peaks. The origin of this discrepancy is unclear.

As the temperature is lowered we see that the single broad peak is resolved or split into four distinct peaks at helium temperature. There are several possible explanations for this. It has been suggested that a superlattice is formed due to the excess cation [5.44,46] in Ag β-aluminas the unit cell, in this model, would consist of four cations; three on BR sites and one on an aBR site. However, for K β-alumina, the four cations would be on three BR sites but with one doubly oc- cupied. Such structures would surely give rise to additional modes. Although sharp transitions have not been observed in the transport properties, gradual shifting of site occupancy and a freezing out of a superlattice may occur. The broad, closely spaced peaks do not permit an accurate temperature dependence study to resolve the issue.

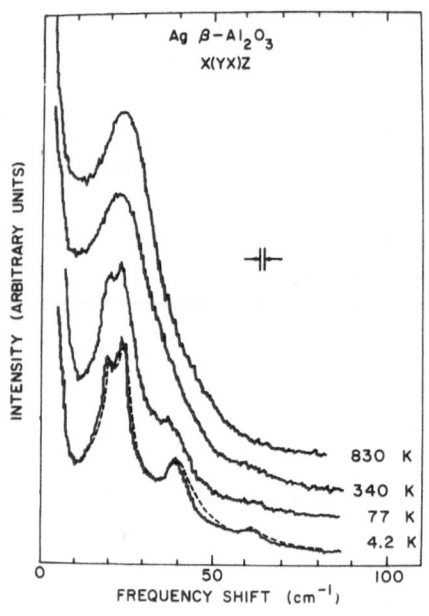

Fig.5.5. Raman spectra of Ag β-Al₂O₃. The dashed line is the calculated spectrum of HSU [5.51]

In an interesting experiment in which ammonium ions were diffused into Na β-alumina at various concentrations [5.50], a new broad peak appeared between 130 and 180 cm^{-1} above the Na peak and well between the two β-alumina peaks at 100 and 258 cm^{-1}. This feature is assigned to the translational in-plane mode of the NH_4 ion. At low concentrations the ammonium ions are assumed to reside at BR sites. However, at high concentrations there is a noticeable splitting of the peak at low temperatures suggesting that there may be double occupancy at these sites. It was also noted, as expected, that with increasing NH_4 concentration the intensity of the Na line at 68 cm^{-1} increased. Substitution of deuterated ammonium yielded the expected frequency shift for a translational mode.

Calculations of the phonon spectrum of the β-aluminas by HSU [5.51] are able to identify the dominant structure in the Raman (see Fig.5.5) and infrared data. Three cases are discussed: the ideal stoichiometric structure, the extended super-lattice, and a defect or nonideal structure. In the model, only the cations in one layer are considered, with the oxygen as a uniform negative background. Elastic and Coulombic potentials are included in the Hamiltonian and the equations of motion are solved in the harmonic approximation. A single peak is found for the stoichiometric case which is due to oscillations at the BR sites. When defects are added, the spectra acquire finite lifetimes and lose symmetry selection rules thus giving rise to a close resemblance between the Raman and infrared spectral density. When excess cations are added one obtains two additional peaks — an in-phase mode below and an out-of-phase mode above the peak in the ideal case — which are due to the superlattice structure as described above. The distinction between the Ag and K superlattices is also found.

We have seen in this section that the Raman spectrum due to the cations in the β-aluminas, which consists of several peaks at low temperature, is well explained by the superlattice model. As in the metal halides the Raman spectrum does not directly exhibit the dynamics of the cations but rather the oscillatory motion at their sites.

The Raman spectra of the sodium β- and β"-gallates have been measured by BURNS et al. [5.52]. The spectra have the same features as the β-aluminas, with the cation-related spectral density separated from that of the spinel blocks.

5.1.5 Anion Conducting Fluorites

The superionic conductors of the fluorite class, PbF_2, $SrCl_2$, etc., have been studied extensively in several works [5.11,53,54]. The most recent of which, by ELLIOTT et al. [5.54] involves both extensive Raman experiments and detailed calculations of the spectral density. The fluorite structure anion conductors have very different properties from those of either the metal halides or β-aluminas. Frenkel defects, interstitial anions and lattice vacancies are the primary mechanism for high anion conductivity [5.55]. The transition to the superionic phase differs from the silver halides in that it is a gradual disorder occurring over a broad temperature range ($\Delta T \simeq 50-100$ K) [5.56]. A more detailed description of this phase change is given in Chap.7.

In crystals with the fluorite structure, space group T_d^2, there is only one allowed Raman active mode, which has T_{2g} character. At room temperature the Raman spectra of the various fluorites indeed show one spectral feature but in most cases it is extremely broad ($\Gamma \sim 50$ cm^{-1}) and very asymmetric. As the temperature increases and passes through the transition to the high conductivity phase this asymmetry grows more apparent, see Fig.5.6. At high temperature the intensity on the low frequency side becomes comparable to the T_{2g} phonon mode itself. This extra scattering is attributed to the anion disorder at high temperatures.

The Raman spectral density of the fluorites has been calculated by ELLIOTT et al. The asymmetry of the T_{2g} peak at ambient temperatures is attributed to the anharmonicity of the lattice. Calculations of the spectral density including third- and fourth-order anharmonic terms in the expansion of the potential give very good fits to the data at room temperature. The strong frequency dependence of the damping arises because the maximum in the two-phonon density of states — the phonon decay involves two zone-boundary phonons — does not coincide with the one-phonon T_{2g} peak.

By considering the effect of the interstitial anions and vacancies on the polarizability, ELLIOTT et al. were able to calculate the observed excess scattering at high temperatures. The Frenkel defects relax momentum conservation and allow observation of the one phonon density of states. The effect is calculated by adding a defect term to the polarizability. Using standard Green's function formalism,

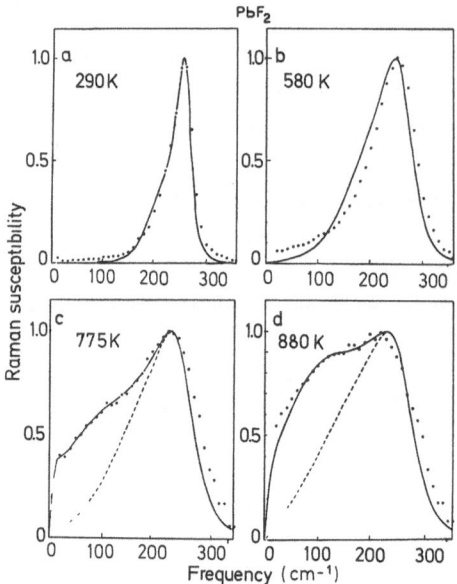

Fig.5.6. Raman spectral density of PbF_2. Dots indicated experimental data, solid line the calculated results, and dashed line the contribution of the anharmonicity. Note that these spectra have the Bose thermal occupation factor divided out [5.54]

the spectral density is then calculated. The only explicitly temperature-dependent parameter is the concentration of defects, which follows the appropriate trends. Accurate estimates cannot be obtained due to approximations made in the calculation. Although the low-frequency shoulder of the spectrum is an artifact of the anion disorder, it is not a consequence of the anion diffusion but rather due to stationary defects. We shall discuss the very low frequency spectrum in Sect.5.2.2. In Fig. 5.6c,d we see excellent fits to the observed spectra when both disorder and anharmonicity are considered.

5.2 Low-Frequency Raman and Brillouin Scattering

In this section we discuss low-frequency light scattering and standard Brillouin scattering. We consider first several theoretical arguments involving the direct effect of the mobile ions on the light scattering spectrum and briefly review normal Brillouin scattering from acoustic phonons. In light of this discussion we then look at the experimental results.

5.2.1 Theoretical Considerations

We first look, in a direct manner, at the effect of diffusing ions of the polarizability of a crystal. Recall from Sect.5.1 that the scattered intensity is directly proportional to the spatial and temporal Fourier transform of the polarizability autocorrelation function $\langle\alpha(\underline{r}',0)\alpha(\underline{r},t)\rangle$. It is necessary then to postulate a model for the variations in $\alpha(\underline{r},t)$ and perform the necessary calculations. In this section we shall review several such approaches.

KLEIN [5.57] and FIELD et al. [5.58,59] formulated a model to explain their light scattering results in $RbAg_4I_5$. The polarizability is defined by

$$\alpha(t) = \sum_{i\ell} \alpha_i N_{i\ell}(t) \quad , \tag{5.3}$$

where α_i is the polarizability of the ith site and $N_{i\ell}(t)$ is the occupation number of that site whose mean value is the site occupation probability P_i. A rate equation is written in terms of the correlations of occupation numbers

$$\frac{d}{dt}\Big\langle N_{i\ell}(t)N_{j\ell'}(0)\Big\rangle = \sum_{ij} C_{ij}(\ell - \ell')\Big\langle N_{i\ell}(t)N_{j\ell'}(0)\Big\rangle \tag{5.4}$$

where $C_{i\ell}(\ell - \ell')$ is the hopping probability. Appropriate boundary conditions are defined and (5.4) is solved for the correlation function of $N_{i\ell}(t)$. This then used to obtain the spectral density for $q = 0$, which has the form

$$S(\omega) = \frac{N}{\pi} \sum_\lambda \frac{\nu(0\lambda)\alpha(\lambda)}{[\nu(0\lambda)]^2 + \omega^2} \quad , \tag{5.5}$$

where $\alpha(\lambda) = \sum_{i=1}^{r} \sqrt{P_i}E_i(0\lambda)\alpha_i$. Here $\nu(0\lambda)$ is the eigenvalue of the relaxation mode, $E_i(0\lambda)$ is the eigenvector, and r is the number of sites. We see that the spectral density is a sum of Lorentzians centered at $\omega = 0$. It has been calculated by GALLAGHER [5.36] that for $RbAg_4I_5$ there should be 56 relaxation modes with $q = 0$ of which 13 are Raman active. With so many overlapping Lorentzians of similar width it is not possible to obtain quantitative results that can be related to diffusion experiments. However, it has been shown [5.58] that it is possible to relate a mean relaxation rate to the diffusion constant through the relation $\bar{\nu} = 6D/a_0^2$, where a_0 is the hopping distance. One must assume that all the jump distances are equal, that there are no jumps from identical sites in different unit cells, and that all sites have the same diagonal elements in the jump-rate matrix. The first two approximations are reasonable but the third is somewhat questionable. We shall see in the next section, however, that the value obtained for $RbAg_4I_5$ is close to the observed experimental value.

A similar model which yields nearly identical results but uses very different calculational methods has been developed by GEISEL in several recent articles [5.60,61] and which is summarized in Chap.8 of this book. We shall briefly review it here for comparison to the above model. The polarizability is defined by

$$\alpha_{if}(\underline{r},t) = \alpha_{if}(\underline{r}) \sum_j \delta[\underline{r} - \underline{r}_j(t)] \quad . \tag{5.6}$$

where the polarizability of a particular site is written explicitly $\alpha_{if}(r)$ and the movement of the jth cation is included through the delta function of position and time. The scattered intensity is defined in the same manner, and one obtains

$$I(\underline{q},\omega) \propto N \sum_{\underline{k},\underline{k}'} \alpha_{if}(\underline{k})\alpha_{if}(\underline{k}')S(\underline{k}' - \underline{q}, \underline{k} + \underline{q},\omega) \quad , \tag{5.7}$$

where

$$S(\underline{Q}',\underline{Q},\omega) = \frac{1}{2\pi N} \int dt \exp(-i\omega t) \sum_{j\ell} \left\langle \exp[i\underline{Q}' \cdot \underline{r}_\ell(0)] \exp[i\underline{Q} \cdot \underline{r}_j(t)] \right\rangle \quad . \tag{5.8}$$

It is then necessary to formulate a model for the particle motion, solve for $\underline{r}_j(t)$ and calculate the correlation function. This is done in a model using Brownian motion in a periodic potential described by the Langevin equation. We refer the reader to the references for further details of the calculation. In the large-friction regime, applicable to α-AgI, a solution of the form

$$S(Q',Q,\omega) = \frac{1}{\pi} \sum_n M_n(0)M_n^*(Q') \frac{\lambda_n}{\omega^2 + \lambda_n^2} \tag{5.9}$$

is obtained with

$$M_n(Q) = \int dx \ e^{iQx} \psi_n(x)\psi_0(x) \quad . \tag{5.10}$$

The $\psi_n(x)$ are effectively eigenfunctions defining the relaxation modes, e.g., $\psi_0(x) \propto \exp[-V(x)/2k_BT]$, and the λ_n are the corresponding relaxation rates. In the case of the two potential minima per unit cell and a polarizability which is even about each site one obtains, in one dimension, a narrow Lorentzian centered at zero frequency corresponding to a low-order relaxation mode. This model has been successfully applied to α-AgI as will be seen in the next section.

We turn briefly to a discussion of two light scattering theories which calculate fluctuations in the dielectric constant in a hydrodynamic formulism. SUBBASWAMY [5.62] treats superionic conductors in a charged cage-fluid approximation. In this theory the cage is expressed in terms of a displacement field while the velocity field and density are the appropriate dynamical variables for describing the fluid (the mobile cations). The theory considers both propagating and non-propagating density fluctuations in the fluid with relaxation modes introduced by a frequency-dependent viscosity coefficient. The strain fluctuations of the case and the density fluctuations of the fluid influence each other; thus, coupling is built into the formulism. The spectral density is calculated in the usual manner.

There are two propagating waves; the longitudinal mode of the cage, and the fluid moving in and out of phase. There is also a nonpropagating fluctuation, corresponding to the entropy fluctuations and internal relaxation modes, and a transverse vibration of the cage which is decoupled from the fluid. In the next section we compare these results to the experimental data.

HUBERMAN and MARTIN [5.63] have developed a theory of the dynamics of the coupled cage-fluid system in similar to [5.62]. Their model differs in that the charged density fluctuations of the liquid obey a driven diffusion equation and there exist fluctuations in the local site populations. Three internal couplings are considered: phonon coupling to the site-population density fluctuations, longitudinal phonon coupling to density fluctuations, and the transverse phonon coupled to the viscous ion current via inertial forces. The spectral density shows a phonon doublet and a Rayleigh component of width Dq^2, where D is the diffusion constant. This model is discussed further in Chap.8.

In Brillouin scattering experiments, light scattering from acoustic lattice phonons, the frequency of the acoustic mode will, for a given phonon wave vector, allow determination of the velocity of sound in the material. From this information the elastic constants of the crystal can be determined. In the following section, we shall look at the effect of the superionic phase on the elastic properties of the crystals. For a complete review of Brillouin scattering, see [5.1-4].

5.2.2 Silver Halides

WINTERLING et al. [5.64] have made low-frequency Raman and Brillouin measurements on AgI at several temperatures in the α phase. As we have seen in the Raman spectrum, in Sect.5.1.1, the intensity increases rapidly at low frequency. A distinct change of slope and increase in intensity is noted below 15 cm^{-1}. To isolate the low-frequency structure, WINTERLING has fitted a broad Lorentzian with $\Gamma = 32$ cm^{-1} to his data and substracted this out to leave the low-frequency structure. This resulting peak, Fig.5.7, was then fitted with a Lorentzian with $\Gamma = 3.8$ cm^{-1}. The normal Brillouin modes are observed at still lower frequencies ($\omega \leq 0.35$ cm^{-1}). These acoustic modes are very strong and are on top of the broader Lorentzian structure. There is no indication of structure at 1 cm^{-1} which was expected from microwave conductivity measurements [5.26].

The narrow Lorentzian peak, $\Gamma = 3.8$ cm^{-1}, is attributed to the jump diffusion of the cations between nearest-neighbor tetrahedral sites. We have seen in the previous section that the theories of Geisel, in Chap.8, and KLEIN [5.57,58] predict this spectral shape. If we only consider the tetrahedral d sites of the group O_h in α-AgI, six per unit cell, there are three Raman active relaxation modes: $A_{1g} + E_g + T_{1u}$. Geisel, however, has found that only the E_g model has a nonvanishing matrix element with eigenvalue $\lambda = 3/2\tau$. Using the relation

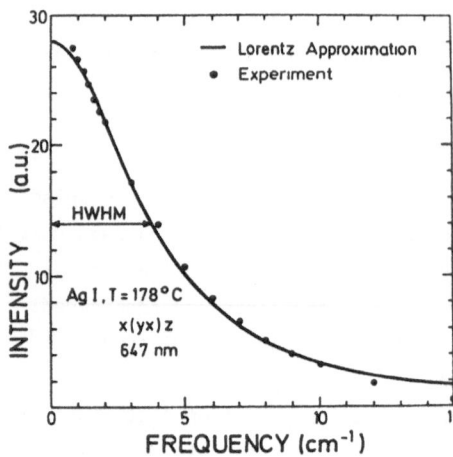

Fig.5.7. Low-frequency spectrum of α-AgI. Dots are experimental points and solid line is the Lorentzian set to the data. [5.64]

$$\tau = a_0^2/6D \qquad\qquad (5.11)$$

with a_0 = 1.78 Å and D = 2.14 × 10^{-5} cm^2/s we find Γ = 3.2 cm^{-1} which is very close to the measured value of 3.8 cm^{-1}. We must note that cation correlations have been neglected and that the cations may not have the exact symmetry of the tetrahedral site [5.21]. The experimental value of Γ can also be considered approximate because the functional form of the scattering at higher frequencies, discussed in terms of attempt frequencies, one and two phonon density of states, is not well understood and will most likely not have a purely Lorentzian shape that was used by WINTERLING to obtain the diffusional peak.

In $RbAg_4I_5$ there are similar relaxation modes; however, the situation is greatly complicated by the crystal structure. As in α-AgI there is a sharp rise in the scattering intensity at low frequency, in this case below 10 cm^{-1}. FIELDS et al. [5.58] have fitted this region with two Lorentzians of width 0.3 and 1.3 cm^{-1}, Fig.5.8a and b, respectively. The Brillouin peaks are very intense, narrow (Γ < 0.3 GHz), and consistent with measured sound velocities. A very narrow dynamic central peak of width approximately 350 MHz was also observed when the Rayleigh peak (Γ < 10 MHz) was removed with an iodine filter.

GALLAGHER [5.36] has determined that for $RbAg_4I_5$ there should be 13 Raman active relaxation modes with symmetry: $2A_1 + 4E + 7T_2$. The matrix elements for these modes have not been calculated; however, we can use the approximation for the average frequency and obtain a width

$$\Gamma = \frac{\bar{\nu}}{2\pi} = D/\pi a^2 = 0.36\ cm^{-1} \qquad . \qquad\qquad (5.12)$$

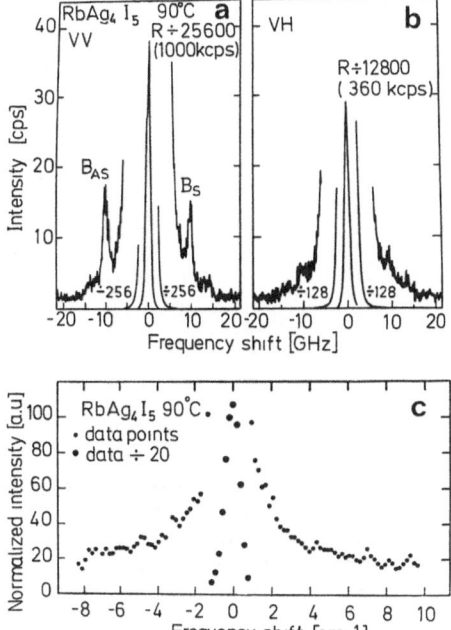

Fig.5.8a-c. Low-frequency spectra of
α-RbAg$_4$I$_5$. (a) and (b) Brillouin peaks
and narrow central peak with $\Gamma = 0.3$ cm^{-1},
(c) Raman spectrum taken with iodine
filter of the broader central peak with
$\Gamma = 1.3$ cm^{-1}

The narrower Lorentzian is therefore identified with the diffusive motion. The
origin of the broader Lorentzian is less clear. It has been suggested that it
is due to the time of flight between sites [5.58]. The fact that the diffusion
time and the time of flight are of the same order of magnitude complements this
view.

The intense, narrow Brillouin lines are not, contrary to the theoretical predictions of SUBBASWAMY [5.62], strongly coupled to the broad central peak
(relaxation modes). The very narrow dynamic central peak can be associated with
the nonpropagating entropy fluctuations predicted in the model. It does not appear, however, that the hydrodynamic model is currently appropriate to describe
the observed spectra.

5.2.3 Other Superionic Conductors

The Brillouin spectra of PbF$_2$ and SrCl$_2$ have been measured over a wide temperature
range in both normal and superionic phases by HARLEY et al. [5.11]. The elastic
constants were determined and it was found that C_{11} decreased by 30% above the
phase transition: see Fig.5.9a for PbF$_2$. This effect was also observed in SrCl$_2$.
We note that C_{44} is unaffected by the transition. The elastic constant C_{11} appears
to couple to the order parameter for the transition to the superionic state.
C_{11} is linear, decreasing with increasing temperature as expected, in both normal
and superionic phases; however, in the transit region there is an anomalous de-

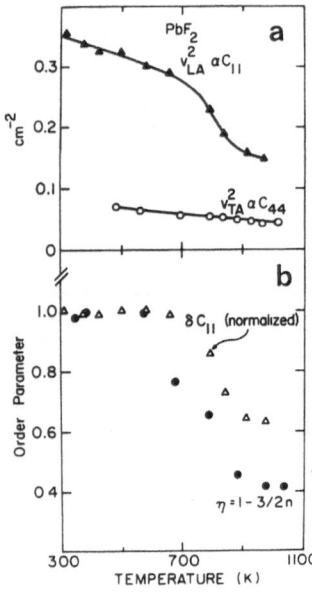

Fig.5.9. (a) Elastic constants of PbF_2 obtained from Brillouin scattering experiment [5.53]. (b) Order parameter obtained from neutron experiments [5.56] plotted against the normalized change in C_{11}

crease. In Fig.5.9b we have plotted the order parameter $\eta = 1 - 3/2\, n$, where n is the fraction of interstitial anions as determined from neutron scattering measurements [5.65], against the normalized change in C_{11}. No evidence of relaxation modes is reported in these results.

In the β-aluminas the acoustic modes have been observed in both Brillouin measurements by HAO [5.47] and CHASE [5.66] and in neutron experiments [5.49]. In the work of HAO a broad Lorentzian-like structure was observed at low frequency, as in the silver halides, that is also interpreted as due to relaxation modes although analysis in this direction has not been completed.

5.3 Infrared Absorption and Frequency Dependent Conductivity

There have been considerable efforts, both experimental and theoretical, directed at determining the frequency-dependent conductivity $\sigma(\omega)$ of superionic conductors [5.26,67-70]. Experimentally, in the frequency region of interest to us, $\sigma(\omega)$ is determined by microwave and infrared absorption and reflectivity. Measurements above $2\ cm^{-1}$ involve reflectivity and absorption of infrared radiation to determine the complex dielectric constant. The frequency-dependent conductivity is determined by the relation

$$\sigma(\omega) = 4\pi\omega\, Im\{\varepsilon(\omega)\} \quad . \tag{5.12}$$

Below $2\ cm^{-1}$ microwave absorption techniques are employed.

The infrared absorption spectrum of crystals will show a sharp feature at the transverse optical-phonon frequencies. These features then correspond to peaks in $\sigma(\omega)$ as found in the spectrum of α-AgI [5.26], Fig.5.10b. In most superionic conductors, however, $\sigma(\omega)$ has very broad peaks of the order of tens of wavenumbers in the optical mode region. As an example, the peak in $\sigma(\omega)$ in α-AgI, Fig.5.10a, at 110 cm^{-1} has $\Gamma \sim 80$ cm^{-1} [5.26] and Na β-alumina at 60 cm^{-1} has $\Gamma \sim 60$ cm^{-1} [5.48].

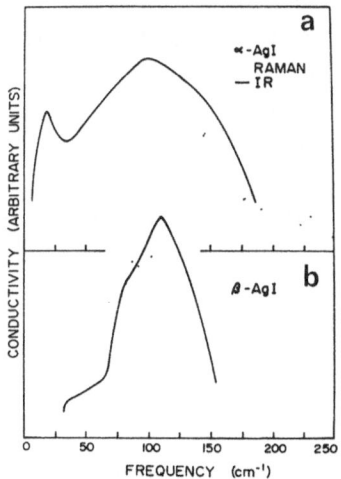

Fig.5.10a and b. Frequency-dependent conductivity of silver iodide: (a) α phase, and (b) β phase. The solid line is the frequency-dependent conductivity obtained from infrared absorption experiments [5.26], and the dots are the conductivity calculated from the Raman spectral density [5.17]

In earlier sections of this chapter we have compared the Raman spectra with $\sigma(\omega)$. Phonons may or may not be both Raman and infrared active depending on the symmetry of the mode [5.1,71]. The Raman spectrum can be related to $\sigma(\omega)$ through the fluctuation-dissipation theorem [5.72],

$$I(\omega) \propto [n(\omega) + 1]Im\{\epsilon(\omega)\} \quad , \qquad (5.13)$$

where $n(\omega)$ is the Bose thermal occupation factor. In the high-temperature limit we obtain

$$\sigma(\omega) \propto \omega^2 I(\omega) \quad . \qquad (5.14)$$

This technique has been used in the past to compare the Raman spectra of α-AgI and RbAg$_4$I$_5$ [5.12,15] to $\sigma(\omega)$. The spectra are very similar over a wide frequency range, when compared by this method, as seen in Fig.5.10a for α-AgI. This technique should be used with caution; with broad, complicated spectra such as are found in α-AgI. The relation should only apply for transverse optical modes at q = 0; however, multiphonon processes will effect both the absorption and light scattering spectra. An example is the case of AgBr and AgCl [5.73] where the multiphonon absorption appears to broaden the reststrahl absorption.

As we have noted in previous sections, the peak in $\sigma(\omega)$ for several silver iodide based compounds is due to the transverse optical breathing mode of the iodine tetrahedra. It has been suggested, in agreement with our discussion of Raman data, that the low-frequency peak in $\sigma(\omega)$ is related to the cation attempt frequency.

The Raman spectra and $\sigma(\omega)$ of the β-aluminas are also very similar. This is expected due to the weak interaction between cations in different planes. In several copper and silver halides a peak in $\sigma(\omega)$, determined by microwave absorption, is seen between 0.1 and 1 cm^{-1} and is explained in terms of cation jump effects [5.67,68]. These peaks are not seen in the Raman spectra were relaxation modes are related to the cation diffusion.

There has recently been much theoretical work aimed at calculating the frequency-dependent conductivity of superionic conductors, and we refer the reader to Chaps.4 and 8 for reviews of this.

5.4 Conclusion

In this chapter we have reviewed a collection of light scattering experiments on superionic conductors and found that, while in many cases the interpretation of the observed spectra was both complicated and speculative, a definite distinction exists between dynamic and static effects. At low frequencies, overlapping both Brillouin and low-frequency Raman scattering, the motion of the mobile ions manifests itself in terms of relaxation modes. In Sect.5.2, we saw that these relaxation modes are Lorentzian in shape and centered at zero frequency. The width of the Lorentzian is related directly to the residence (or dwell) time of the mobile ion and hence the diffusion constant as we found in Sect.5.2.2 for α-AgI and RbAg$_4$I$_5$. The relation can become complicated if there are several available sites per unit cell as well as different site occupancy. Aside from relaxation modes arising from the residence time of the mobile ion there may also exist relaxation modes originating from the time of flight of the ion. From simple arguments we saw that the residence time of the ion is greater than, but of the same order of magnitude as, the time between sites. Relaxation modes due to time-of-flight effects would have larger widths as proposed for RbAg$_4$I$_5$.

In Sect.5.1 we found that the Raman spectra of the superionic conductors are very broad and structureless. The effect of the static disorder brought about by the distribution of mobile ions over several possible sublattice sites is responsible for the spectral shape. The disorder manifests itself in the spectral density through several different mechanisms. The anharmonicity of the lattice causes very strong, asymmetric damping of the optical phonons. This effect is observed in both normal and superionic phases, and has a particularly strong influence on the spectra of the fluorites as noted in Sect.5.1.4. The disorder of the mobile ions in the rigid host lattice removes the momentum-conservation selection rule, allowing direct observation of the one-phonon density of states. This accounts in part for the strong Raman scattering, due to zone-boundary acoustic phonons, below the optical-phonon frequencies in the metal halides and fluorites. Dispersionless low-frequency "Einstein oscillation" modes due to the diffusion-related oscillatory movements of the cations at the sublattice sites have been proposed to exist in the metal halides and β-aluminas. The attempt-frequency peaks observed in the Raman spectra of the β-aluminas (Sect.5.1.3) are easily identifiable due to the separation from the spectral density due to the host lattice. In the metal halides the low-lying attempt modes are proposed to explain the strong scattering above the relaxation modes. We also saw that the anharmonicity of the lattice has made second-order scattering processes of importance in the metal halides. The totality of these effects, as seen particularly in the metal halides, gives the Raman spectra a liquidlike appearance. A description of the transition to the superionic phase as a sublattice melting is appropriate; the static meltlike disorder is reflected in the Raman spectrum while the dynamics of the ionic motion are more directly seen in the low-frequency relaxation modes.

References

5.1 R. Loudon: Adv. Phys. *13*, 423 (1964)
5.2 M. Cardona (ed.): *Light Scattering in Solids*, Topics in Applied Physics, Vol.8 (Springer, Berlin, Heidelberg, New York 1975)
5.3 B.J. Berne, R. Pecora: *Dynamic Light Scattering* (Wiley and Sons, New York 1976) p.27
5.4 A.A. Maradudin, E.W. Montroll, G.H. Weiss, I.P. Ipatova: *Theory of Lattice Dynamics in the Harmonic Approximation* (Academic Press, New York 1971)
5.5 M. Born, K. Huang: *Dynamical Theory of Crystal Lattices* (Oxford University Press, Oxford 1954)
5.6 G.B. Benedek: "Thermal Fluctuations and the Scattering of Light," in Brandeis Summer Inst. in Theor. Phys., ed. by M. Chretien, E.P. Gross, S. Deser (Gordon an Breach, New York 1966)
5.7 J.A. Bucaro, T.A. Litovitz: J. Chem. Phys. *54*, 3846 (1971)
5.8 W. Jost: *Diffusion in Solids, Liquids and Gases* (Academic Press, New York 1960)
5.9 P. Vashishta, A. Rahman: Phys. Rev. Lett. *40*, 1337 (1978)
5.10 R.C. Hanson, T.A. Fjeldy, H.D. Hochheimer: Phys. Status Solidi (b) *70*, 567 (1975)
5.11 R.T. Harley, W. Hayes, A.J. Rushworth, J.F. Ryan: J. Phys. C. *8*, 1350 (1975)
5.12 M.J. Delaney, S. Ushioda: Solid State Commun. *19*, 297 (1976)
5.13 G. Burns, F.H. Dacol, M.W. Shafer: Solid State Commun. *19*, 287 (1976)
5.14 M.J. Delaney, S. Ushioda: Phys. Rev. B *16*, 1410 (1977); and in *Proc. Intern. Conf. on Lattice Dynamics*, ed. by M. Balkanski (Flammarion, Paris 1978)

138

5.15 G. Burns, F.H. Dacol, M.W. Shafer: Phys. Rev. B *16*, 1416 (1977)
5.16 R. Alben, G. Burns: Phys. Rev. B *16*, 3746 (1977)
5.17 M.J. Delaney: Ph.D. dissertation (University of California, Irvine 1977)
5.18 G.L. Bottger, C.V. Damsgard: J. Chem. Phys. *57*, 1215 (1972)
5.19 S. Hoshino: J. Phys. Soc. Jpn. *12*, 315 (1957)
5.20 W. Buhrer, P. Bruesch: Solid State Commun. *16*, 155 (1975)
5.21 J.B. Boyce, T.M. Hayes, W. Stutius, J.C. Mikkelsen: Phys. Rev. Lett. *38*, 1362 (1977)
5.22 A. Kvist, R. Tarneberg: Z. Naturforsch. *25*a, 257 (1970)
5.23 Samuelsen, Andersen, Feder (eds.): *Structural Phase Transitions and Soft Modes* (Universitetsforlaget, Oslo 1971)
5.24 W. Buhrer, R.M. Nickow, P. Bruesch: Phys. Rev. B *17*, 3362 (1978)
5.25 S.M. Shapiro, D. Semmingsen, M. Salamon: *Proc. Intern. Conf. on Lattice Dynamins*, ed. by M. Balkanski (Flammarion, Paris 1978)
5.26 K. Funke, A. Jost: Ber. Bunsenges. Phys. Chem. *75*, 436 (1971)
5.27 G.L. Bottger, C.V. Damsgard: Solid State Commun. *9*, 1277 (1971)
5.28 T. Fukumoto, S. Namashima, K. Tabuchi, A. Mitsuishi: Phys. Status Solids (b) *73*, 341 (1976
     B. Prevot, M. Sieskind: Phys. Status Solidi B *59*, 133 (1973)
5.29 A.J. Barker: J. Phys. C *5*, 2276 (1972)
5.30 J.P. McTaque, P.A. Fleury, D.B. DuPre: Phys. Rev. *188*, 303 (1969)
5.31 S. Geller: Science *157*, 310 (1967)
5.32 S. Geller: Phys. Rev. B *14*, 4345 (1976)
5.33 G. Burns, F.H. Dacol, M.W. Shafer: Phys. Lett. *58*a, 229 (1976); Solid State Commun. *19*, 291 (1976)
5.34 D. Gallagher, M.V. Klein: J. Phys. C *9*, L687 (1976)
5.35 D. Gallagher, M.V. Klein: Phys. Rev. B (April 1979)
5.36 D. Gallagher: Ph.D. dissertation (University of Illinois, Urbana-Champaign 1978)
5.37 I.D. Brown, H.E. Howard-Lock, M. Natarajan: Can. J. Chem. *55*, 1511 (1977)
5.38 L.E. Topol, B.B. Owens: J. Phys. Chem. *72*, 2106 (1968)
5.39 R. Nemanich: Private Communication
5.40 G. Burns, F.H. Dacol, M.W. Shafer, R. Alber: Solid State Commun. *24*, 753 (1977)
5.41 J.H. Kennedy: In *Solid Electrolytes*, ed. by S. Geller, Topics in Applied Physics, Vol.21 (Springer, Berlin, Heidelberg, New York 1977) p.105
5.42 J.C. Wang, M. Gaffari, S. Choi: J. Chem. Phys. *63*, 772 (1975)
5.43 C.R. Peters, M. Bettman, J.W. Moore, M.D. Glick: Acta Crystallogr. B *27*, 1826 (1971)
5.44 B.D. McWhan, S.J. Allen, J.P. Remeika, P.D. Dernier: Phys. Rev. Lett. *35*, 953 (1975)
5.45 L.L. Chase, C.H. Hao, G.H. Mahan: Solid State Commun. *18*, 401 (1976)
5.46 C.H. Hao, L.L. Chase, G.D. Mahan: Phys. Rev. B *13*, 4306 (1976)
5.47 C.H. Hao: Ph.D. thesis (University of Indiana 1978)
5.48 S.J. Allen, J.P. Remeika: Phys. Rev. Lett. *33*, 1478 (1974)
5.49 D.B. McWhan, S.M. Shapiro, J.P. Remeika, G. Shirane: J. Phys. C *8*, L487 (1975)
5.50 J.B. Bates, T. Kaneda, J.C. Wang: Solid State Commun. *25*, 629 (1978)
5.51 W.Y. Hsu: Phys. Rev. B *14*, 5161 (1976)
5.52 G. Burns, G.V. Chandrashekhar, F.H. Dacol, L.M. Foster: Solid State Commun. *21*, 1057 (1977)
5.53 W. Hayes, A.J. Rushworth, J.F. Ryan, R.J. Elliott, W.G. Kleppmann: J. Phys. C *10*, L111 (1977)
5.54 R.J. Elliott, W. Hayes, W.G. Kleppmann, A.J. Rushworth, J.F. Ryan: Proc. R. Soc. London *360*, 317 (1978)
5.55 C.E. Derrington, M.O'Keefe: Nature (London) Phys. Sci. *246*, 44 (1973)
5.56 M.O'Keefe: In *Superionic Conductors*, ed. by G.D. Mahan, W.L. Roth (Plenum Press, New York 1976)
5.57 M.V. Klein: In *Light Scattering in Solids*, ed. by M. Balkanski, R.C.C. Leite, S.P.S. Porto (Flammarion, Paris 1973)
5.58 R.A. Field, D. Gallagher, M.V. Klein: Phys. Rev. B *18*, 2995 (1978)
5.59 R. Field: Ph.D. dissertation (University of Illinois, Urbana-Champaign, 1978)
5.60 T. Geisel: Solid State Commun. *24*, 155 (1977)

5.61 W. Dieterich, T. Geisel, I. Peschel: Z. Phys. B *29*, 5 (1978)
5.62 K.R. Subbaswamy: Solid State Commun. *19*, 1157 (1976)
5.63 B.A. Huberman, R.M. Martin: Phys. Rev. B *13*, 1498 (1976)
5.64 G. Winterling, W. Senn, M. Grimsditch, R. Katiyar: In *Proc. Intern. Conf. on Lattice Dynamics*, ed. by M. Balkanski (Flammarion, Paris 1977)
5.65 S. Shapiro: In *Superionic Conductors*, ed. by G.D. Mahan, W. Roth (Plenum Press, New York 1976)
5.66 L.L. Chase: Private communication
5.67 G. Echold, K. Funke: Z. Naturforsch. *28*a, 1042 (1973)
5.68 D. Clemen, K. Funke: Ber. Bunsenges. Phys. Chem. *79*, 1119 (1975)
5.69 H.R. Chandrasekhar, G. Burns, G.V. Chandrashekhar: Solid State Commun. *25*, 547 (1978)
5.70 P. Bruesch, S. Strassler, H.R. Zeller: Phys. Status Solidi (a) *31*, 217 (1975)
5.71 M. Tinkham: *Group Theory and Quantum Mechanics* (McGraw-Hill, New York 1964)
5.72 A.S. Barker, R. Loudon: Rev. Mod. Phys. *44*, 18 (1972)
5.73 G.L. Bottger, A.L. Geddes: J. Chem. Phys. *46*, 3000 (1967)

6. Magnetic Resonance in Superionic Conductors

P. M. Richards

With 10 Figures

Nuclear magnetic resonance can be a powerful tool for studying motions of ions and therefore has had extensive application [6.1] in the field of superionic conductors. Being concerned here with basic physics, we make the following distinction between superionic (SI) and low-ionic-conductivity materials: in the latter only a small number of independent particles or vacancies hop whereas large numbers are involved in the former. Thus correlated motion is a major aspect of the problem, and it is of interest to see what can be learned about this from nuclear magnetic resonance (NMR). To understand the features of the dynamics which can be delineated and how NMR might be useful in distinguishing correlated from independent particle motion, we consider this technique in conjunction with bulk conductivity and infrared or Raman spectroscopy (Chap.5).

The most important property of superionic conductors, at least for technical applications, is the dc ionic conductivity σ which may be expressed as

$$\sigma = \frac{\nu}{6} \left(\frac{N}{V}\right) e^2 \ell^2 / k_B T = \left(\frac{N}{V}\right) e^2 D / k_B T \quad , \qquad (6.1)$$

where N is the number of mobile ions of charge e contained in volume V, T is the absolute temperature, and ν^{-1} is the average time for the ion to hop an average distance ℓ. The first equality in (6.1) assumes a specific discrete hopping model and may require revisions —or redefinitions of ν and ℓ as effective quantities— if there are departures from the simple, uncorrelated hopping model which we use for the present. The second, Einstein-Nernst, equality, which relates σ to a bulk diffusion coefficient is quite general, however [6.2].

If the number of carriers is known, then the conductivity measures the product $\nu \ell^2$ and can thus determine the hopping frequency ν only if the jump mechanisms are sufficiently well understood that the mean jump distance ℓ is known. The hop may be to nearest-neighbor sites only, and such determination can be made, but this is not always the case.

The advantage of NMR is that it can often provide a direct estimate of ν. The reason, as treated in more detail in Sect.6.1, is that the NMR relaxation process is governed by the rate at which an interaction, such as coupling to an electric field gradient or to the spin of a nearby ion, fluctuates. This generally depends only on how long the nucleus remains at a given site and not on how far it travels from the site once it jumps. Conduction, on the contrary, clearly depends both on how frequent and how far the jumps are.

For processes where an Arrhenius relation

$$\nu = \nu_0 \exp(-\Delta/k_B T) \tag{6.2}$$

is obeyed, the activation energy Δ and "attempt frequency" ν_0 often can both be extracted from NMR relaxation data. Δ is obtained from the overall temperature dependence, and once it is known, ν_0 can be inferred from the temperature at which ν reaches a critical value associated with a maximum spin-lattice relaxation rate ($\nu \approx \omega_0$, the NMR frequency) or the onset of motional narrowing of the linewidth ($\nu \approx 1/T_{20}$, where T_{20} is the rigid lattice spin-spin relaxation time). A fundamental question is whether ν_0 is related to a vibrational frequency $\tilde{\nu}$ of the mobile ion residing in a potential well between hops. Simple classical diffusion theory [6.3] predicts that $\nu_0 = \tilde{\nu}$, which can be tested by combined NMR and infrared or Raman measurements since the latter methods probe vibrational frequencies. Hopping produces a broadening but, at least at sufficiently low temperatures, should not interfere with the spectroscopic determination of $\tilde{\nu}$. The complementary value of NMR here is that the spin relaxation is insensitive to the rapid vibration $\tilde{\nu}$, so that the vibrations between hops normally do not interfere with measurements of ν as long as $\nu \sim \omega_0 \ll \tilde{\nu}$.

In the simplest case of discrete, nearest-neighbor, classical hopping, the activation energy and attempt frequency inferred from conductivity and NMR should agree, and the attempt frequency should correspond to a measured optical phonon mode. A fascinating aspect of the physics of SI conductors is that there are often vast discrepancies among the three sets of measurements. These may result from the correlated hopping of many ions but could also originate (at least in the case of NMR) in an incorrect relationship between a given experimental parameter and the hopping process. One goal of this chapter is to determine the extent to which such discrepancies may be ascribed to correlation effects as opposed to an improper application of simple hopping ideas to the theory of NMR relaxation.

That correlated hopping may cause a discrepancy can be seen by noting that the ν of (6.1) from the measured conductivity is the true, correlated frequency which describes the collective motion of all the ions. But ν_{NMR}, inferred from NMR relaxation, is related to the motion of a single, distinguishable (by its spin label) ion. If correlation effects are important, these frequencies can be quite different. A gross effect occurs in the one-dimensional diffusion of particles which cannot hop to occupied sites [6.4-6]. Here, the bulk diffusion coefficient D of all the particles is well defined and the second equality of (6.1) holds. However, the motion of a single labeled particle is not diffusive (its mean

square displacement x^2 is proportional to $t^{\frac{1}{2}}$ rather than to t for long times t).
One would not, therefore, expect NMR-inferred dynamics to agree with those obtained
from bulk conductivity. In less extreme, three-dimensional examples, where a distinguishable particle diffusion coefficient D^* can be defined (and measured by NMR
techniques [6.7]) D^* is known [6.8] to differ from the bulk D. Thus the conductivity ν and ν_{NMR} will differ even if the jump distance ℓ is the same for collective
and single-particle motion.

Another example in which correlations are important is the inequivalent site
model as shown in Fig.6.1. A particle jumps rapidly from a b-type to an unoccupied
a-type well, but the a → b process is much slower. If, as shown, the neighboring
a wells are initially occupied, a nucleus will hop back and forth between the
nearest b sites until one of the other a wells is vacated. This motion makes no
contribution to the conductivity yet can be effective in relaxing the nuclear spin
if it experiences different interactions at a and b sites.

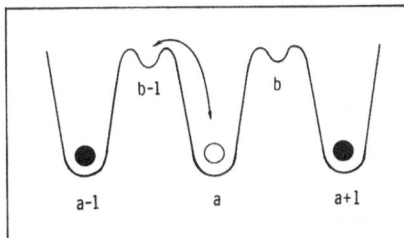

Fig.6.1. Inequivalent site model with deep
a-type and shallow b-type wells. Rapid hopping
back and forth of nucleus among sites b - 1,
a, b gives rise to a high frequency component
in NMR relaxation; but there is no net conduction unless particles are displaced from a - 1
or a + 1

In the above cases there is a true difference between distinguishable-particle
and total dynamics. As will be discussed in detail later, there can also be situations in which the difference is only apparent. That is, a simple hopping picture is valid but care has to be taken in interpreting the NMR results. Low dimensionality, a common occurrence for the channel- or planar-structured SI materials,
can produce apparent anomalies in the attempt frequency and activation energy.
Quadrupole effects and coupling to paramagnetic impurities may alter the motional
narrowing condition $\nu T_{20} \sim 1$, thereby leading to an incorrect attempt frequency.

Electron paramagnetic resonance (EPR) has been used less than NMR in studying
superionic conductors. The monovalent mobile ions do not have unpaired electron
spins, making direct EPR studies of diffusing ions impossible. The mobile ions can
influence the EPR spectrum, however, through their modulation of crystal-field
splittings and hyperfine interactions. EPR is also a good monitor of local site
symmetry and can be used to obtain information about the surroundings which influence the ionic motion. Because of the strong spin-spin interaction between paramagnetic ions, information of this type generally requires small concentrations of
EPR species as substitutional impurities.

6.1 Theory of NMR Relaxation of and by Rapidly Diffusing Ions

NMR relaxation is characterized by the longitudinal and transverse relaxation times T_1 and T_2, respectively. Longitudinal and transverse refer to components of nuclear spin parallel and perpendicular to the applied static field. T_1 processes require a change in nuclear Zeeman energy $\hbar\omega_0$ accompanied by a compensating change in lattice energy in order to conserve energy. In the language of the semiclassical theory of relaxation by time-dependent interactions [6.9], the fluctuation spectrum of the interaction must have appreciable spectral density at the frequency ω_0. T_2 processes lead to line broadening and can take place without the exchange of Zeeman energy. As a consequence the spectral density at zero frequency can be effective for transverse, but not longitudinal relaxation. Static inhomogeneous broadening, caused by sources such as random strains, contribute to the linewidth but not to T_2 as measured by spin-echo techniques [6.10]. We shall not consider such broadening here; thus T_2 will be assumed to be that measured by spin echos or that obtained from the linewidth for a system in which inhomogeneous contributions are negligible. A third relaxation time [6.11], the "rotating frame T_1" or T_1 may be measured by suitable pulses with a strong rf field H_1. It has the feature of probing spectral density at low nonzero frequency $\omega_1 \ll \omega_0$ where ω_1 is H_1 expressed in frequency units.

6.1.1 General Correlation Functions and Interactions

The general theory of magnetic relaxation [6.9,12] expresses relaxation rates in terms of the spectral densities

$$J_{\alpha\beta}(\omega) = \mathrm{Re}\left\{\int_0^\infty e^{-i\omega t}[G_{\alpha\beta}(t) + G_{-\alpha\beta}(t)]dt\right\} \tag{6.3}$$

and the correlation functions

$$G_{\alpha\beta}(t) = \frac{\hbar^{-2}<[\mathscr{H}_\alpha'(t),I_\beta][I_{-\beta},\mathscr{H}_{-\alpha}'(0)]>}{<I_\beta I_{-\beta}>} \tag{6.4}$$

where $\mathscr{H}_\alpha'(t)$ is that part of the perturbing interaction which induces a change α in Zeeman quantum number ($\alpha = 0, \pm 1, \pm 2$ for the important mechanisms), and I_β is the component of total nuclear spin which gives a change β in Zeeman quantum number; $\beta = 0$ corresponds to I_z, $\beta = \pm 1$ to $I_\pm = I_x \pm iI_y$. Angular brackets indicate thermal averaging.

The interaction can be one of three types. (I) Spin-spin coupling between like nuclear spins, which we denote n-n and for which

---

$\hbar = h/2\pi$ (normalized Planck's constant)

$$\mathcal{H}_\alpha'(t) = \hbar \sum_{i \neq j} \sum_{\alpha'} A_{ij}^{\alpha;\alpha'}(t) I_i^{\alpha'} I_j^{\alpha - \alpha'} \quad , \tag{6.5}$$

where $A_{ij}^{\alpha;\alpha'}$ is the classical dipole coupling between spins i and j. It is time dependent because the position vector $\underline{r}_{ij}$ between i and j changes both in magnitude and direction due to ion motion.

(II) Coupling of the quadrupole moment to electric field gradients, Q-coupling:

$$\mathcal{H}_\alpha'(t) = \hbar \sum_i B_i^\alpha(t) \mathcal{O}_i^\alpha \quad , \tag{6.6}$$

where $\mathcal{O}_i^\alpha$ is that quadratic combination of spin operators which produces the change α in the Zeman quantum number $[\mathcal{O}_i^0 = I_i^{z^2} - \frac{1}{3} I(I + 1), \quad \mathcal{O}_i^{\pm 1} = I_i^z I_i^\pm + I_i^\pm I_i^z,$ $\mathcal{O}_i^{\pm 2} = I_i^\pm I_i^\pm]$ and $B_i^\alpha(t)$ is proportional to the quadrupole moment and electric field gradient (efg). As the environment experienced by a nucleus changes, the efg changes, making $B_i^\alpha(t)$ time dependent.

(III) Spin-spin coupling between unlike spins I_j and S_i, e-n coupling

$$\mathcal{H}_\alpha'(t) = \hbar \sum_{ij} \sum_{\alpha'} C_{ij}^{\alpha;\alpha'}(t) S_i^{\alpha'}(t) I_j^{\alpha - \alpha'} \quad . \tag{6.7}$$

The distinction between (6.5) and (6.7) is that in the latter case the spins S_i are assumed to have resonance frequencies far removed from the observed resonance of I at ω_0, so that their time dependence may be considered independently. The case of practical importance is when S_i is the electron spin of a paramagnetic impurity. As before, $C_{ij}^{\alpha\alpha'}(t)$ is time dependent since the electronic-nuclear spin-spin (e-n) interaction varies as the nucleus diffuses away from the impurity. Transferred hyperfine coupling may equal or even exceed classical dipole coupling in determining the magnitude of $C_{ij}^{\alpha\alpha'}(t)$. Electron-spin precession and relaxation make $S_i^{\alpha'}(t)$ time dependent.

In general the various correlation functions are complicated and involve several aspects of the dynamics which may be difficult to disentangle. Our procedure is first to obtain the complete expressions and then, in certain limiting cases of simple interactions and lattice geometry, to reduce them to forms which relate directly to single, distinguishable particle motion or to site occupancy (vacancy motion). We assume that the hopping is essentially instantaneous so that a nucleus can be regarded as occupying a given lattice site at any time. Occupation numbers are defined by

$$P_\mu = \begin{cases} 1 \text{ if site } \mu \text{ is occupied by } any \text{ nucleus} \\ \\ 0 \text{ otherwise} \end{cases} \quad , \tag{6.8}$$

$$n_{i\mu} = \begin{cases} 1 \text{ if site } \mu \text{ is occupied by the } i^{th} \text{ nucleus, whose spin is } I_i \\ \\ 0 \text{ otherwise} \end{cases} \quad . \quad (6.9)$$

P_μ and $n_{i\mu}$ are thus site- and distinguishable-particle functions, respectively, whose average values are the respective occupation probabilities. It is evident that

$$P_\mu = \sum_i n_{i\mu} \quad . \quad (6.10)$$

With greek subscripts indicating site indices; Latin ones denoting particles, the coefficients of (6.5,6) and (6.7) may be expressed as

$$A_{ij}^{\alpha,\alpha'}(t) = \sum_{\mu\lambda} n_{i\mu}(t)n_{j\lambda}(t)A_{\mu\lambda}^{\alpha,\alpha'} \quad (6.11a)$$

$$B_i^\alpha(t) = \sum_\mu n_{i\mu}(t)B_\mu^\alpha(t) \quad , \quad (6.11b)$$

$$C_{ij}^{\alpha,\alpha'}(t)S_i^{\alpha'}(t) = \sum_\mu n_{j\mu}(t)C_{\mu_0\mu}S_{\mu_0}(t) \quad , \quad (6.11c)$$

where we have assumed that all time dependence of the n-n and e-n couplings arises from the nuclear motion, and thus we have neglected lattice vibrations. The time dependence of $S_{\mu_0}(t)$, an electron spin fixed at site μ_0, is, however, governed by sources other than the hopping (the indices i and μ_0 are interchangeable since the ith electron spin is always at site μ_0). The quadrupole coupling B_μ^α is still time dependent

$$B_\mu^\alpha(t) = B_\mu^\alpha\left[P_{\mu'\neq\mu}(t)\right] \quad (6.12)$$

since the efg for a nucleus at site μ is influenced by the occupancy of other sites μ'. The form (6.12), as (6.11a), neglects any vibrational or lattice-relaxation influence on the time dependence.

Performing the high-temperature averages over nuclear spin coordinates ($<I_i^\alpha I_j^{\alpha'}> = 0$ unless $i = j$ and $\alpha' = -\alpha$) in (6.4) results in

$$G_{n-n}(t) \propto N^{-1} \sum_{\mu\lambda} \sum_{\mu'\lambda'} \overline{A_{\mu\lambda}A_{\mu'\lambda'}} \sum_{ij} \left\langle n_{i\mu}(t)n_{j\lambda}(t)n_{i\mu'}(0)n_{j\lambda'}(0)\right\rangle \quad , \quad (6.13)$$

$$G_Q(t) \propto N^{-1} \sum_{\mu\mu'} \left\langle \overline{B_\mu(t)B_{\mu'}(0)} \sum_i n_{i\mu}(t)n_{i\mu'}(0)\right\rangle \quad , \quad (6.14)$$

$$G_{e-n}(t) \propto N^{-1} \sum_{\mu_0,\mu,\mu'} \overline{C_{\mu_0\mu}C_{\mu_0'\mu'}} \left\langle S_{\mu_0}^\alpha(t)S_{\mu_0}^{-\alpha}(0)\right\rangle \sum_j \left\langle n_{j\mu}(t)n_{j\mu'}(0)\right\rangle \quad , \quad (6.15)$$

where, apart from the electron-spin correlation, the Zeeman quantum number indices have been suppressed since they are not germane to the discussion. In (6.15) it has also been assumed that there is no correlation between impurity spins on different sites. The bars over the A's, B's, and C's indicate an angular average for the case of polycrystalline samples. By virtue of (6.12) there are generally correlations between $B_\mu(t)B_{\mu'}(0)$ and $n_{i\mu}(t)n_{i\mu'}(0)$, thus accounting for the positioning of the brackets in (6.14).

Of the three forms, only the e-n coupling (6.15) can be related to a simple hopping correlation functions without further approximations. Since $\langle n_{j\mu}(t)n_{j\mu'}(0)\rangle$ is independent of j for identical ions, the e-n relaxation is specified by the time-dependent probability that a distinguishable ion is at site μ at time t given that it was at μ' at $t = 0$. A particularly simple case results for a dominant nearest-neighbor transferred hyperfine interaction at high temperature where the hopping time is fast compared with the electronic relaxation time. Then $\langle S_{\mu 0}^z(t)S_{\mu 0}^z(0)\rangle$ may be replaced by its $t = 0$ value $1/3\ S(S + 1)$ and the correlation $\langle n_{j\mu}(t)n_{j\mu'}(0)\rangle$ need be considered only for μ and μ' neighbors of the same impurity. We shall argue in Sect.6.2.4 that Mn-doped PbF_2 satisfies these conditions [6.13].

The nuclear spin-spin coupling (6.13) involves a complex two-distinguishable-particle correlation function. It can be simplified to $\langle n_{i\mu}(t)n_{i\mu}(0)n_{j\lambda}(t)n_{j\lambda}(0)\rangle$, the joint autocorrelation function that i remain at μ and j remain at λ, on the assumption that $\overline{A_{\mu\lambda}A_{\mu'\lambda'}} \approx \overline{|A_{\mu\lambda}|^2}\ \delta_{\mu,\mu'}\delta_{\lambda,\lambda'}$, which is not unreasonable because of the angular average.

The strong efg coupling for nuclei with quadrupole moments generally makes the n-n interaction an important T_1 mechanism only for $I = \frac{1}{2}$. As a result of (6.12) the correlation functions (6.14) in $G_Q(t)$ can be quite involved unless simplifying assumptions are made. First we note that ionic motion can result in n-n or Q relaxation for both stationary and mobile nuclei since the dipole field or efg experienced by a stationary nucleus fluctuates due to the motion of nearby ions. When nucleus i is stationary (6.14) reduces to $G_Q(t) \propto \langle B_\mu(t)B_\mu(0)\rangle$ if all sites μ of the stationary resonating nuclei are equivalent. Further, assume that $B_\mu(t)$ may be written as

$$B_\mu(t) = B_\mu^0 + \sum_{\mu'} \Delta B_{\mu\mu'}[P_{\mu'}(t) - \langle P_{\mu'}\rangle]\ , \tag{6.16}$$

where B_μ^0 reflects the average efg when neighboring sites μ' have the average occupation $\langle P_{\mu'}\rangle$ and $\Delta B_{\mu\mu'}$ relates to the fluctuation in efg at μ as ions hop on and off sites μ'. Keeping only the time-dependent part of $G_Q(t)$ that can lead to a T_1 process for the nucleus at site μ, we find

$$G_Q(t) \propto \sum_{\mu'\mu''} \overline{\Delta B_{\mu\mu'}\Delta B_{\mu\mu''}}\langle \delta P_{\mu'}(t)\delta P_{\mu''}(0)\rangle\ , \tag{6.17}$$

where $\delta P_{\mu'}(t) = P_{\mu'}(t) - \langle P_{\mu'}\rangle\ .$

The feature of note in (6.17) is the rather obvious fact that Q relaxation of a stationary nucleus depends on site occupation only, since the spin label which makes the particles distinguishable does not influence the efg. This contrasts with n-n relaxation of a stationary nucleus, since the dipole field depends on the spin orientation as well as the distance between particles.

Quadrupole relaxation of a mobile nucleus involves correlations among both distinguishable particle and site functions as seen from (6.14). If $B_\mu^0 \gg \Delta B_{\mu\mu'}$, in which case B_μ may be taken as time independent, (6.14) simplifies to

$$G_Q(t) \propto \sum_{\mu\mu'} \overline{B_\mu^0 B_{\mu'}^0} \left\langle n_{i\mu}(t) n_{i\mu'}(0) \right\rangle \quad . \tag{6.18}$$

This is of the same form as (6.17) except that a distinguishable-particle correlation function is involved. Suppose in particular that there are two inequivalent sites a_μ, b_μ for the mobile ion in the μth unit cell for which B_μ has values B_a and B_b, respectively. Then (6.18) becomes

$$G_Q(t) \propto (B_a - B_b)[B_a \left\langle n_a(t) n_a(0) \right\rangle - B_b \left\langle n_b(t) n_b(0) \right\rangle] \quad , \tag{6.19}$$

where $<n_a(t) n_a(0)>$ is the probability that a *particular* nucleus is on *any* A site at time t given it was on *any* A site at t = 0, i.e., $n_a(t) = \sum_{a_\mu} n_{i\mu}$, with a similar expression for $<n_b(t) n_b(0)>$. We have used $n_a(t) + n_b(t) = 1$ and ignored time-independent terms in arriving at (6.19). For the case of a stationary nucleus in the μth unit cell we postulate that the fluctuating part ΔB is determined entirely by occupation of sites a_μ and b_μ and thus obtain from (6.17)

$$G_Q(t) = (\Delta B_a - \Delta B_b)[\Delta B_a \left\langle \delta P_a(t) \delta P_a(0) \right\rangle - \Delta B_b \left\langle \delta P_b(t) \delta P_b(0) \right\rangle] \quad , \tag{6.20}$$

where $\Delta B_a \equiv \Delta B_{\mu,a_\mu}$, $\Delta B_b \equiv \Delta B_{\mu,b_\mu}$ and $\delta P_{a(b)} \equiv P_{a(b)} - <P_{a(b)}>$ with $P_{a(b)}$ the probability that *any* ion is on a particular a(b) site. Equations (6.19) and (6.20) differ only in that the former is for a distinguishable particle and the latter for a distinguishable site.

The above examples illustrate the types of particle-particle, site-site, or site-particle time correlation functions which determine relaxation rates for the various coupling mechanisms. Since it is desirable to relate NMR relaxation to conductivity studies, we also quote the correlation function which gives σ. According to KUBO [6.2],

$$\sigma \propto \int_0^\infty dt \left\langle V(t) V(0) \right\rangle = \int_0^\infty dt \sum_{\mu\mu'} \left\langle x_\mu x_{\mu'} \dot{P}_\mu(t) \dot{P}_{\mu'}(0) \right\rangle \quad , \tag{6.21}$$

where

$$V(t) = \sum_i \dot{x}_i(t) = \sum_{i,\mu} x_\mu \dot{n}_{i\mu}(t) = \sum_\mu x_\mu \dot{P}_\mu(t)$$

is the net velocity in the x direction. It is evident that detailed hopping models are required to make definite statements about whether, for example, the activation energy for σ should be the same as that for the NMR T_1.

6.1.2 Calculation of T_1 and T_2 from Correlation Functions

We next discuss calculation of T_1 and T_2 once the difficult problem of finding $G_{\alpha\beta}(t)$ has been solved. Results obtained for T_1 and T_2 in various limits from Sects.6.1.2-5 are summarized in Tables 6.1 and 6.2 together with some expressions not explicitly derived in the text. We introduce a characteristic time τ_c associated with decay of $G_{\alpha\beta}(t)$ which, if much less than the observation time (of the order of the relaxation time) permits the application of standard second-order, time-dependent perturbation theory. This gives the motional narrowing expressions

$$1/T_1 = J_{10}(\omega_0) + J_{20}(2\omega_0) \quad , \tag{6.22}$$

$$1/T_2 = J_{01}(0) + J_{11}(\omega_0) + J_{21}(2\omega_0) \quad , \tag{6.23}$$

for either n-n or Q processes. An expression similar to (6.23) exists for $T_{1\rho}$ with the first term on the right side replaced by a spectral density at ω_1. With a simple exponential decay of $G_{\alpha\beta}(t)$, T_1 calculated by (6.22) is consistent with the condition $T_1 \gg \tau_c$ as long as $G_{\alpha 0}(0)\tau_c^2/(1 + \omega_0^2\tau_c^2) \ll 1$, which will be true as long as the line is narrow enough to be observable. For e-n relaxation the situation is complicated by the nonuniformity of the interaction. The average value $\langle J(\omega_0)\rangle_{av} \sim cJ'(\omega_0)$ where $J'(\omega_0)$ is the value when the nucleus is on a nearest-neighbor site of an impurity, and c is the concentration of impurities satisfies $T_1 \gg \tau_c$ when used in (6.22). The local relaxation rate [6.9] $J'(\omega_0)$, however, gives $T_1 < \tau_c$ except at high temperatures. The more stringent requirement, $J'(\omega_0)\tau_c \ll 1$, must hold before (6.22) applies with $J(\omega_0)$ replaced by $\langle J(\omega_0)\rangle_{av}$. Care must also be taken in defining τ_c since both particle motion and electron-spin relaxation contribute, as seen from (6.15). The subject of e-n relaxation has recently been studied by the author [6.14].

The expression (6.23) for T_2 holds only in the motional-narrowing regime where $J_{01}(0)\tau_c \ll 1$, or $G_{01}(0)\tau_c^2 \ll 1$. This condition is violated at low temperatures. In the limit $G_{01}(0)\tau_c^2 \gg 1$ one obtains the "rigid lattice" result

$$\frac{1}{T_{20}} = [G_{01}(0)]^{\frac{1}{2}} = (\Delta\omega^2)^{\frac{1}{2}} \quad , \tag{6.24}$$

Table 6.1. Limiting behavior of T_1 for independent particle diffusion as a function of hop time τ_c, frequency ω_0 and dimensionality d. Apparent activation energy $d(\ln T_1)/d(1/k_B T)$ is derived by assuming an activated $\tau_c = \tau_0 \exp(\Delta/k_B T)$

Conditions		T_1(proportional to)	$d(\ln T_1)/d(1/k_B T)$
$\omega_0\tau_c \gg 1$, all d		$\omega_0^2\tau_c$	Δ
$\omega_0\tau_c \ll 1$	d = 3	$1/\tau_c$	$-\Delta$
	d = 2	$[\tau_c \ln(1/\omega_0\tau_c)]^{-1}$	$-\Delta[1 + (\Delta/k_B T + \ln \omega_0\tau_0)^{-1}]$
	d = 1	$(\omega_0/\tau_c)^{\frac{1}{2}}$	$-\Delta/2$
$\omega_0\tau_c \ll 1$, quasi 1d with interchannel hop time $\tau_\perp$ and activation energy $\Delta_\perp$.	$\omega_0\tau_\perp \gg 1$	$(\omega_0/\tau_c)^{\frac{1}{2}}$	$-\Delta/2$
	$\omega_0\tau_\perp \ll 1$	$(\tau_c\tau_\perp)^{-\frac{1}{2}}$	$-(\Delta +\Delta_\perp)/2$

Table 6.2. Limiting behavior of T_2 for independent particle diffusion as a function hop time τ_c, dipolar second moment $\langle\Delta\omega^2\rangle$ and dimensionality d. Apparent activation energy is derived as in Table 6.1

Conditions		T_2	$d(\ln T_2/d(1/k_B T)$
$\langle\Delta\omega^2\rangle\tau_c^2 \gg 1$, all d		$\langle\Delta\omega^2\rangle^{-\frac{1}{2}}$	0
$\langle\Delta\omega^2\rangle\tau_c^2 \ll 1$, $\omega_0\tau_c \gg 1$	d = 3	$\langle\Delta\omega^2\rangle^{-1}\tau_c^{-1}$	$-\Delta$
	d = 2	$\langle\Delta\omega^2\rangle^{-1}\tau_c^{-1}\left[\ln\left(\frac{1}{\langle\Delta\omega^2\rangle\tau_c^2}\right)\right]^{-1}$	$-\Delta\left\{1 +\left(\frac{\Delta}{k_B T}+\frac{1}{2} \ln[\langle\Delta\omega^2\rangle\tau_0^2]\right)\right\}$
	quasi-1d,		
	$\langle\Delta\omega^2\rangle^{2/3}\tau_c^{1/3}\tau_\perp \gg 1$	$\langle\Delta\omega^2\rangle^{-2/3}\tau_c^{-1/3}$	$-\Delta/3$
	$\langle\Delta\omega^2\rangle^{2/3}\tau_c^{1/3}\tau_\perp \ll 1$	$\langle\Delta\omega^2\rangle^{-1}(\tau_c\tau_\perp)^{-\frac{1}{2}}$	$-(\Delta + \Delta_\perp)/2$
$\omega_0\tau_c \ll 1$	d = 3	$T_2 = T_1$ [a]	
	d = 1 or 2	$T_2 < T_1$ [b]	

[a] Assumes $I = \frac{1}{2}$ or negligible quadrupole splitting. See (6.27,28).

[b] This occurs because of the anomalously large secular (independent of ω_0) contribution to T_2 for low dimensions. For quasi-1d $T_2 = T_1$ for $\omega_0\tau_\perp \ll 1$, i.e., when the system acts like 3d.

where we have replaced $G_{01}(0)$ by the more familiar notation for the second moment $<\Delta\omega^2>$. For intermediate values of $G_{01}(0)\tau_c^2$ one can use the general Kubo-Tomita expression [6.12] for the relaxation function $\phi(t)$

$$\phi(t) = \exp\left[-\int_0^t (t - \tau)G_{01}(\tau)d\tau\right] \quad , \tag{6.25}$$

and define T_2 by $\phi(T_2) = 1$. The Bloembergen-Purcell-Pound (BPP) expression [6.9,15]

$$\frac{1}{T_2} \sim \int_0^{T_2} G_{01}(\tau)d\tau \quad , \tag{6.26}$$

leads to a similar dependence of T_2 on τ_c. Note for future discussion that T_2 increases from its rigid lattice value T_{20} when motion becomes rapid enough that $T_{20}/\tau_c \sim 1$.

We are now in a position to describe the general features of T_1, $T_{1\rho}$, and T_2 as functions of temperature. As shown in Fig.6.2 T_1^{-1} will go through a maximum at $\omega_0 \sim 1/\tau_c$, where $1/\tau_c$ is the frequency at which $J(\omega)$ is maximal, corresponding to the characteristic decay time of $G(t)$. If τ_c is further identified as the hopping time, it is evident that, for an activated process, T_1^{-1} will have a maximum as a function of temperature T and vary rapidly on either side. $T_{1\rho}^{-1}$ has a maximum at a lower temperature due to its being proportional to $J(\omega_1)$; ω_1, the precession frequency in the rf field, is much less than the NMR frequency ω_0. T_2^{-1} will decrease nearly monotonically with increasing temperature since the narrowing becomes more complete with decreasing τ_c. There is the possibility of a "glitch" in the T_2^{-1} versus T^{-1} curve near the point at which T_1^{-1} has a maximum since the nonsecular components at ω_0 and $2\omega_0$ in (6.23) are also maximal there. These features of a T_1 minimum and motionally narrowed T_2 are nearly universally observed in superionic conductors. [Exceptions do occur when the T_1 minimum is masked by effects of a phase transition (Chap.7).] It is in the frequency dependence and quantitative values of τ_c that wide variations occur.

6.1.3 T_1/T_2 Ratio

The high-temperature ratio T_1/T_2 is of interest since it gives an indication of the dominant relaxation process for $I > \frac{1}{2}$. When $\omega_0\tau_c \ll 1$, the spectral density is independent of frequency and T_1 and T_2 processes occur with equal facility. Further, in a powdered sample where the interactions favor no particular direction symmetry would seem to dictate $T_1 = T_2$.

An exception occurs for half-integer $I > \frac{1}{2}$ in the presence of a static efg coupling large enough to give resolved quadrupole splitting. Then only the central $\frac{1}{2} \rightarrow -\frac{1}{2}$ transition, which is unaffected to first order in the efg, is observed in the main resonance. Thus, in a T_2 calculation, rather than a sum over all spin states

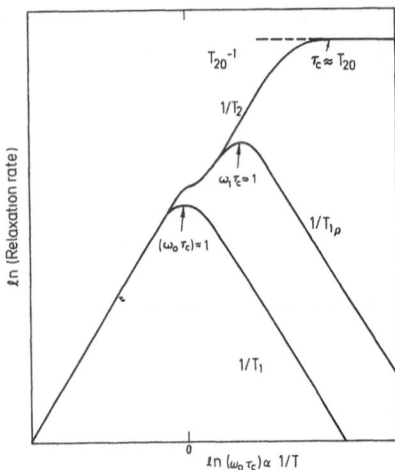

Fig.6.2. Sketch of BPP temperature dependence of relaxation rates

implied in (6.3), one is restricted to those which contribute to the ½→ -½ transition. This has the effect of enhancing $G_{\alpha 1}$ for $\alpha \neq 0$ while leaving G_{01}, the longitudinal term, unchanged. It is assumed that T_1 still involves all spin states and thus is unaffected. This is because in a typical T_1 experiment numerous rf pulses are applied of sufficient strength and duration to saturate all the transitions, not just the ½ → -½, such that recovery of the magnetization may be described by a spin temperature. Under these conditions it has been shown [6.16] that for $\omega_0 \tau_c \ll 1$ and a powder average,

$$\left(\frac{T_1}{T_2}\right)_{e-n} = (I + \tfrac{1}{2})^2 \tag{6.27}$$

for e-n coupling. The calculation is more difficult for efg coupling to the quadrupole moment, but for all the spin values $I > \tfrac{1}{2}$ considered [6.17] the result is consistent with

$$\left(\frac{T_1}{T_2}\right)_Q = \tfrac{1}{4}(I + \tfrac{1}{2})^2 \tag{6.28}$$

under the same conditions. Expressions have also been derived [6.17] for $(T_1/T_2)_{n-n}$ when only the ½ → -½ transition is observed, but the n-n mechanism is generally so much weaker than the Q or e-n ones that $(T_1/T_2)_{n-n}$ is only of academic interest. Equations (6.27) and (6.28) are used in Sect.6.3 to distinguish between relaxation mechanisms, particularly for $I = 3/2$ [7]Li-ion conductors.

6.1.4 Simple Random-Walk Values

Noninteracting ions which undergo hopping motion may be described by a random-walk model. It is necessary to determine T_1 and T_2 in this model before one can attempt to discuss possible deviations produced by correlated motion. Correlation functions beyond the simple random-walk calculation are discussed in Chap.4. The random-walk equation for a single particle is

$$\frac{dP_\mu}{dt} = \sum_\lambda (- W_{\mu\lambda}P_\mu + W_{\lambda\mu}P_\lambda) \quad , \tag{6.29}$$

where $W_{\mu\lambda}$ is the transition probability per unit time for hopping from site μ to site λ. Since the lattice contains only one particle there is obviously no difference between the equation for site occupancy P_μ and the distinguishable particle n_{i_μ}. Equations (6.29) are readily solved for a crystalline lattice by a spatial Fourier transformation to the wave vector normal modes P_k which decay exponentially with time constants τ_k. For nearest-neighbor hopping in a lattice with one mobile ion site per unit cell, we have

$$\tau_k = \tau_c'/(1 - \gamma_k) \tag{6.30}$$

where $\tau_c'^{-1}$ = zW with z the number of nearest neighbors to which the hopping rate is $W_{\mu\lambda}$ = W and $\gamma_k = \frac{1}{z} \sum' \exp(i\underline{k}\cdot\underline{r}_{\mu\lambda})$ with the sum restricted to the z nearest-neighbor sites. The spectral density for single-particle random-walk contribution to relaxation can then be written

$$J(\omega_0) = \int d\underline{k} \; F(\underline{k}) \; \frac{\tau_k}{1+\omega_0^2\tau_k^2} \tag{6.31}$$

where $F(\underline{k})$ is related to the spatial Fourier transform of whichever interaction is responsible for the relaxation.

Although (6.31) can involve complicated numerical integration, two points are evident regarding the frequency dependence: for $\omega_0\tau_c' \ll 1$ and a three-dimensional (3D) lattice where $d\underline{k} \sim k^2 dk$, $J(\omega_0) \propto \tau_c'$ since the integral converges at $\omega_0 = 0$ as long as $F(\underline{k})$ does not diverge at $k \to 0$; and for $\omega_0\tau_c' \gg 1$ and any lattice $J(\omega_0) \propto 1/\omega_0^2\tau_c'$. The commonly-used BPP approximation [6.15]

$$J(\omega_0) = \frac{C\tau_c}{1+\omega_0^2\tau_c^2} \quad , \tag{6.32}$$

where C is an appropriate constant, basically assumes that $k^2F(\underline{k})$ is peaked at some value k_0 and takes $\tau_c = \tau_{k_0}$. The basic features are quite insensitive to the ap-

154

proximation and, unless details of the lattice hopping are well understood, it may
be a fruitless exercise to try to go beyond the BPP formula (6.32). Evaluations of
(6.31) for dipole-dipole interaction and cubic lattices have been given [6.18]. In
Fig.6.3 we compare the BPP expression with (6.31) evaluated for F($\underline{k}$) independent of
$\underline{k}$ in a simple cubic lattice. A $\underline{k}$-independent F($\underline{k}$) would result, for example, from
nearest-neighbor transferred hyperfine coupling to a paramagnetic impurity with
only one nearest-neighbor mobile-ion site.

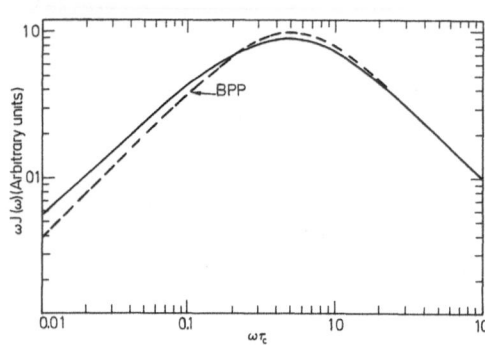

Fig.6.3. Comparison of BPP (6.32) with
$\tau_C = \tau'_C$ and (6.30) and (6.31) for F($\underline{k}$)
constant in a simple cubic lattice

In the region where $\langle\Delta\omega^2\rangle\tau_C^2 \ll 1$, and $\omega_0\tau_C \gg 1$ the motional narrowing expression (6.32) describes $1/T_2$ with the zero-frequency term $J_{01}(0)$ dominant. As a
consequence BPP predicts

$$\frac{1}{T_2} \sim \langle\Delta\omega^2\rangle\tau_C \quad . \tag{6.33}$$

6.1.5 Diffusion in Lower Dimensions

Ionic motion confined to one or two dimensions (1D or 2D) is a common occurrence in
superionic conductors because of structural features. Prime examples of 1D and 2D,
respectively, are β-eucryptite (LiAlSiO$_4$) [6.19] and β-alumina [6.20]. The spectral
density J(ω_0) given by the diffusion result (6.31) diverges as $\omega_0 \to 0$ in 1D or 2D.
This divergence has profound effects on both NMR and EPR relaxation in low-dimensional
concentrated magnetic systems [6.21]. Similar anomalies should occur from ionic
motion, and at least one such case has been documented in 2D [6.22]. In the high-
temperature region where $\omega_0\tau_C \ll 1$, (6.31) and (6.22) predict

$$\frac{1}{T_1} \propto \omega_0^{-\frac{1}{2}} \quad , \quad \ln(1/\omega_0\tau_C) \text{ for 1D, 2D} \quad , \tag{6.34}$$

and thus, in contrast to 3D, there is a frequency dependence on the high-temperature
side of the T_1 minimum. The relation $T_1 \propto \omega_0^2$ on the low-temperature side of the
minimum is independent of dimensionality.

Real systems are only quasi-1D or -2D. There is a characteristic time $\tau_\perp \gg \tau'_c$ for interchannel or interplane hopping. The frequency dependence of (6.34) will be observed only when $\omega_0 \tau_\perp \gg 1$ and $\omega_0 \tau_c \ll 1$ hold; for frequencies lower then $1/\tau_\perp$, the long-time ultimate 3D behavior is sampled. Therefore a very large ratio $\tau_\perp/\tau_c$ is required to detect the predicted low-dimensional frequency variation.

The effect on T_2 may be more apparent due to a delayed onset of motional narrowing with increasing temperature. Equation (6.25) or (6.26) has to be used instead of (6.22) since $J_{01}(0)$ diverges. Because $G_{01}(\tau) \approx \langle \Delta\omega^2 \rangle (\tau/\tau_c)^{-d/2}$ for diffusion in d-dimensions, one obtains in 1D,

$$\frac{1}{T_2} \sim \left\langle \Delta\omega^2 \right\rangle^{2/3} \tau_c^{1/3} \tag{6.35}$$

for $T_2 < \tau_\perp$, and

$$\frac{1}{T_2} \sim \left\langle \Delta\omega^2 \right\rangle \tau_c^{\frac{1}{2}} \tau_\perp^{\frac{1}{2}} \tag{6.36}$$

for $T_2 > \tau_\perp$. If $\tau_\perp$ is long enough for (6.35) to hold in the initial narrowing portion of the T_2 versus T curve, one will see an apparent activation energy Δ' one-third that of τ_c. This can further lead to an anomalously low prefactor τ'_0 if one assumes that $\tau_0 \exp[\Delta'/k_B T] = \langle \Delta\omega^2 \rangle^{-\frac{1}{2}}$ at the lowest temperature at which motional narrowing is observed. Further details are contained in a recent paper [6.23].

6.1.6 Effects of Correlated Hopping

Hopping of a distinguishable particle is said to be correlated if it has a "memory" of its previous hop, i.e., if the probability of hopping to a given site on the nth jump depends on the site it hopped from on the $(n-1)^{st}$ jump. A well known example [6.8] occurs for vacancy diffusion in the limit of very small vacancy concentration. Since a tracer particle in this case can jump only into a vacant site, it is most likely to hop back to the unoccupied site from whence it came. As a result the effective time τ_p for migration of a *labeled particle* away from a given site is longer than the hopping time τ_v which characterizes vacancy diffusion. Vacancy migration is a random process so long as the tracer is indistinguishable from the other particles in regard to the transition probability for jumping. Further, in such a system the conductivity is governed by vacancy motion since this is the only mechanism for variation of charge density in the lattice.

If the tracer is labeled by a nuclear spin whose resonance is observed it follows that

$$\tau_{NMR} > \tau_{conductivity} \tag{6.37}$$

if we make the identifications $\tau_{NMR} = \tau_p$ and $\tau_{conductivity} = \tau_v$. Here τ_{NMR} is the effective correlation time of the perturbation which produces NMR relaxation and is the same as the particle migration time τ_p for the important cases of coupling to efg's or paramagnetic spins of fixed ions. Equation (6.37) shows in principle that the importance of correlated hopping could be detected by careful comparison of NMR results with other measurements. Similar information is obtainable from comparison of tracer and bulk diffusion coefficients, D^* and D, respectively, since, in this picture, $\tau_{conductivity}/\tau_{NMR}$ is the same as the Haven ratio D^*/D [6.8]. If τ_v is known, the amount of correlation can be estimated from the condition $\omega_0\tau_{NMR} \approx 1$ at the T_1 minimum. This translates to $\omega_0\tau_v \approx f < 1$ where $f = \tau_v/\tau_{NMR}$.

The form of the spectral densities $J(\omega)$ of (6.3) as well as their magnitudes (magnitude of τ_{NMR}) can be affected by correlated hopping. It has been shown by WOLF [6.24], for example, that for relaxation by n-n dipole coupling there is a 10% asymmetry in the $J(\omega)$ curve for a simple cubic lattice with small vacancy concentration (monovacancy limit) as compared with the symmetric BPP curve of (6.32). Asymmetry is defined [6.24] essentially as the relative difference between the magnitudes of the slope of $\ln J(\omega)$ versus $\ln(\omega\tau_{NMR})$ for $\omega\tau_{NMR} \ll 1$ and $\omega\tau_{NMR} \gg 1$. The usefulness of this technique is limited by the fact that the normal random-walk diffusion of (6.31) also produces asymmetry. Indeed WOLF found that (6.31) gives an even greater 19% asymmetry for the simple cubic lattice. The situation is better in fcc crystals where monovacancy calculation has a 14% asymmetry compared with 4% for the random walk. The curve of Fig.6.3 has 22% asymmetry. A similar problem arises in the above relation $\omega_0\tau_v \equiv f$ at the T_1 minimum since f can be identified precisely as the correlation factor only if $\omega_0\tau_{NMR} = 1$ exactly at that point. The results do show $\omega_0\tau_{NMR} = 1$ and thence give $f = 0.54$ for the fcc lattice in the limit of very small vacancy concentration.

A second type of correlation occurs when groups of particles tend to move together. It has been proposed [6.25] that significant lowering of the activation energy can be achieved by such correlated group movement. This can have a dramatic effect on NMR relaxation processes which involve coupling between mobile ions. As long as the position vector $\underline{r}_{ij}$ between two particles remains unchanged, there is no fluctuation and an infinite correlation time. In general such motion would have $\underline{r}_{ij}$ changing more slowly than $\underline{r}_i + \underline{r}_j$ which describes position relative to a fixed origin and thus (6.37) should be applicable.

Lattice relaxation, although not treated in this chapter, must also be mentioned as a possible source of correlation. As distinct from the above mechanisms, this can produce non-random-walk behavior even for a single particle with a large number of vacant sites available. The point is that if the lattice relaxation time associated with the surrounding "stationary" ions is longer than the hopping time, the site from which the ion last came will still be in a different state than other neighboring sites when the ion is again ready to jump, and the probability for jumping to this site will be different. Such correlation affects both τ_{NMR} and $\tau_{conductivity}$ so that their relation is not necessarily altered as in the previous examples.

Correlation is most dramatic in 1D. It is impossible in this restricted geometry for an ion to move a distance na without pushing of the order of ρn particles ahead of it. (a is the jump distance and ρ is the particle density). As a consequence [6.4] single-particle motion is not diffusive although bulk motion of all the particles is diffusive. The effective time for a labeled particle to migrate from, say, a paramagnetic impurity has a larger activation energy than $\tau_{conductivity}$ and thus the inequality in (6.37) becomes extreme at low temperatures.

6.2 Comparison with Experiment

In this section we discuss the extent to which various features mentioned in Sect.6.1 have been confirmed—or not confirmed—by NMR experiments. It is not the intention to give a complete survey of the literature, but rather to illustrate with a few examples the basic concepts. A fairly complete list of the pre-1977 literature on NMR experiments is contained in the review by WHITTINGHAM and SILBERNAGEL [6.1].

6.2.1 Thermal Activation

Probably the best and most easily verified feature of NMR relaxation is the activation energy. As discussed in connection with (6.31) and (6.32), one expects $T_1 \propto \tau_c$ on the low-temperature ($\omega_0\tau_c \gg 1$) side of the T_1 minimum. This translates to $T_1^{-1} \propto \exp(-\Delta/k_BT)$ for processes characterized by an activation energy Δ. When, however, $\omega_0\tau_c$ is many orders of magnitude greater than unity the hopping motion is too slow to make a significant contribution to the spectral density for relaxation. Lattice vibrations or the spin-fluctuation of paramagnetic impurities therefore become important at low temperatures and lead to a weaker-than-activated temperature dependence of T_1. Nonetheless, there are examples where $T_1 \propto \exp(-\Delta/k_BT)$ over two or more decades on the low temperature side of the minimum. Noteworthy are the ^{19}F resonance [6.26] in PbF_2 and ^7Li in [6.27] β-eucryptite ($LiAlSiO_4$).

Line narrowing described by (6.33) has been used frequently to obtain activation energies. It is difficult to satisfy the relation $\langle\Delta\omega^2\rangle^{1/2} \ll 1/\tau_c \ll \omega_0$, so that (6.33) is seldom applicable over more then one decade. Nonetheless activation energies extracted from T_2 data generally agree with those from T_1 data and these correlate well with measurements of dc conductivity. Table 6.3 contains a comparison of activation energies obtained by NMR(Δ_{NMR}) and conductivity (Δ_σ). In several cases the NMR data do not cover a sufficient range to establish an exponential dependence

Table 6.3. Activation energies determined by NMR and conductivity (σ) measurements in same temperature range

Compound	Δ_{NMR}[eV]	Δ_{σ}[eV]
PbF_2	0.7 [a]	0.7 [b]
$LiAlSiO_4$	0.8 [c]	0.7 [d]
CuI	0.5 [e]	0.6 [f]
$Li_{4.2}Si_{0.8}Al_{0.2}O_4$	0.2 [g]	0.4 [g]

[a] [6.26]
[b] R.W. Bonne, J. Schoonman: J. Electrochem. Soc. 124, 28 (1977
[c] [6.27]
[d] [6.19]
[e] [6.41]
[f] J.B. Wagner, C. Wagner: J. Chem. Phys. 26, 1597 (1957)
[g] R.D. Shannon, B.E. Taylor, A.D. English, T. Berzins: Electrochim. Acta 22, 783 (1977)

beyond doubt; but the straight line portion of the ln T_1 versus $1/T$ curve does seem to yield a number which corresponds well with Δ_{σ} in most instances.

The simple BPP expression (6.32) predicts the same activation energy on either side of the T_1 minimum. Evidence for this in the nonmetallic substances normally classified as superionic conductors is rather scant. It is often difficult to obtain data over a broad enough temperature range, and there are definite instances, one of which is cited below, where activation energies differ on either side of the minimum. Proton diffusion in metal hydrides, however, does provide a good example of the classic BPP behavior. This is illustrated in Fig.6.4 for the proton T_1 in $TiH_{1.7}$ [6.28]. Note that the BPP-like curve is obtained only after a Korringa rate $(T_1)_e^{-1} \propto T$ due to conduction electrons is subtracted. This illustrates how other mechanisms can influence the interpretation and have to be accounted for if meaningful analyses are to be performed. (Poor electronic conductors will of course have no Korringa term, but phonon contributions proportional to T^2 can be important.)

Figure 6.5 shows T_1 in Na β-alumina [6.29] where there is a substantial difference between activation energies inferred by using the BPP formula on either side of the minimum. The discrepancy has been interpreted [6.30] as being due to a distribution of activation energies and reasonable fits to the data have been obtained for a number of Na β-alumina samples grown by different techniques. The general expressions (6.30) and (6.31) illustrate how a diffusion process involving several hopping times can lead to dissimilar slopes above and below T_{min}. At high temperature, T_1^{-1} is given by an average of τ_k, whereas at low temperature it is given by an average of $1/\tau_k$. If the activation energy of τ_k is k dependent, these two averages will not yield the same effective activation energy. To be more specific, the high temperature T_1^{-1} is weighted in favor of the longer τ_k's, which have the higher activation energies, while shorter, lower activation energy, τ_k's dominate at low

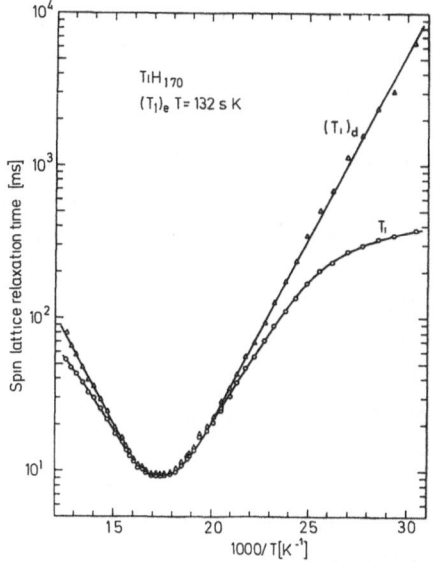

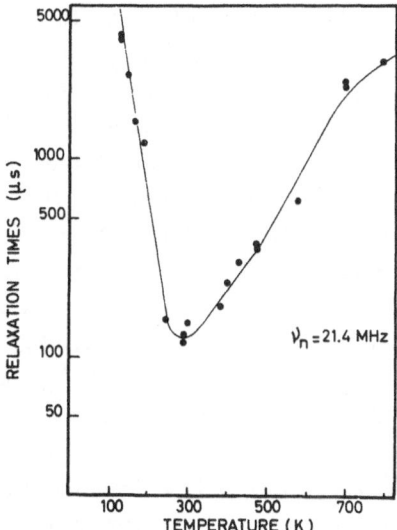

Fig.6.4. Proton T_1 of TiH$_{1.7}$. Diffusion part $(T_1)_d$ is related to observed T_1 by $(T_1)_d^{-1} = T_1^{-1} - (T_1)_e^{-1}$ where $(T_1)_e^{-1}$ is Korringa rate obtained from low-temperature data [6.28]

Fig.6.5. T_1 of ^{23}Na in Na β-alumina (only the long time constant is shown where a double-exponential decay was observed [6.29]

temperature. Thus the larger apparent activation energy on the high-temperature side of the T_1 minimum is accounted for. Implicit in this argument is the assumption that a form such as (6.31) holds in the more general case of inequivalent sites with the indices k now regarded as the normal modes of the system and τ_k the relaxation time of the kth normal mode.

6.2.2 Frequency Dependence

The BPP formula (6.32) predicts $T_1 \propto \omega_0^2$ on the low-temperature side of the T_1 minimum and $T_1 \propto \omega_0$ at the minimum. The more general (6.31) gives the same prediction for a single characteristic time τ_c'. The fact that there are some SI materials where this dependence is not observed remains an unexplained and challenging problem. We consider first, however, cases where the dependence is verified. If one includes data on the "rotating frame" $T_{1\rho}$, it can be argued that results of BOYCE et al. [6.26] for the ^{19}F relaxation in PbF$_2$ agree with BPP frequency dependence for frequencies between 60 kHz and 27 MHz. (The T_1 data covered 8 MHz to 27 MHz and the $T_{1\rho}$ data probed spectral densities at $2\omega_1/2\pi = 60$ kHz and 100 kHz where ω_1 is the rf field in frequency units.) The data of WALSTEDT et al. [6.30] on β-alumina do not extend far enough below T_{min} to confirm an ω_0^2 dependence, but $T_1 \propto \omega_0$ is verified at the minimum for the two frequencies employed.

Much attention has been paid to the absence of frequency dependence in β-eucryptite (LiAlSiO$_4$) [6.27] and Li$_2$Ti$_3$O$_7$ [6.31,32]. These systems both display T$_1$ minima and apparent thermal activation. Further, both the mobile [7]Li and stationary [27]Al have been studied in β-eucryptite and both show a frequency-independent T$_1$ minimum. T$_1$ can be frequency independent if the spectral density J(ω) contains a characteristic frequency ω$_c$ greater than the NMR frequency ω$_0$. But one cannot at the same time have J(ω$_0$) ~ 1/ω$_c$, which would be the case for a simple BPP type spectral density, since T$_1^{-1}$ must decrease as temperature is lowered and (presumably) ω$_c$ decreases. A spectral density was proposed by FOLLSTAEDT and RICHARDS [6.27] for β-eucryptite as sketched in Fig.6.6. This is based on the structure below 450°C [6.33], where, as in Fig.6.1, the Li$^+$ ions occupy preferred a sites and occasionally hop to neighboring b sites. The fast rate ω$_c$ in this case is identified as the b → a transition rate W$_{ba}$, and the spectral density in the flat region is of the order of W$_{ab}$/W$_{ba}^2$ since only a fraction W$_{ab}$/W$_{ba}$ << 1 of the decay is characterized by this rapid process. If the slow a → b transition rate W$_{ab}$ has a much larger activation energy than W$_{ba}$, the desired effect of a frequency-independent T$_1$ which increases with decreasing T is achieved.

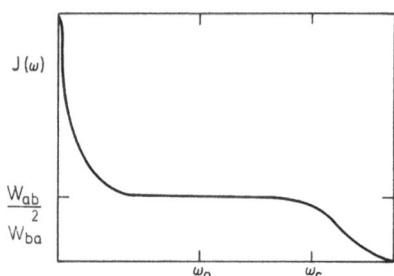

Fig.6.6. Proposed spectral density to explain frequency-independent low-temperature T$_1$ in β-eucryptite [6.27]

The T$_1$ minimum in this model has nothing to do with ω$_0$ becoming equal to a characteristic frequency. Rather it is assumed to result from an order-disorder transition which is believed [6.33] to occur at about the same temperature, 450°C, as T$_{min}$. The nature of this transition is not well known and it does not seem to affect the dc conductivity, although it has been detected in the microwave conductivity [6.34].

Validity of the FOLLSTAEDT and RICHARDS model [6.27] in the low-temperature region requires that the initial motion of an ion at site a$_0$ is confined to a$_0$ and the neighboring b wells. Once other ions move out of the next-nearest-neighbor a type wells, the ion in question can migrate away and this induces an additional time dependence. It was assumed [6.27] that this slower decay makes a negligible contribution to J(ω$_0$), but this has been questioned recently in the light of analytical calculations [6.35] of hopping in a 1D model appropriate to β-eucryptite. In light of this and uncertainties regarding the order-disorder transition, the

lack of frequency dependence may still be an unresolved problem. An obvious
question is whether $Li_2Ti_3O_7$ also undergoes a transition at the temperature
(220°C) of the T_1 minimum. This is not known and the conductivity paths even at
room temperature are uncertain [6.36]. $T_{1\rho}$ measurements could be helpful in the
materials which show no frequency dependence of T_1 since at low enough frequency
there should be some variation, as suggested by Fig.6.6.

The resonance of ^7Li in Li_5AlO_4 provides another puzzling frequency dependence
[6.37]. In the low-temperature (< 780°C) α phase, a T_1 minimum is observed which
satisfies $T_1 \propto \omega_0$ in accord with BPP, but this linear dependence continues well
below T_{min} rather than switching to $T_1 \propto \omega_0^2$. In the metastable β phase, by contrast, there is negligible frequency dependence of T_1 at the minimum, but a near-
BPP variation $T_1 \propto \omega_0^{1.8}$ at low temperatures. The relaxation is believed to be
dominated by paramagnetic impurities, a fact the author has used [6.14] to explain
the weak, low-temperature frequency dependence in the α phase. But, it is difficult to see how this model alone can account simultaneously for both phases over
the whole temperature range. Consideration of details of different hopping processes in the two phases — there are two inequivalent Li^+ sites in the β phase and
five in the α phase — may be required to resolve this dilemma.

Diffusion in 1D or 2D is expected to produce a frequency dependence on the high
temperature side of T_{min} as well [see (6.34) and Table 6.1]. To our knowledge this
has not yet been observed for ionic diffusion in a 1D system, but it has been
reported for 2D, where a logarithmic variation of T_1 with ω_0 was seen by SILBER-
NAGEL and GAMBLE [6.1,22] for proton resonance in the intercalation compound
$TaS_2(NH_3)$. As mentioned in connection with (6.34), interchannel hopping can mask
the predicted pure 1D dependence, and the weak 2D logarithmic dependence obviously
cannot be detected unless a broad range of frequencies is used (ω_0 was varied by
more than a factor of 10 in [6.22]). Thus it may not be too surprising that more
examples have not been encountered; the experimentalist has to take some pains
to seek out the low-dimensional features.

6.2.3 Prefactors

Table 6.4 contains values of the prefactor or "attempt frequency" ν_0 in (6.2)
obtained by NMR and, where data are available, by other means. Its measurement
requires knowledge of the precise magnitude of the hopping frequency ν at some
temperature and thus is more complex than that of the activation energy. If the
frequency dependence is BPP-like, one can then assert with reasonable confidence
that $\nu_0 \exp(-\Delta/k_B T_{min}) = \omega_0$ at T_{min} and thereby obtain ν_0 from the known slopes
(Δ) away from the minimum. But if there is no frequency dependence, as for some
of the above examples, it is not reasonable to make this identification. An
alternative approach is to apply (6.33) to the initial increase of T_2 or line
narrowing as temperature is increased. There are several examples of this method

Table 6.4. Prefactor (attempt frequency) ν_0 measured by NMR and, where data are available, by conducting σ. Vibration frequency $\tilde{\nu}$ measured by infrared or Raman spectroscopy also given where data are available

Compound	ν_0(NMR[s^{-1}])	$\nu_0(\sigma[s^{-1}])$	$\tilde{\nu}[s^{-1}]$	Remarks
$LaNi_5H_6$ [a]	2×10^{12}			line narrowing and $T_{1\rho}$
H_xWO_3 [b,c,d]	$2 \times 10^7 - 3 \times 10^{11}$			line narrowing, T_1 and $T_{1\rho}$
$LiAlSiO_4$	5×10^7 [e]	5×10^{14} [f]	5×10^{12} [g]	line narrowing
$\gamma-Li_{1.12}V_3O_{7.9}$ [h]	$4 \times 10^{10} - 3 \times 10^{11}$			line narrowing
$Li_{2.15}B_{0.15}C_{0.85}O_3$	10^8 [i]	$\geq 10^{12}$ [i]	4×10^{12} [j]	line narrowing, ir for Li_2CO_3
$Li_2Ti_3O_7$ [k]	4×10^7			line narrowing
Li_5AlO_4	2×10^{12} [l]	10^{13} [m]		T_1 minimum
NH_4NO_3	5×10^{13} [n]		6×10^{12} [j]	line narrowing, T_1 and $T_{1\rho}$
$(NH_4)_{0.33}WO_3$ [o]	5×10^9			T_1
$(NH_4)_4Fe(CN)_6 \cdot H_2O$ [p]	$10^9 - 10^{10}$			
PbF_2	9×10^{14} [q]	5×10^{12} [r]	6×10^{12} [s]	T_1, large NMR prefactor attributed in [6.26] to defect formation entropy
Na β-alumina	2×10^{11} [t]	5×10^{11} [u]	2×10^{12} [u]	T_1
CuI	5×10^{11} [v]	3×10^{13} [w]	4×10^{12} [x]	line narrowing

[a] T.K. Halstead, N.A. Abood, K.H.J. Buschow: Solid State Commun. 19, 425 (1976)
[b] M.A. Vannice, M. Boudart, J.J. Fripiat: J. Catal. 17, 359 (1970)
[c] P.G. Dickens, D.J. Murphy, T.K. Halsted: J. Sol. State Chem. 6, 370 (1973)
[d] K. Nishimura: Solid State Commun. 20, 523 (1976)
[e] D.M. Follstaedt: Unpublished measurement
[f] [6.19]
[g] [6.39]
[h] T.K. Halstead, W.U. Benesh, R.D. Gulliver II, R.A. Huggins: J. Chem. Phys. 58, 3530 (1973)
[i] R.D. Shannon, B.E. Taylor, A.D. English, T. Berzins: Electrochim. Acta 22, 783 (1977)
[j] R.A. Nyquist, R.O. Kagel: Infrared Spectra of Inorganic Compounds (Academic Press, New York 1971)
[k] [6.32]

[l] [6.37]
[m] R.M. Biefeld, R.T. Johnson: J. Electrochem. Soc. 126, 1 (1979)
[n] M.T. Riggin, R.R. Knispel, M.M. Pinter: J. Chem. Phys. 56, 2911 (1972)
[o] L.D. Clark, M.S. Whittingham, R.A. Huggins: J. Sol. State. Chem. 5, 487 (1972)
[p] [6.45]
[q] [6.26]
[r] R.W. Bonne, J. Schoonman: J. Electrochem. Soc. 124, 28 (1977)
[s] R.T. Harley, W. Hayes, A.J. Rushworth, J.F. Ryan: J. Phys. C 8, L530 (1975)
[t] [6.30]
[u] S.J. Allen, L.C. Feldman, D.B. McWhan, J.P. Remeika, R.E. Walstedt: In [6.20]
[v] [6.41]
[w] J.B. Wagner, C. Wagner: J. Chem. Phys. 26, 1597 (1957). We have extrapolated their zinc-blende (γ) phase data to obtain σ_0 and used (6.1)
[x] G. Burns, F.H. Dacol, M.W. Shafer: Solid State Commun. 24, 753 (1977)

in the literature, but it may also be suspect for low dimensionality $\Big($see (6.35,36), Table 6.2, and [6.23]$\Big)$ or spin $I > \frac{1}{2}$ as discussed below.

A major question is wheter ν_0 corresponds to an optical phonon frequency $\tilde{\nu}$ associated with motion of the ion within its cage between infrequent hops. Since ν_0 contains an entropy for vacancy formation [6.38], it is reasonable to expect such a correspondence only for systems with a temperature-independent number of vacancies. Thus there is a large apparent discrepancy between ν_0 and $\tilde{\nu}$ in PbF_2, in spite of the BPP behavior, which is resolved [6.26] upon realizing that the number of vacancies is activated at low temperature. In addition to the entropy S which enters the free energy as -TS, any mechanism which produces an effective barrier height which has a linear temperature dependence will alter the prefactor and possibly confuse the interpretation.

Consider the α phase of Li_5AlO_4, where FOLLSTAEDT and BIEFELD [6.37] observed BPP frequency dependence at the T_1 minimum. The prefactor was estimated to be $\nu_0 = 2 \times 10^{12} s^{-1}$ from a BPP fit, which is within the acceptable range for an optical phonon frequency. Although Li_5AlO_4 has not been studied by infrared or Raman techniques, Raman studies [6.39] of $LiAlSiO_4$ show a mode at $5 \times 10^{12} s^{-1}$ which was tentatively identified with Li^+ ion motion.

What would happen if ν_0 were inferred from T_2 data? This is shown in Fig.6.7 where the solid curve is the rate predicted by the author [6.14] using the same parameters used to obtain a fit to T_1. The data show slight motional broadening before narrowing sets in, and that at a considerably higher temperature than expected. If we use (6.33), together with the observed activation energy, we obtain $\nu_0 = 1.9 \times 10^{10} s^{-1}$, a discrepancy of two orders of magnitude. It is likely that this discrepancy comes from an incomplete understanding of the onset of line narrowing as temperature increases, particularly in view of the small, but experimentally significant, increase in $1/T_2$ seen at first. FOLLSTAEDT's data on the α phase [6.37] of Li_5AlO_4 and on β-eucryptite [6.40] show similar behavior. If ν_0 were extracted from the β-eucryptite T_2, one would infer $\nu_0 = 4.8 \times 10^7 s^{-1}$. Unfortunately, ν cannot be obtained from T_1 in this compound because of the lack of frequency dependence.

We do not know the origin of the initial broadening or delayed onset of narrowing, but speculate that it may be due to features of the quadrupole-split spectrum in these materials, which lack cubic symmetry. It is believed that only the central, unperturbed $\frac{1}{2} \rightarrow -\frac{1}{2}$ transition is seen in the $I = 3/2$ ^7Li resonance. Motion might cause some overlap of the noncentral transitions with the main line and thus lead to broadening or at least inhibit the narrowing. It may be noteworthy in this respect that a more reasonable $\nu_0 = 5 \times 10^{11} s^{-1}$ is obtained [6.41] from T_2 measurements on $I = 3/2$ ^{63}Cu in the cubic zinc-blende phase of CuI where no quadrupole splitting is expected. This is still less than the Raman value $\tilde{\nu} = 4 \times 10^{12} s^{-1}$ [6.42], however.

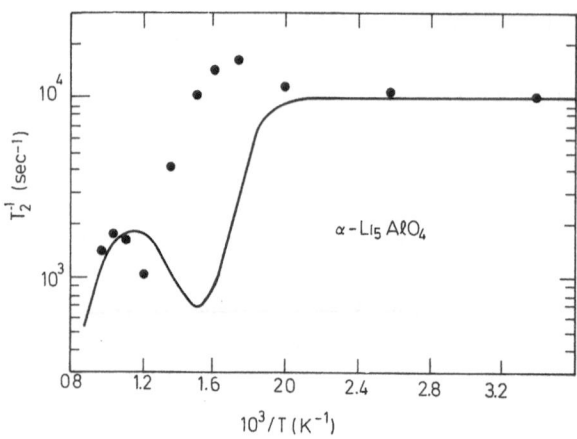

Fig.6.7. Relaxation rate T_2^{-1} for ^7Li in α phase of Li_5AlO_4. Dots are experimental points of [6.37]; solid line is theory of [6.14] for same parameters used to fit T_1 data

$Li_2Ti_3O_7$, like β-eucryptite, has a value of $\nu_0 = 4 \times 10^7 s^{-1}$ [6.32] that is orders of magnitude too small. This was interpreted by BOYCE and HUBERMAN [6.32] as a "real" effect, i.e., not caused by peculiarities of quadrupole couplings or low dimensionality affecting the T_2 processes. It was argued that energy transfer is inefficient when an ion is at the top of its well, so there is a high probability that it stays in a "gas phase" instead of getting trapped in an adjacent well. This leads to a reduction in hopping rate which can make the effective ν_0 orders of magnitude less than $\tilde{\nu}$. The further claim was made that this "breakdown of absolute rate theory" should be more pronounced in complex molecular structures, and this is why ν_0 is observed to be of the order of $\tilde{\nu}$ in the relatively simple PbF_2 and CuI. There may be truth to this argument, but in view of the above-mentioned problems with interpretation of T_2, some care may have to be taken in accepting it solely on the basis of existing data.

We have pointed out [6.23] that a number of compounds which exhibit large prefactor anomalies are low dimensional and this can cause problems as discussed in connection with (6.36). Certainly β-eucryptite falls in this category since the conductivity has been shown [6.19] to be highly 1D as suggested by the channel structure. It is also curious to note that the prefactor deduced from applying (6.1) to the conductivity data [6.19] is $\nu_0 \approx 5 \times 10^{14} s^{-1}$, so the anomaly is in the other direction. We claimed in [6.23] that $Li_2Ti_3O_7$ was likely to be 1D and therein lay the cause of the small prefactor. This is now in doubt because the single-crystal dc conductivity [6.43] and microwave dielectric constant [6.44] show no strong anisotropy, and the structure [6.36] is inconclusive as to whether Li^+ ions hop along the 1D channels or into off-channel vacancies.

It appears that many more experiments are needed before the last word can be said about prefactors, particularly their relation to optical phonon frequencies.

6.2.4 Coupling to Paramagnetic Impurities

An indication of relaxation by paramagnetic impurities is a maximum in $1/T_2$ as
evidenced by Figs.6.7,8. This arises because the paramagnetic impurity contribution
$(1/T_2)$ in the low-temperature region is limited by the rate $1/\tau_c$ at which the nucleus can diffuse up to an impurity, while at high temperatures, where the nucleus
encounters many impurities before relaxing, the limiting factor is the time τ_c
spent in vicinity of an impurity. HOGG et al. [6.13] demonstrated this feature for
controlled amounts of paramagnetic impurities on the ^{19}F resonance in Mn-doped
PbF_2. An interesting and potentially useful feature is that no effect due to impurities was seen on the ^{207}Pb resonance, even though this resonance shows motional
narrowing because of interaction with the hopping F^- ions. The interpretation was
that paramagnetic impurities provide an effective relaxation mechanism only if the
nuclei can diffuse to their vicinity. (This diffusion can be accomplished by mutual
spin flips for stationary nuclei, but spin diffusion is a much less effective relaxation mechanism than particle diffusion at high temperatures.) Thus one could
determine that F^-, not Pb^{++}, is the mobile species in PbF_2. This is hardly surprising for such a crystal, but there are other compounds where identity of the
mobile ion is not so obvious, and for these the "magnetic tagging" could provide
valuable information. A case in point is ammonium ferrocyanide [6.45] where it is
not known whether the whole ammonium ion or just the protons are mobile. Comparison of H and N resonance might resolve this.

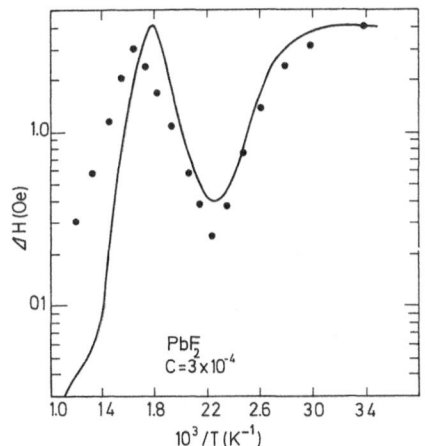

Fig.6.8. Comparison between theory [6.14]
and experiment [6.13] for ^{19}F linewidth in
PbF_2:Mn

It was pointed out by HOGG et al. [6.13] that very small ($\sim$3 ppm) paramagnetic
impurity concentrations could influence the relaxation at high temperatures and that
this might explain the anomalous increase of $1/T_2$ with temperature above 300°C observed by BOYCE et al. [6.26] in "pure" PbF_2. Evidence for the importance of un-

controlled paramagnetic impurities may be seen from the high-temperature T_1/T_2 ratios for the [7]Li resonance in β-eucryptite, α- and β-phases of Li_5AlO_4, and $Li_2Ti_3O_7$. Data from [6.32] for $Li_2Ti_3O_7$ are shown in Fig.6.9. Instead of approaching unity for $T > T_{min}$, the T_1/T_2 ratio attains a constant value nearly equal to $4 = (I + \frac{1}{2})^2$ for $I = 3/2$. $T_1/T_2 \approx 4$ is also found in the other Li^+ conductors above the T_1 minimum. Reference to (6.27) and (6.28) shows that this is the expected ratio for coupling to paramagnetic impurities and definitely not for efg coupling. Two other pieces of information tend to substantiate this. First, room-temperature EPR measurements [6.46] on "pure" Li_5AlO_4 reveal ~100 ppm paramagnetic impurities per Al^{3+} ion, which is about the amount needed to explain [6.14] the observed relaxation rates. Second, for a case where the relaxation is thought to be via quadrupole interactions (Q) (6.28) seems to hold: the [27]Al resonance observed by FOLLSTAEDT [6.40] in β-eucryptite has $T_1/T_2 \approx 2$, which agrees with (6.28) for $I = 5/2$. As mentioned above, one does not expect stationary nuclei to be as strongly influenced by paramagnetic impurities and thus relaxation of [27]Al should be via the Q mechanism. The relaxation in CuI was verified by BOYCE and HUBERMAN [6.41] to be Q type by the fact that the rates of [63]Cu and [65]Cu (both $I = 3/2$) scale as the square of their quadrupole moments; and $T_1 = T_2$ is observed at high temperature, consistent with (6.25). However, in view of the remark made about CuI in Sect.6.2.3, this may not be conclusive since all —rather than just the $\frac{1}{2} \rightarrow -\frac{1}{2}$— transitions may be contributing.

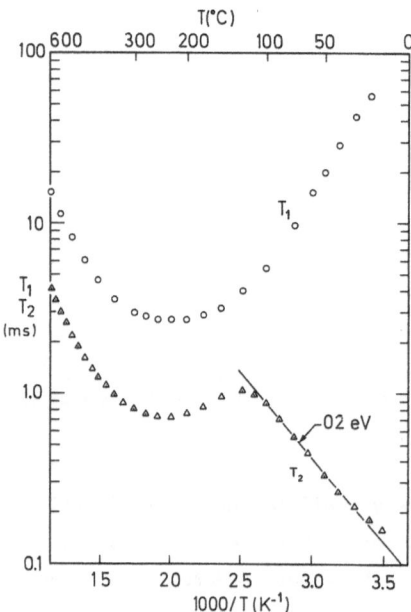

Fig.6.9. T_1 and T_2 for [7]Li in $Li_2Ti_3O_7$ [6.32]. Note $T_1 \approx 4T_2$ at high temperature

It thus appears that half-integer $I > \frac{1}{2}$ nuclei for which only the $\frac{1}{2} \rightarrow -\frac{1}{2}$ transition is detected might possess an advantage over $I = \frac{1}{2}$ nuclei such as ^{19}F in that the T_1/T_2 ratio can be used as an additional handle on the relaxation mechanism. Future experiments are needed on the Li^+ conductors in which (I) the dopant is controlled and (II) a level of purity is reached such that $T_1/T_2 \approx 1$ can be confirmed in the absence of paramagnetic ions.

Intentional doping with paramagnetic ions may provide the best means of sorting out microscopic particle dynamics as stated in connection with (6.15). Elusive features of correlated motion might be tractable. In particular *if* the Mn-F transferred hyperfine interaction and Mn concentration are as quoted by HOGG et al. [6.13] then the discrepancy of a factor of 7 in the high temperature region of Fig.6.8 might be taken seriously enough to suggest correlated hopping as in (6.37). The solid curve is the theory [6.14] using parameters stated in [6.13] and $\tau_{conductivity}$ inferred by BOYCE et al. [6.26]. In the high-temperature region the theory should be least susceptible to doubt and leads to the expression given in [6.13]. Since $1/T_2 \propto \tau_c$ in this range, if the correlated hopping time for NMR is greater than $\tau_{conductivity}$, the discrepancy could be explained [6.47].

Final note should be made of the anomalous, high-temperature T_1 seen [6.26] in PbF_2. The original interpretation [6.26] of this as evidence for correlated motion is in doubt because of the paramagnetic impurity problem. However, although e-n coupling can likely explain the T_2 data, the situation for T_1 may not be that clear cut, since T_1 is independent of frequency in a region where it decreases with increasing temperature. This occurs in the theory only if the nuclear spin stays in vicinity of an impurity long enough to be relaxed in a single encounter. But, $\tau_{conductivity} \sim 3 \times 10^{-11}s$ at the temperature where T_1 starts its anomalous decrease implies much too short a dwell time to allow complete relaxation by any reasonable coupling mechanism.

6.3 Electron Paramagnetic Resonance

Divalent and trivalent paramagnetic ions, whose resonances are commonly studied [6.48], are not mobile in the SI conductors. Rather they form part of the semirigid lattice through which the mobile ions move. As a consequence less information about motion has been obtained from EPR than from NMR. Although it is possible to see diffusion effects on the NMR of stationary nuclei, e.g., ^{207}Pb in PbF_2 and ^{27}Al in $LiALSiO_4$ as mentioned in Sect.6.2, it is a more difficult problem in EPR. Mobile ions modulate the crystal field and could thus produce spin-lattice relaxation. However, the characteristic frequencies are better suited for NMR. Spin-lattice relaxation requires spectral density at the EPR angular frequency ω_e, typically

$6 \times 10^{10} s^{-1}$, which is larger than ionic hopping frequencies except at the highest temperatures. Coupling to high-frequency acoustic phonons is normally a more efficient relaxation mechanism.

In spite of these difficulties, two EPR studies have purported to show the effects of ionic motion. The work of EVORA and JACCARINO [6.49] on Mn^{++} in PbF_2 is noteworthy for showing both a narrowing and broadening effect. Figure 6.10 shows Mn^{++}EPR spectra at $-196^{\circ}C$ (77 K) and $387^{\circ}C$. The six main lines are due to hyperfine interaction with the $I = 5/2$ ^{55}Mn nucleus. At low temperature each of these lines shows structure because of further transferred hyperfine interaction with ^{19}F, which is regarded as the main source of broadening for the S-state Mn^{++} ion in a cubic crystal field. As the F's become mobile the structure disappears and the lines narrow. The temperature ($220^{\circ}C$) of the onset of motional narrowing (see [Ref.6.49, Fig.3]) is consistent with the condition $A^{19} \tau_{conductivity} \sim 1$ where A^{19} is the Mn-F transferred hyperfine angular frequency. These results provide a check on the NMR determinations.

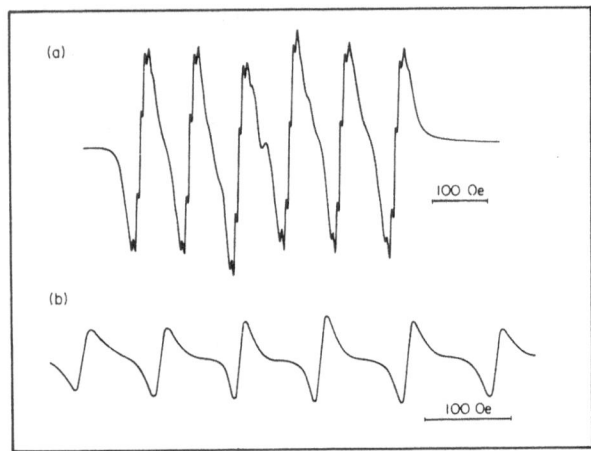

Fig.6.10. EPR spectra of Mn^{++} in PbF_2 at $-196^{\circ}C$ (a) and $387^{\circ}C$ (b), from [6.49]

Above about $400^{\circ}C$ the lines start to broaden in a manner which depends on the ^{55}Mn nuclear spin quantum number m_I associated with each line, and which was attributed to a fluctuating crystal field. The m_I dependence results from the off-diagonal terms $\frac{\hbar}{2} A(I^+S^- + I^-S^+)$ of the ^{55}Mn hyperfine interaction $\hbar A \underline{I} \cdot \underline{S}$. In their absence the frequency of an $m_S \rightarrow m_S + 1$ electronic transition, within a given hyperfine line of nuclear quantum number m_I, is independent of the electronic quantum number m_S. The off-diagonal terms lift this degeneracy and produce a frequency spread [6.49,50] $A^2/\omega_e(2S - 1)m_I$ in the $2S + 1$ possible electronic transitions at the nuclear state m_I. An interesting feature of this effect is that it leads to different dependences of linewidth on m_I depending on whether the rate τ_c^{-1} of crystal-field fluctuations is fast or slow compared with the electronic frequency ω_e. For rapid modulation all electronic transitions (within a given m_I) are excited and a term of the order of $(A^2 m_I/\omega_e)^2 \tau_c$ adds to the broadening in the usual way so that linewidth increases with m_I^2, as observed. For slow modulation, however, the noncentral transitions are very much broader than the central $m_S = -\frac{1}{2} \rightarrow \frac{1}{2}$ transition so that only this latter resonance is observed. It is noteworthy that the condition for dominance of the $-\frac{1}{2} \rightarrow \frac{1}{2}$ transition is $\omega_e \tau_c \gg 1$ rather than the more stringent (for short τ_c) $(A^2/\omega_e)\tau_c \gg 1$ [6.51]. If $(A^2/\omega_e)\tau_c \leq 1$, the satellite transitions are not resolved, but their width remains much greater than that of the central transition as long as $\omega_e \tau_c \gg 1$. At the temperatures where m_I-dependent linewidth was observed by EVORA and JACCARINO [6.49] we have $\omega_e \tau_c \gg 1$ and $(A^2/\omega_e)\tau_c \ll 1$; τ_c is taken as the hop time for F^- diffusion.

Because the random crystal field produces no secular (static, $\Delta m_S = 0$) broadening of a pure $m_S = -\frac{1}{2} \rightarrow \frac{1}{2}$ transition, the central component is broadened for $\omega_e \tau_c \gg 1$ only to the extent that there is admixture to $m_S \neq \pm\frac{1}{2}$ states through the off-diagonal terms in the hyperfine interaction. The first-order wave function is

$$\Psi(m_S, m_I) = |m_S, m_I\rangle$$

$$- \frac{A}{\omega_e} [(S - m_S)(S + m_S + 1)(I + m_I)(I - m_I + 1)]^{\frac{1}{2}} |m_S + 1, m_I - 1\rangle$$

$$+ \frac{A}{\omega_e} [(S + m_S)(S - m_S + 1)(I - m_I)(I + m_I + 1)]^{\frac{1}{2}} |m_S - 1, m_I + 1\rangle \quad ,$$

where $|m_S, m_I\rangle$ is the unperturbed wave function with the electronic and nuclear spins individually quantized. The mean square amount of admixture for large S is proportional to $\frac{1}{2}(I + m_I)(I - m_I + 1) + \frac{1}{2}(I - m_I)(I + m_I + 1) = I(I + 1) - m_I^2$, a *decreasing* function of m_I^2. The linewidth is expected to decrease with m_I^2 in this opposite limit of slow modulation.

Based on their observation that the width increased with m_I^2, EVORA and JACCARINO deduced that the spectral density of the fluctuating crystal field was not consistent with hopping of the F^- ion, but rather, required considerably more weight at ω_e and $2\omega_e$ than could be accounted for by (6.32) [with ω_0 replaced by $\omega_e \approx 6 \times 10^{10}\text{s}^{-1}$] in the temperature range where $\omega_e^2 \tau_c^2 \gg 1$. The required high-frequency spectral density was interpreted as coming from acoustic phonons, whose interaction was enhanced by the sublattice melting. That the phonon coupling is greatly enhanced by the rapid ionic diffusion is inferred from the fact that Mn^{++} has lines as narrow as 0.2 gauss (a cubic-field splitting of this order was observed in [6.52]) at high temperature in a nonconducting cubic material such as ZnS.

A second observation of motional effects in EPR involves Cu^{2+} in single-crystal β-sodium gallate [6.53] which has the β-alumina structure. As opposed to the S-state Mn^{2+} ion, Cu^{2+} has an anisotropic hyperfine interaction and g tensor. If the local symmetry fluctuates via motion of either Cu^{2+} or the surrounding ligands, the hyperfine splitting will be washed out and only an average, angularly independent, g factor will be seen. This was the case at room temperature from which TITLE and CHANDRASHEKHAR [6.53] deduced an effective correlation time on the order of 10^{-11} s. They claimed this represented motion of the Cu^{2+} or the Cu-O complex itself since motion of the Na^{+} ions would not destroy the ligand-field symmetry. If true, this is a remarkable result since it would imply more rapid motion for Cu^{2+} than for Na^{+} whose hopping time is of the order of 10^{-9} s at room temperature [6.29,30]. (The conductivity of Na β-gallate is nearly the same as that of Na β-alumina [6.54].) The correlation time of 10^{-11} s suggests high-frequency acoustic phonons as employed by EVORA and JACCARINO to explain their data [6.49], and it may be possible that rapid fluctuations of the bridging O^{2-} ions with respect to Cu^{2+}, triggered by the rapid Na^{+} diffusion could account for the isotropic, structureless line. Structure was observed at the low temperature of 35 K and at room temperature in a crystal which had absorbed water and thereby suffered a loss in conductivity. Thus, a motional effect which correlates with ionic diffusion does seem to be involved. No motional effects were observed for Cu^{2+} in Na β-alumina.

6.4 Structure Determination

The main emphasis of this chapter is on the use of magnetic resonance to elucidate ionic motion. For completeness, however, we also mention that it can be a good tool for determining structural properties at low temperatures before features of the spectra are washed out by motional averaging. The single crystal spectrum of nuclei with quadrupole moments can be used to deduce the site symmetry and thereby locate the ions which become mobile at higher temperatures [6.55]. This is especially helpful for light ions such as Li^{+} whose position is difficult to obtain by X-ray diffraction. For example, it was found from the quadrupole splittings that there are two inequivalent Li^{+} sites in β-eucryptite [6.56]. Further, the magnitude of the splitting gives the efg at the site which could in principle be used as a check on model calculations of the potential well and static barrier the ion must overcome.

The site symmetry of a paramagnetic impurity can be deduced from the crystal field split single-crystal EPR spectrum. Such information is particularly important in nonstoichiometric defect structures such as β-alumina where the conductivity can be affected by impurities which block favored hopping paths. One could assess the likely effect of an impurity by knowing its location. Such studies have been carried out in β-alumina [6.57].

6.5 Summary and Conclusions

This chapter has presented a survey of theory and experiment pertaining to magnetic resonance in superionic conductors. Most attention was paid to NMR relaxation since this is the most direct magnetic resonance handle on the ionic motion. Review of the general theory showed that the NMR longitudinal (T_1), transverse (T_2) and rotating-frame ($T_{1\rho}$) relaxation times are related to time correlation functions of the particle motion. These are fairly complex and can involve single-site, two-particle, or combinations of site and distinguishable particle correlations depending on whether the coupling is dipole-dipole or quadrupole and whether the observed nucleus is mobile or stationary. Only in the case of coupling to paramagnetic impurities is the correlation function directly related to the hopping of a single distinguishable particle.

The much-used BPP model predicts a minimum T_1 at $\omega_0\tau \approx 1$ and $T_1 \propto \tau^{-1}(\omega_0^2\tau)$ on the high- (low-) temperature side of the T_1 minimum T_{min} where τ is the hopping time. This permits determination of both the activation energy Δ and attempt frequency ν_0 in the expression $1/\tau = \nu_0 \exp(-\Delta/k_BT)$. Because of the above complexities some care has to be exercised in assuming that the τ thus measured is the true, single-particle hopping time. A more complete diffusion relation modifies the BPP expression somewhat, but the basic form is maintained in three dimensions. If one is only interested in factor-of-two determination of τ, BPP suffices for 3D diffusion processes. However, much more serious discrepancies occur in 1D or 2D systems, and these are important considerations for the channel or planar superionic conductors. Both the activation energy and prefactor can be misinterpreted. Correlated hopping in general produces the result $\tau_{NMR} > \tau_{conductivity}$, where τ_{NMR} is the hopping time inferred from an NMR experiment and $\tau_{conductivity}$ that from dc conductivity ($\tau_{conductivity} = \tau_v$, the vacancy hopping time, if diffusion is via vacancies). One must make sure that an apparent difference between τ_{NMR} and $\tau_{conductivity}$ is not due to inapplicability of BPP to the problem before one can make statements about how correlated the hopping is.

A T_1 which goes through a minimum and is activated on either side of T_{min} is observed in a number of compounds. The activation energy Δ is in accord with that measured by dc conductivity in most cases. Line narrowing or T_2 measurements also show activation and generally agree with the T_1 determination of Δ. The BPP frequency dependence $T_1 \propto \omega_0^2$ on the low-temperature side of T_{min} is seen in some but by no means all materials. A predicted $T_1^{-1} \propto \ln(1/\omega_0)$ for 2D conductors on the high-temperature side of T_{min} has been reported, but a similar $T_1^{-1} \propto \omega_0^{-\frac{1}{2}}$ for 1 D has not yet been observed. The attempt frequency tends to agree roughly with an optical phonon frequency as measured by infrared or Raman spectroscopy when it is obtained in 3D compounds by application of the BPP formula to T_1 data. Orders-of-magnitude discrepancy have been reported, however, when it is inferred from T_2 measurements. Many of the compounds which display these anomalously low prefactors are either low-dimensional or display curious "motional broadening" which may be due to quadrupole effects. It remains, in our opinion, an open question as to whether the "true" attempt frequency is really much lower than the frequency with which an ion vibrates near the bottom of its well.

Paramagnetic impurities have been shown to be a major source of relaxation in SI conductors because of the ability of a mobile nucleus to diffuse to their vincinity. This has been demonstrated both by intentional doping and after-the-fact deductions from observed T_1 and T_2 behavior in "pure" samples. Future NMR studies with controlled amounts and types of paramagnetic ions appear to be a promising means of monitoring distinguishable particle diffusion because of the relatively simple correlation functions involved.

Motional effects have been detected in some EPR experiments. The correlation times associated with observed broadening are too short to be attributed to mobile ion hopping and, at least in one instance, are believed to reflect coupling to high-frequency acoustic phonons, a coupling which is greatly enhanced by the liquidlike ionic motion. In most instances, EPR of dilute paramagnetic impurities does not show motional effects but rather has structure which reflects the local site symmetry. This has been used to advantage in determining the location of impurity atoms in defect structures. Similarly, NMR with resolved quadrupole structure at low temperature has been used to pinpoint equilibrium sites of the mobile species and the magnitude of electric field gradients.

The subject is by no means closed. Several fundamental questions remain, particularly in regard to prefactors and correlated hopping. The latter is particularly important since it should be a main feature of the motion of large numbers of ions in these materials. However, it is not yet clear to what extent correlations have been shown experimentally to produce marked differences from simple BPP behavior. Superionic conductors tend to have rather complex structures and this has hindered progress. One would like to be able to probe systems where the correlation functions have the simplest possible form such as given in some of the "special case" examples of Sect.6.1.1. Studies which could make a clear-cut distinction between site and particle correlations would be useful. This might be accomplished by careful comparison in the same compound between quadrupole relaxation of a stationary nucleus, which monitors only site correlations, and paramagnetic impurity relaxation of a mobile nucleus, which gives the distinguishable-particle hopping. It is hoped that this review serves more as a stimulus to future work than as an authoratative final word on the subject.

Acknowledgement. This work was supported by the US Department of Energy under Contract AT (29-1) 789.

References

6.1 M.S. Whittingham, B.G. Silbernagel: In *Solid Electrolytes: General Principles, Characterization, Materials, Applications*, ed. by P. Hagenmuller, W. van Gool (Academic Press, New York 1977)
6.2 R. Kubo: J. Phys. Soc. Jpn. *12*, 570 (1957)
6.3 C. Zener: In *Imperfections in Nearly Perfect Crystals*, ed. by W. Shockley, J.H. Hollomon, R. Maurer, F. Seitz (Wiley and Sons, New York 1950)
6.4 P.M. Richards: Phys. Rev. B *16*, 1393 (1977)
6.5 P.A. Fedders: Phys. Rev. B *17*, 40 (1978)
6.6 S. Alexander, P. Pincus: Phys. Rev. B *18*, 2011 (1978)
6.7 A.N. Garroway, R.M. Cotts: Phys. Rev. A *7*, 635 (1973)
6.8 A.D. Le Claire: In *Fast Ion Transport in Solids*, ed. by W. van Gool (North-Holland, New York 1973)
6.9 A. Abragam: *The Principles of Nuclear Magnetism* (Oxford University Press, New York 1961) Chap.8
     C.P. Slichter: *Principles of Magnetic Resonance*, 2nd ed. Springer Series in Solid-State Sciences, Vol.1 (Springer, Berlin, Heidelberg, New York 1978)
6.10 Ref.6.9, pp.58-63
6.11 D.C. Look, I.J. Lowe: J. Chem. Phys. *44*, 2995 (1966)
6.12 R. Kubo, K. Tomita: J. Phys. Soc. Jpn. *9*, 888 (1954)
6.13 R.D. Hogg, S.P. Vernon, V. Jaccarino: Phys. Rev. Lett. *39*, 481 (1977)
6.14 P.M. Richards: Phys. Rev. B *18*, 6358 (1978)
6.15 N. Bloembergen, E.M. Purcell, R.V. Pound: Phys. Rev. *73*, 679 (1948)
6.16 R.E. Walstedt: Phys. Rev. Lett. *19*, 146, 816 (1967)
6.17 P.A. Fedders: Phys. Rev. B *13*, 2768 (1976)
6.18 D. Wolf: J. Mag. Res. *17*, 1 (1975)
6.19 U. v. Alpen, H. Schulz, G.H. Talat, H. Bohm: Solid State Commun. *23*, 911 (1977)
6.20 W.L. Roth, F. Reidinger, S. La Placa: In *Superionic Conductors*, ed. by G.D. Mahan, W.L. Roth (Plenum Press, New York 1976) and references therein
6.21 P.M. Richards, F. Borsa: In *Local Properties at Phase Transitions*, Proceedings of the International School of Physics "Enrico Fermi", Course LIX, ed. by K.A. Müller, A. Rigamonti (North-Holland, New York 1976)
6.22 B.G. Silbernagel, F.R. Gamble: Phys. Rev. Lett. *32*, 1436 (1974)
6.23 P.M. Richards: Solid State Commun. *25*, 1019 (1978)
6.24 D. Wolf: J. Phys. C *10*, 3545 (1977)
6.25 J.C. Wang, M. Gaffari, S. Choi: J. Chem. Phys. *63*, 772 (1975)
6.26 J.B. Boyce, J.C. Mikkelsen, M. O'Keeffe: Solid State Commun. *21*, 955 (1977)
6.27 D.M. Follstaedt, P.M. Richards: Phys. Rev. Lett. *37*, 1571 (1976)
6.28 C. Korn, D. Zamir: J. Phys. Chem. Solids *31*, 489 (1970)
6.29 D. Jerome, J.P. Boilot: J. Phys. (Paris) *35*, L-129 (1974)
6.30 R.E. Walstedt, R. Dupree, J.P. Remeika, A. Rodriguez: Phys. Rev. B *15*, 3442 (1977)
6.31 J.B. Boyce, J.C. Mikkelsen: Bull. Am. Phys. Soc. *21*, 285 (1976)
6.32 B.A. Huberman, J.B. Boyce: Solid State Commun. *25*, 759 (1978)
6.33 H. Schulz, V. Tscherry: Acta Crystallogr. B *28*, 2168 (1972)
6.34 P.M. Richards: Phys. Lett. *69* A, 58 (1978)
6.35 P.A. Fedders: Phys. Rev. B *17*, 2098 (1978)
6.36 B. Morosin, J.C. Mikkelsen: Acta Cryst. B *35* (1979, in press)
6.37 D.M. Follstaedt, R.M. Biefeld: Phys. Rev. B. *18*, 5928 (1978)
6.38 F.G. Brown: *The Physics of Solids* (Benjamin, New York 1967) Chap.10
6.39 B. Morosin, P.S. Peercy: Phys. Lett. *53* A, 147 (1975)
6.40 D.M. Follstaedt: Unpublished

6.41 K.D. Becker, G.W. Herzog, D. Kanne, H. Richtering, E. Stadler: Ber. Bunsenges.
     Phys. Chem. *14*, 527 (1970)
     J.B. Boyce, B.A. Huberman: Solid State Commun. *21*, 31 (1977)
6.42 G. Burns, F.H. Dacol, M.W. Shafer, R. Alben: Solid State Commun. *24*, 753 (1977)
6.43 J.B. Boyce: Private communication
6.44 P.M. Richards: Unpublished
6.45 M.S. Whittingham, P.S. Connell, R.A. Huggins: J. Solid State Chem. *5*, 321 (1972)
6.46 E.L. Venturini, J.B. Boyce: Private communication
6.47 V. Jaccarino: Private communication
6.48 A. Abragam, B. Bleaney: *Electron Paramagnetic Resonance of Transition Ions*
     (Oxford University Press, New York 1970)
6.49 C. Evora, V. Jaccarino: Phys. Rev. Lett. *39*, 1554 (1977)
6.50 G.R. Luckhurst: In *Electron Spin Relaxation in Liquids*, ed. by L.T. Muus, P.W.
     Atkins (Plenum Press, New York 1972) p.313
6.51 A. Carrington, G.R. Luckhurst: Mol. Phys. *8*, 125 (1964)
6.52 W.M. Walsh, J. Jeener, N. Bloembergen: Phys. Rev. *139*, A 1338 (1965)
6.53 R.S. Title, G.V. Chandrashekhar: Solid State Commun. *20*, 405 (1976)
6.54 J.P. Boilot, J. Thery, R. Collonges: Mater. Res. Bull. *8*, 1143 (1973)
6.55 H.S. Story, W.C. Bailey, I. Chung, W.L. Roth: In Ref.20
6.56 L.V. Dmitrieva, Z.N. Zonn, G.M. Shakhdinarov: Sov. Phys.-Solid State *12*, 32
     (1970)
6.57 R.C. Barklie, K. O'Donnell, A. Murtagh: J. Phys. C *10*, 4815 (1977)
     J.P. Boilot, A. Kahn, J. Thery, R. Collongues, J. Antoine, D. Viven, C.
     Chevrette, D. Gourier: Electrochim. Acta *22*, 741 (1977)

7. Phase Transitions in Ionic Conductors

M. B. Salamon

With 12 Figures

Upon melting, the ionic conductivity σ of a salt generally shows a sharp increase, typically several orders of magnitude, to a liquid value of $\sigma \sim 10^2 \ \Omega^{-1}$. Crystals with superionic [7.1] or solid electrolyte phases, however, pass, either gradually or through a series of phase transitions, from normal ionic conductivity ($\sigma < 0.1 \ \Omega^{-1} \mathrm{m}^{-1}$) to liquidlike values while still solid. The purpose of this chapter is to examine the nature of the phase transitions between low- and high-conductivity phases in the light of modern understanding of critical phenomena. We show that analysis of the critical behavior near the transitions between the high and low ionic conductivity phases of these materials contributes to our understanding of the nature of ion-ion interactions, the role of lattice strain, the importance of phonon coupling to the ionic motion, and the degree of collectiveness involved in superionic conduction.

It has occasionally been stated that the superionic phase is a result of "sublattice melting," [7.2]; that is, atoms on a certain set of lattice positions become mobile, almost liquidlike, while the remaining atoms retain their normal lattice positions. This is, of course, an oversimplification, but it cannot be disputed that materials with superionic phases melt in stages. This is, in some ways, analogous to the better understood case of liquid crystals in which the crystal melts into an isotropic liquid by means of a series of phase transitions, each of which marks the melting of another degree of freedom of the molecules in the crystal [7.3]. In the present case, several ionic species are present, and one or more of the species, located on one or more sets of equivalent sites, disorders and becomes mobile at each transition.

In the liquid crystal case, the various phase transitions have been thoroughly classified [7.3]. Some attempt has been made to categorize the transitions in solid electrolytes as well, but the situation is less clear. O'KEEFE [7.2] divided ionic solids into three classes characterized by their phase transitions. Class I contains "normally melting" salts such as alkali halides which show large discontinuities in ionic conductivity at the melting point, with low conductivities and large activation energies (~ 1 eV) in the solid phase. Crystals showing first-order solid-electrolyte transitions belong to Class II. In a subclass IIa, O'KEEFE places those materials showing a large, discontinuous change in ionic conductivity and a change

in the lattice symmetry of both mobile and nonmobile sublattices, typified by the
$\alpha - \beta$ transition in AgI. Into IIb are placed those crystals for which the changes
in the immobile sublattice are minor. We shall return later to what "minor" might
mean. Finally, Class III includes materials in which the change from low to high
ionic conductivity is spread over a wide temperature range and which have no change
in the symmetry of the underlying nonconducting lattice. PARDEE and MAHAN [7.4]
have attempted to classify ionic conductors according to the order of the phase
transition. Thus, they make no distinction between first-order transitions of type
IIa and IIb, but add a category of second-order transitions, characterized by a
smooth change in ionic conductivity (but with a discontinuous temperature derivative)
and a lambda anomaly in the specific heat. Class III remains a separate category.

In Fig.7.1 we show the ionic conductivity of AgI (Class IIa), Ag_2HgI_4 (Class IIb)
[7.5] and $PyAg_5I_6$ [Py $\equiv$ (C_5H_5NH)] (Class III) [7.6]. Also shown is the transition
in $NH_4Ag_4I_5$ [7.7] which does not fit O'Keefe's scheme. Class III is sometimes called
a Faraday transition [7.8].

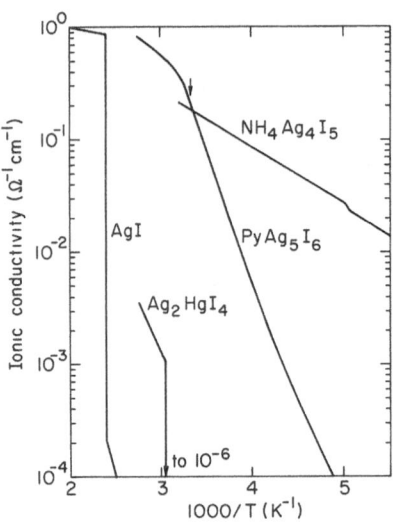

Fig.7.1. The ionic conductivity of several conductors near their conducting-insulating transitions, shown with an arrow

These attempts to organize the various effects observed are not entirely adequate
for our present discussion. General rules for classifying phase transitions were
established by LANDAU [7.8,9], and are based on the symmetry changes which occur at
the transition. The most fundamental of these rules requires that the space group
G of the low-temperature phase of the crystal be a subgroup of the high-temperature
space group G_0. The remaining rules determine whether the transitions will be of
first or second order. The distinction between major and minor changes of lattice
symmetry alluded to by O'KEEFE [7.2] in his subclasses IIa and IIb is, in each case,
that Class IIb transitions satisfy the first Landau rule while Class IIa transitions

do not. Thus we shall distinguish, in a way analogous to O'KEEFE, between non-Landau transitions such as the hexagonal-cubic transitions in AgI, and Landau transitions such as the rhombohedral-cubic transition in $RbAg_4I_5$ [7.10]. Further analysis, required within the Landau theory to determine the nature of the transition, will be the subject of the remainder of this chapter.

Class III transitions pose more of a challenge, since no symmetry breaking occurs. We shall argue that the situation is similar to that of a ferromagnet in an applied field, where the spontaneous symmetry breaking of the phase transition is removed by the external field.

In Sect.7.1, we summarize the Landau concept of phase transitions and the definition of an order parameter. This will be used to discuss the features of modern phase-transition theory which are useful tools for the study of ionic conductors — the concepts of universality, dimensionality and the number of degrees of freedom of the order parameter [7.11]. These ideas will be applied to models proposed for phase transitions in ionic conductors, and an explicit example of setting up a Ginzburg-Landau-Wilson Hamiltonian [7.12] will be presented in Sect.7.2. In Sect.7.3, experimental aspects will be reviewed in the light of the theoretical discussion, and Sect.7.4 will conclude in terms of the information obtained from the critical point studies.

7.1 Modern Theory of Phase Transitions

The starting point for most thermodynamic calculations, especially those involving a phase transition, is the determination of the partition function Z [7.11,12]. Computation of Z requires an integration over all the degrees of freedom of the system (a trace in quantum problems), those which are immaterial to the phase transition as well as those which show critical behavior. Thus, the first task in such a calculation is the separation of critical and noncritical parts of Z. This was first attempted by LANDAU [7.8], who introduced a free energy functional $F_L[p,T,\rho(\underline{x})]$ in which $\rho(\underline{x})$ represents the local particle density, or some other local density such as the magnetization. The density function can be generalized to several components or to vector densities.

The separation of critical and noncritical parts of Z is most conveniently achieved by expanding $F_L[p,T,\rho(\underline{x})]$ about the high-temperature limit of the density [7.9]. This results in a form generally known as a Ginzburg-Landau free energy functional or Hamiltonian. Consider the density function far from the critical point in the disordered phase. This function, which we term $\rho_0(\underline{x})$, is invariant under the group of rotations and translations G_0 which constitute the space group of the crystal in the high-temperature phase. Deviations of $\rho(\underline{x})$ from its high-temperature value are not uniform, and they may be written in terms of functions $\phi_i(\underline{x})$ which transform among themselves according to some n-dimensional irreducible representation of G_0. Intrinsic to the Ginzburg-Landau approach is the assumption that only one such representation will contribute to the critical behavior. The functions $\phi_i(\underline{x})$ are generally taken to be the components of an n-dimensional vector $\vec{\phi}(\underline{x})$, which is called the order parameter of the phase transition [7.11,12].

The partition function is, of course, invariant under rotations and translations of the space group G_0 and, consequently, so is $F_L(p,T,\rho)$. Thus, if we expand $F_L(p,T,\rho)$ around $\rho_0(\underline{x})$, in powers of the deviations $\phi_i(\underline{x})$, only those combinations of $\phi_i(\underline{x})$ may appear which are invariant under G_0. What is more, if we expect the low-temperature phase to have a different symmetry than the high-temperature phase, we must demand that the $\phi_i(\underline{x})$ be not themselves invariants; that is, they must transform according to some representation of G_0 other than the identity. Explicitly, we may write

$$F_L[p,T,\rho(\underline{x})] = F_L[p,T,\rho_0(\underline{x})] + \frac{1}{2} r \sum_i \phi_i^2 + \sum_p v_p \sum_{ijk} \alpha_{ijk}^p \phi_i \phi_j \phi_k + \cdots \quad . \quad (7.1)$$

The sum $\sum_p$ is over all sets of third-order invariants in G_0 and, by convention, second- , third- , and fourth-order coefficients are labelled r, v, and u respectively. The expansion may also include invariant powers of $\partial \phi_i / \partial x_k$ and combinations of these with the ϕ_i. The portions of (7.1) containing components of the order parameter constitute the Ginzburg-Landau Hamiltonian, which we shall denote by $\delta F_L[p,T,\vec{\phi}(\underline{x})]$.

The partition function may now be written as a functional integral over the ϕ_i:

$$Z = Z_0 \int \prod_{i=1}^{n} d\phi_i \exp\{-\beta \int \delta F_L[p,T,\vec{\phi}(\underline{x})]d^dx\} \quad . \quad (7.2)$$

The components of $\vec{\phi}(\underline{x})$ are defined over the d-dimensional physical space of the crystal. A partition function of this type is said to be a d-dimensional, n-vector model, with n the number of "degrees of freedom" of the order parameter [7.11,12]. The factor Z_0 is the result of integrating over all degrees of freedom in the problem except those associated with the order parameter.

The central problem in the statistical mechanics of phase transitions is the evaluation of the integral (7.2). Non-Gaussian terms take on essential importance near T_c, the temperature at which r vanishes, and lead to the singularities which characterize critical behavior. In Landau theory [7.8,11], the integral is simply replaced by its saddle-point value; that is, its value at the minimum of the functional $\delta F_L[p,T,\vec{\phi}(\underline{x})]$ with respect to the components of $\vec{\phi}(\underline{x})$. Then, since the Helmholtz potential is F = -kT ln Z, F contains the minimal value of F_L. As is well known in the Bragg-Williams theory of order-disorder transitions [7.13], the Ginzburg-Landau theory of superconductivity, and other so-called mean-field theories [7.14], a phase transition occurs at $r = 0$, characterized by a uniform order parameter which is zero above T_c and increases as $r^{\frac{1}{2}}$, where $r = (T_c - T)$ below.

The power of $(T - T_c)$ which describes the behavior of the order parameter, a critical exponent, is denoted by β. Other critical exponents which describe the power-law dependence of properties near T_c are listed in Table 7.1 along with their Landau-theory values [7.14]. We shall be mainly concerned with β, with the specific heat exponent α, and with the susceptibility exponent γ. The values of these exponents from Landau theory and the position of the critical temperature are not in agreement with experiment, reflecting the approximate nature of the calculation. However, if the free energy functional F_L and the order parameter $\vec{\phi}(\underline{x})$ have been properly chosen, the qualitative behavior will generally be correct.

7.1.1 Landau Criteria

For the Landau theory to be valid [7.9], three criteria must be satisfied:

(I) The space group G of the ordered phase must be a subgroup of the space group G_0 of the disordered phase and some combination of the $\phi_i(\underline{x})$ must be invariant under G.

(II) No third-order terms should appear in (7.1).

(III) No terms of the form $\phi_i \partial \phi_i / \partial x_k$ may appear in (7.1).

Of these criteria, (I) is necessary for the entire Ginzburg-Landau concept to apply, while (II) and (III) guarantee that the transition which occurs is second order. Thus, transitions of the Landau type, whether first or second order, form a distinct class apart from those lacking a link between phases through the broken symmetry of the disordered-phase space group. It is this distinction to which O'KEEFE [7.2] was referring in dividing Class II solid electrolyte transitions into two subclasses.

The failure of the Landau theory to predict quantitatively the behavior at the critical point is due to its failure to take proper account of fluctuations. An improvement is made by the inclusion of terms of the form $(\partial \phi_i / \partial x_k)^2$ in (7.1), but the difficulty is more fundamental. More recent work, based on WILSON's [7.15] renormalization group (RNG) theory, shows that fluctuations play an essential role below some critical value of the lattice dimensionality d. For dimensions greater than the critical value d_c ($d_c = 4$ for the systems of interest here), fluctuations are not important and the Landau theory is exact. Below d_c, the role of fluctuations is essential, and the system cannot be treated within the saddle-point method. The interested reader is referred to the recent book by PFEUTY and TOULOUSE [7.12] for a more detailed discussion of the RNG.

7.1.2 Renormalization Group

Briefly, in the RNG method one first converts the integral over spatially defined values of the order paramter to one involving their Fourier components. In that form, the integral can be easily partitioned between fluctuations which occur on a scale much shorter than the distance over which the order parameter is correlated (the correlation length ξ) and those on a longer length scale. The former fluctuations are not of importance near T_c and may be integrated out [7.15]. Near $d = d_c$ one next rescales the problem to be identical with the original, but with renormalized values of r, u, and ϕ_i. The process is repeated until no further change in r, u, ... occurs. These values, called r^*, u^*, ... are termed the "fixed point." The nature of the fixed point determines the behavior of the phase transition. In particular, if there exists a direction in u-v space along which successive applications of the change-of-scale trick carries r and u to r^* and u^*, the fixed point is said to be stable. While there is no rigorous proof of the importance of a stable fixed point to the existence of a second-order transition, such a condition has been proposed by MUKAMEL and KRINSKY [7.16] as a fourth Landau criterion:

(IV) The free energy functional (7.1) must possess a stable fixed point.

This last criterion is required to explain why some phase transitions which satisfy the first three criteria are, nonetheless, first order. A prime reason for the failure to find a stable fixed point, it turns out, has to do with a too-large value of n, the number of components of the order parameter [7.16]. For ionic conductors, however, this consideration does not appear to be important.

The power of the Ginzburg-Landau-Wilson approach lies in its predictions for the relationship between n and d and the values of critical exponents. Any Ginzburg-Landau functional having the same n and d and the same general form will lead to the same fixed point, and thus to the same exponents. This result, often termed *universality*, demonstrates that the basic symmetry of the problem underlies its critical behavior, regardless of microscopic details [7.11]. Thus, we may use the values of critical exponents to work back through the argument and deduce the nature of the order parameter. This is of considerable importance in the superionic case, where the nature of the order parameter is not obvious, and is, in fact, one of the central questions. In Table 7.1, we have listed various values of the critical exponents as predicted from the RNG theory [7.11].

Table 7.1. Critical exponents for several model systems and $RbAg_4I_5$. (n,d) refers to the d-dimensional n-vector model

Exponent	Landau	RNG (1,3)	RNG (2,3)	RNG (3,3)	$RbAg_4I_5$ [a]
β	1/2	0.34	0.36	0.376	0.35 ± 0.02
γ	1	1.24	1.30	1.35	1.25 ± 0.1
α	0(step)	0.08	~.02	-0.10	0.14 ± 0.02

[a][7.51]

7.2 Models for Critical Behavior in Superionic Conductors

In this section we review several models which have been proposed to explain the
phase transitions in superionic conductors. The effort to date has focussed on the
formulation of Hamiltonians and Ginzburg-Landau free energy functionals for the
problem, which are then treated in Landau theory or the random-phase approximation.
Finding the correct Landau functional is already a difficult problem since the nature of the order parameter for these transitions is not obvious. The usual choice
has been to take the fractional site occupancy as a starting point and to generate
a free energy functional or interaction Hamiltonian based on either the average
occupancy or on number operators.

7.2.1 Quasi-Chemical Models

The earliest attempts at constructing models for the insulator-electrolyte transition
were based on the quasi-chemical or Bragg-Williams-type model. WIEDERSICH and JOHNSTON
[7.17] presented a detailed model of this type for $RbAg_4I_5$, and more general for-
mulations, differing only in the amount of detail, have been made by RICE, STRÄSSLER,
and TOOMBS (RST) [7.18], by HUBERMAN [7.19], by WELCH and DIENES [7.20], and, most
recently by LAM and BUNDE [7.21]. In these models, the Helmholtz free energy $F = U - TS$
is written down by calculating the entropy S from combinatorics and expanding U in
powers of the occupancy of certain sites. This Helmholtz energy is then treated as a
Ginzburg-Landau functional and is minimized with respect to site occupancies.

In its simplest form [7.18], N mobile cations are postulated to occupy N sites in
a rigid lattice at T = 0 K. These ions may be excited onto Ng interstitial sites
at finite T, the occupancies of which are characterized by a variable c(T). The
entropy is computed from the number of ways of distributing the N ions over the
N(1 + g) sites such that N(1 - c) ions are on the original sites and Nc are on interstitial sites. The energy U is written as a power series in c, but with varying
interpretations as to the underlying physics of the various terms [7.20]. Following
RST [7.18], we may write the resulting functional as

$$\delta F_L(T,c) = U(c) - k_B TS(c)$$

$$\cong c(\varepsilon_0 - \frac{1}{2}\lambda c) - k_B T[c \ln(g) - c \ln(c) - (1 - c) \ln(1 - c)] \quad . \quad (7.3)$$

This form has been studied previously as a Landau functional by STRÄSSLER and
KITTEL [7.22].

Examination of (7.3) shows that it has no minimum at low temperature for $\lambda > 0$,
but develops one at high T, located at

$$c(\infty) = g/(1 + g) \quad . \quad (7.4)$$

At $T = 0$ K, δF_L has a lowest value (not a true minimum) at $c(0) = 0$. The value of δF_L at $c = 1$ becomes lower than its value at $c = 0$ for temperatures above

$$k_B T_0 = \lambda (2\varepsilon_0/\lambda - 1)/2 \ln(g) \quad , \tag{7.5}$$

provided that

$$2\varepsilon_0/\lambda < [1 + \tfrac{1}{2} \ln(g)] \quad . \tag{7.6}$$

This has been widely taken to be a first-order phase transition, sometimes termed a "supertransition" [7.22], since the order parameter c jumps from zero to unity, and then decreases toward its high temperature value. The behavior of this free energy functional is shown in Fig.7.2 for the case $2\varepsilon_0/\lambda = 4/3$ and $g = 3$ at various temperatures.

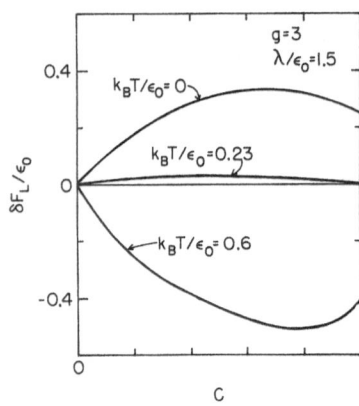

Fig.7.2. Quasi-chemical Ginzburg-Landau free energy at several temperatures. Note that a true minimum is absent at low temperatures

The nonanalytic behavior of (7.3) near $c = 0$ and $c = 1$ makes it very unlikely that the saddle-point evaluation of (7.2) intrinsic in the Landau model will be valid. What is more, the usual Landau expansion predicts a vanishing order parameter in the disordered phase. We show below that the BRAGG-WILLIAMS [7.13] order parameter $w = 2c - 1$ is a more reasonable choice for the study of these quasi-chemical models. Consider the expansion of (7.3) in powers of w:

$$\delta F_L(T,w) = \tfrac{1}{2} [\varepsilon_0 - \tfrac{1}{2} \lambda - k_B T \ln(4g)] + \tfrac{1}{2} [\varepsilon_0 - \tfrac{1}{2} \lambda - k_B T \ln(g)]w$$

$$+ \tfrac{1}{2} k_B [T - \lambda/4k_B]w^2 + (k_B T/12)w^4 + \ldots \quad . \tag{7.7}$$

Equation (7.7) is clearly in the Landau form, and in the absence of the linear term, would have a phase transition at

$$T_c = \lambda/4k_B \quad . \tag{7.8}$$

The term linear in w has the effect of a temperature-dependent magnetic field

$$h = (2k_B T \ln(g) + \lambda - 2\varepsilon_0)/4 \quad . \tag{7.9}$$

Using (7.5) we may rewrite this as

$$h = \frac{1}{2} k_B (T - T_0) \ln(g) \quad , \tag{7.10}$$

so that the first-order transition temperature corresponds to a change in the sign
of the magnetic field term.

From (7.8) and (7.5) it is easy to see that

$$T_0/T_c = [(4\varepsilon_0/\lambda) - 2]/\ln(g) \quad , \tag{7.11}$$

so that $T_0 < T_c$ when (7.6) is satisfied. Thus, the phase transition is smeared by
a temperature-dependent magnetic field term which changes sign below T_c whenever
(7.6) holds. Then we have a first-order transition analogous to the reversal of
the direction of magnetization in a ferromagnet. This is not a "supertransition"
or other unusual change, and occurs after order has already set in. When (7.6) is
not satisfied, the field reversal occurs above T_c. The reversal temperature can be
far above T_c (it is infinite for g = 1) so that h may be large throughout the
region near T_c. Hence, the model, as modified, provides a satisfactory explanation
for Class III transitions, in which the conducting phase arises smoothly, without
a phase transition. Such a model has been used successfully by HIBMA [7.6] to explain the entropy change in $PyAg_5I_6$

Clearly, the model free energy is not complete. The fourth-order term vanishes
with temperature, reflecting the termination of the expansion of U with terms second order in c. Keeping cubic and higher terms would eliminate the field-reversal
effect and stabilize the system. Thus we conclude that (7.3) is not a suitable
Landau free energy functional, but that the modification introduced in (7.7) represents an improvement, at least for cases when $T_0 > T_c$.

7.2.2 Lattice Gas Models

A more microscopic approach to phase transitions in ionic conductors is possible
through the lattice gas model, which is the subject of Chap.4 of this volume. An
early attempt to account for phase transitions in these materials by PARDEE and
MAHAN [7.4] used such a model. The Hamiltonian consists of three terms: the phonons
of the nonconducting sublattice, the ion-ion repulsion,

$$H_c = \sum_{ij} V_{ij} n_i n_j \quad , \tag{7.12}$$

and an ion-phonon term. In (7.12) $n_i = 1$ implies that the ith site is occupied by a mobile ion, while $n_i = 0$ means that the site is empty. The ion-phonon term may be removed by a transformation of the phonon coordinates, which also has the effect of reducing the ion-ion repulsion, so that V_{ij} is replaced by a reduced value U_{ij} in (7.12). Thermodynamic calculations are performed in mean-field theory, using two types of nearest-neighbor sites, denoted by n_i and p_i. Each site then has a mean-field grand canonical potential given in terms of average occupancies $\bar{n}$ and $\bar{p}$ by

$$H_n = n_i(zU\bar{p} - \mu_n) \quad , \tag{7.13}$$

and

$$H_p = p_i(zU\bar{n} - \mu_p) \quad .$$

Here U is the nearest-neighbor value of U_{ij}, z is the number of nearest neighbors, and μ_n and μ_p are the chemical potentials of the two types of site. A straightforward calculation of the order parameter $\phi = (\bar{n} - \bar{p})/(\bar{n} + \bar{p})$ gives

$$\phi = \tanh \left[\frac{1}{2} \beta(zUn_c\phi - \Delta)\right] \quad , \tag{7.14}$$

where $n_c = \bar{n} + \bar{p}$, and $\Delta = \mu_p - \mu_n$.

When $\Delta = 0$, this is the usual Weiss mean-field result [7.14], and the model has the critical behavior of the Landau model, Table 7.1. The quantity Δ plays the same role as the field h in the above model, and tends to smear the phase transition. This is because a chemical potential difference between the sites lifts the site degeneracy just as a magnetic field lifts the degeneracy of up and down spin. In fact, if the total number of carriers is to be held constant, Δ must be temperature dependent, as it is in the quasi-chemical model. Thus, the lattice gas model in this approximation is quite similar to our modified quasi-chemical model, and predicts either a second-order transition or a smeared transition depending on the presence of a difference in site energies.

In the case of three different types of sites [7.21] or when two mobile-ion species may occupy a given site [7.23], the order parameter takes on three values and is similar to a spin-1 problem. The lattice gas Hamiltonian may then be written in terms of a spin operator $S_i = 0, \pm 1$. In the three-level problem [7.21], S_i is associated with the difference between occupancy of the two interstitial sites n_{i2} and n_{i3}, while S_i^2 measures the removal of an ion from the original site, n_{i1}:

$$\begin{align} S_i &= n_{i2} - n_{i3} \\ S_i^2 &= 1 - n_{i1} \end{align} \tag{7.15}$$

In the case of the two-species problem, which has been applied to Ag_2HgI_4 and Cu_2HgI_4 [7.23], a single site may contain an ion of type A, or one of type B, or no ion at all. Then, labeling the number of A ions by n_i^A and B ions by n_i^B, we define

$$S_i = n_i^A - n_i^B$$
$$S_i^2 = n_i^A + n_i^B$$

(7.16)

In either case, the lattice gas Hamiltonian in the absence of distortions of the nonconducting lattice becomes

$$H = -D \sum_i S_i^2 - J \sum_{<ij>} S_i S_j + \text{higher order terms} \quad .$$

(7.17)

This has the form of the antiferromagnetic Blume-Capel model and is, therefore, expected to have a tricritical point. GIRVIN and MAHAN [7.23] have investigated this Hamiltonian for the structure of Ag_2HgI_4 using renormalization techniques and concluded that this compound lies in the first-order region of the tricritical phase diagram. LAM and BUNDE [7.21] have included a term in $S_i^2 S_j^2$ in the three-site problem, which, in Landau theory, leads to two transitions. The latter approach, however, suffers from the same difficulty as the original formulation of the quasi-chemical problem and cannot be viewed as a valid model.

None of these calculations has been concerned with the symmetry requirements on the order parameter, which leads to the inclusion of terms which break the symmetry of the high-temperature phase and preclude a true phase transition. We now turn to a specific example for which the symmetry of the order parameter has been worked out [7.24].

7.2.3 The Order Parameter for $RbAg_4I_5$

The high-temperature α phase of $RbAg_4I_5$ is cubic and belongs to the space group $P4_132(O^7)$. At 208 K the crystal undergoes a transition into a rhombohedral phase [7.10] with the space group $R32(D_3^7)$, a subgroup of $P4_132$. Thus, this transition between two superionic phases satisfies the first Landau criterion. At this transition, the conductivity does not exhibit a jump, but rather a smooth kink, such as in Fig.7.1. At lower temperatures, near 122 K, there is a second transition into a trigonal phase with a larger unit cell, marked by a first-order jump in ionic conductivity; this transition clearly belongs to O'Keefe's Class IIa.

In the cubic phase, the unit cell contains four formula units, but with the 16 Ag ions distributed over 56 interstitial sites [7.25]. These are comprised of two sets of 24 equivalent positions, labeled Ag(II) and Ag(III) by GELLER [7.25], and a set of 8 equivalent positions labeled Ag(c). The Ag(II) sites occur in pairs

which are situated such that both cannot be simultaneously occupied by Ag ions.
They are equivalent only on the average, with the mobile Ag ions instantaneously
on one or the other member of the pair. The Ag(II) sites contain a larger-than-
average fraction of the mobile ions at room temperature and play an important role
in the geometry of conducting channels, since every second conducting hop must
involve one of these sites [7.25].

In the rhombohedral β phase, the nonconducting lattice undergoes only a slight
distortion of the unit cell along a [111] direction with no change in cell
volume [7.10]. The various Ag sites, however, are split into groups containing
no more than 6 equivalent positions, as illustrated in Fig.7.3. One of the four
sets of equivalent positions, Ag(2), derived from the Ag(II) sites of the cubic
phase, is empty while another, Ag(4), is nearly filled. Each of the filled sites
along with its neighboring empty site forms one of the Ag(II) pairs of the high-
temperature phase. We are thus led to use this ordering of the occupancies of the
pair states as an order parameter. There are other choices involving larger com-
plexes of the sites —which must be of importance in accounting for the emptying of
the Ag(III) sites —but we believe the essential physics is contained in this
choice [7.24].

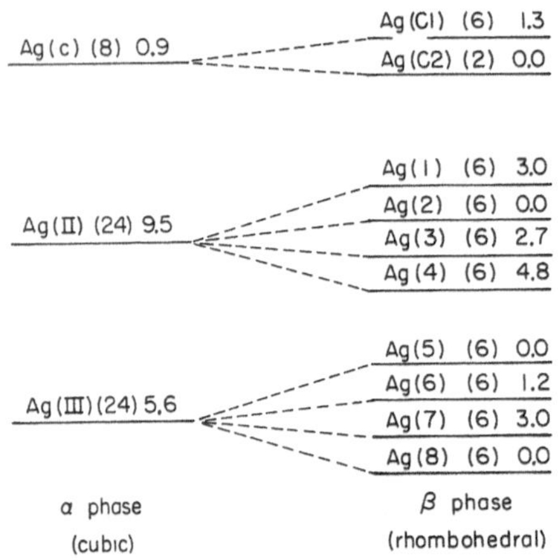

Ag(c) (8) 0.9

Ag(Cl) (6) 1.3
Ag(C2) (2) 0.0

Ag(II) (24) 9.5

Ag(1) (6) 3.0
Ag(2) (6) 0.0
Ag(3) (6) 2.7
Ag(4) (6) 4.8

Ag(III)(24) 5.6

Ag(5) (6) 0.0
Ag(6) (6) 1.2
Ag(7) (6) 3.0
Ag(8) (6) 0.0

α phase
(cubic)

β phase
(rhombohedral)

Fig.7.3. Schematic diagram of the redis-
tribution of Ag occupancies in RbAg$_4$I$_5$.
Numbers in parentheses are the site degen-
eracies while decimal numbers are site occupancies at room temperature (β phase) and
at 130 K (β phase). [7.10]

In order to construct an order parameter we number the Ag(II) sites and pick the
i^{th} site in the jth unit cell of the crystal. We call this an a site and denote its
occupancy $n^a_{ij} = 0,1$. Its pair partner we label b, with occupancy n^b_{ij}. Those sites
obtainable from the first site by the application of the identity element of the
group, the eight 3-fold rotations, and the three 2-fold rotations about the cube
axes are all a sites. The pair partners of these are the b sites. We now define a
pseudospin operator

$$S^z_{ij} = n^a_{ij} - n^b_{ij} \quad , \tag{7.18}$$

and its Fourier components

$$S^z_i(\underline{k}) = N^{-\frac{1}{2}} \sum_j S^z_{ij} \exp(i\underline{k}\cdot\underline{R}_j) \quad . \tag{7.19}$$

Let us examine the effect of the space-group operations on a particular pair in the zeroth unit cell with pseudospin S^z_{10}. The first member of the pair is located at the point (x,y,z) where $x = 0.53$, $y = 0.27$, and $z = 0.80$, in units of the lattice constant a_0 [7.25]. The occupancy of this site is n^a_{10} and of its nearest-neighbor Ag(II) site, located at $(3/4 - y, 3/4 - x, 3/4 - z)$, ~1.8 Å distant, n^b_{10}. Application of a typical space-group operation, such as rotation about the z axis combined with a translation of $\frac{1}{2}$ a_0 along the [101] direction carries the above pair into another pair $(1/2 - x, \bar{y}, 1/2 + z)$, $(3/4 + \dot{y}, 1/4 - x, 1/4 - z)$ whose pseudospin has been labeled S^z_{40}. This translation has assumed each cell equivalent and so corresponds to the k = 0 component of (7.19). Similarly, a two-fold rotation about the [110] axis, coupled with a translation of $\frac{1}{4}$ a_0 along [311] carries the site (x,y,z) into $(3/4 + y, 1/4 - x, 1/4 - z)$, and thus transforms S^z_{10} into $-S^z_{40}$.

Proceeding in a like manner, we can generate a representation of the space group at k = 0 from the 12 pseudospins. However, the space group at k = 0 is isomorphous with the cubic group 432, with maximum dimensionality of 3, indicating that this is a reducible representation. From the character table it is easy to show that this 12-dimensional representation decomposes into $A_2 + E + 2T_1 + T_2$. Lacking a change in unit cell dimension at T_c, $RbAg_4I_5$ has an order parameter associated with the point k = 0 [7.9], which can therefore be constructed from the sum over all cells of each of the 12 pseudospins, equal to $S^z_i(0)$ from (7.19).

Of the possible representations generated by the 12 pseudospins, only T_2 is compatible with the identity representation of the rhombohedral group 32, and the order parameter must have T_2 symmetry. Under the operations of the cubic group, basis functions of T_2 transform as yz, zx, and xy. Thus, we require linear combinations of the $S^z_i(0)$, which we label as ϕ_4, ϕ_5, and ϕ_6 in analogy with elastic strains, which transform among themselves as do yz, etc. Using standard methods [7.9], one finds, for example, that ϕ_4 is proportional to $S^z_1(0) + S^z_2(0) - S^z_3(0) - S_4(0) + S^z_{10}(0) - S^z_9(0) + S^z_{12}(0) - S^z_{11}(0)$. The definitions have been written out completely by SALAMON [7.24]. It is now straightforward to write down a Landau free energy functional in the form (7.1) including not only invariant combinations of the ϕ's, but also combinations involving the lattice strains e_4, e_5, e_6. The expression is

$$\delta F_L = \sum_{m=4}^{6} (\frac{1}{2} r\phi_m^2 + u\phi_m^4 + 2C_{44}e_m^2 + \eta e_m \phi_m) + u'(\phi_4^2\phi_5^2 + \phi_5^2\phi_6^2 + \phi_6^2\phi_4^2) - v\phi_4\phi_5\phi_6 \quad .$$

(7.20)

Characteristic of the rhombohedral distortion, the magnitude of the three components must be equal, but with phases which result in the four possible [111] distortions. Thus, there is only one independent amplitude in the problem, but four different domains are possible. The phase transition which results from (7.20) is first order and occurs at

$$T_c^* = T_c + \eta^2/8r_0C_{44} + v^2/32r_0(u + u') \quad ,$$

(7.21)

where we have assumed $r = r_0(T - T_c)$. The first-order transition arises as a result of the cubic term, which appears in (7.20) in violation of the second Landau criterion. As we note in Sect.7.3, v appears to be small and sensitive to strain so that the first-order transition occurs only close to T_c, a situation often termed "weakly first order."

An important feature of (7.20) is the term bilinear in the order parameter and the strain. This results in a rhombohedral distortion which is proportional to the order parameter rather than the more familiar case in which the lattice coupling is through the pressure dependence of T_c, and thus quadratic in the order parameter. An important test of this Ginzburg-Landau theory is a determination of the temperature dependence of the strain. As discussed in Sect.7.3.4, the birefringence, which measures the distortion, is linear in the order parameter, thus corroborating the above form. In the microscopic Hamiltonian underlying (7.20), a bilinear coupling must exist between local strain and pseudospin variables. As has been pointed out elsewhere by the author [7.24], the Hamiltonian which is deduced resembles in many ways that of cooperative Jahn-Teller transitions, a point which has been elaborated upon by LAM and BUNDE [7.21].

That linear combination of the ϕ's which is invariant in a rhombohedral crystal whose trigonal axis coincides with the original cubic [111] direction is $\phi_4 + \phi_5 + \phi_6$. The combination of pseudospin variables corresponding to this quantity is $\sum_j(S_{2j}^z + S_{6j}^z - S_{3j}^z - S_{7j}^z + S_{10j}^z - S_{11j}^z)$. The order parameter is zero above T_c, but grows to +1 at low temperatures. Thus, pseudospins labelled 2, 6, and 10 tend toward positive values, meaning the a sites are increasingly occupied, while those labelled 3,7, and 11 tend toward negative values with b sites occupied. An explicit comparison of sites shows that the a sites of 2, 6, and 10 and the b sites of 3, 7, and 11 correspond to the set of sites labelled Ag(4) in Fig.7.3. The respective opposite members belong to Ag(2). The order parameter as defined leads naturally therefore to the occupancy of sites inferred from the X-ray refinement of the β phase.

Thus, with a properly chosén order parameter, Ginzburg-Landau models are capable of providing a good qualitative description of the superionic transitions, and are proper starting points for a full RNG calculation.

7.3 Critical Behavior of Physical Properties

In this section we examine experimental results for the behavior of various physical properties near the critical points of ionic conductors. As above, we focus on Class IIa and III transitions, plus second-order Landau transitions, as only these show well defined critical behavior. We do not attempt an exhaustive survey, but rather choose representative examples in each case.

7.3.1 Specific Heat

An anomaly in the specific heat is one of the key indications of a phase transition, especially in cases, as with ionic conductors, where the order parameter is not directly accessible. JOHNSTON et al. [7.26] located the 208 K phase transition in $RbAg_4I_5$ first through their observation of a lambda anomaly in the specific heat. Subsequently, PARDEE and MAHAN [7.27] pointed out the Ising-like nature of the low-temperature side of the specific heat curve. This represented the first use of the phase transition as a tool in elucidating the nature of the partial ordering transition in an ionic conductor.

A detailed study of the specific heat of the MAg_4I_5 family, for M= Rb, K, and NH_4, has been carried out by LEDERMAN et al. [7.28] and by VARGAS et al. [7.7]. The entire family has been shown to have specific heat exponents $\alpha = \alpha' = 0.15 \pm 0.02$ and a ratio of critical amplitudes of $A^+/A^- = 0.6 \pm 0.05$, both in agreement with RNG values for $n = 1$ and $d = 3$ (cf. Table 7.1). In Fig.7.4 we show data for $NH_4Ag_4I_5$ very close to its critical point at 198 K. There is no evidence for a first-order transition which means that the cubic term in (7.20) only becomes important very close to T_c.

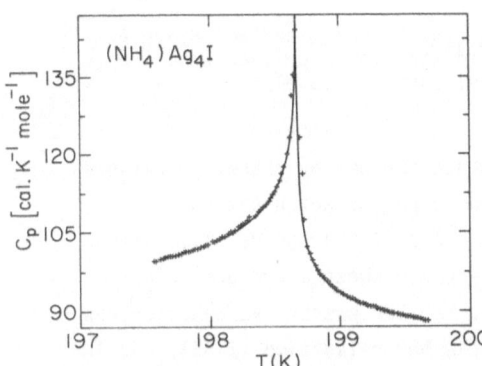

Fig.7.4. Specific heat of $NH_4Ag_4I_5$. Solid line is the fitted curve

These data show quite clearly that the $\approx$ 200 K transitions in MAg_4I_5 are of the Landau type, and have an order parameter with one degree of freedom. Such behavior can be obtained from a Ginzburg-Landau functional of the form (7.20) by means of RNG analysis, and indicates the basic soundness of the approach. As we shall discuss in Sect.7.3.4, there is some evidence for a first-order transition in $RbAg_4I_5$ which does not show up in the C_p data. It may be that either the entropy of transition is extremely small, or the conditions of the specific heat measurements (clamped sample) result in sufficient strain to suppress the first-order effect. This point will require further study.

Specific heat measurements have been carried out on a larger number of Class III, or diffuse transitions. These behave like a magnet in an external field, and the specific heat anomaly is thus rounded and spread over a wide temperature range. DWORKIN and BREDIG [7.29] determined the specific heat accompanying the rise of ionic conductivity in $SrCl_2$, CaF_2 and the anti-isotype, K_2S. All show a specific heat maximum spread over several tens, or even a hundred degrees (up to 10% of T_c). A similar measurement has been reported by DERRINGTON et al. [7.30] for PbF_2. The most complete of these studies was made by HIBMA [7.6] on $PyAg_5I_6$. HIBMA's data for the excess specific heat are shown in Fig.7.5 along with his calculated curve. The theoretical curve was determined from a quasi-chemical model similar to the RST model, but with pair occupancies and single occupancies taken to be independent. This decouples the pair energy from the average occupancy (the order parameter) and facilitates the calculation of C_p. The agreement is satisfactory and yields a site energy difference of 0.065 eV and a pair energy of 0.092 eV, both rather large compared with k_BT_c.

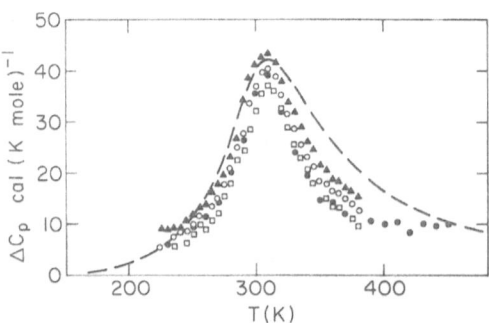

Fig.7.5. Specific heat of $PyAg_5I_6$. Solid line is determined from the quasi-chemical model

An interesting situation arises in CuI, which has two nearby phase transitions [7.31]. The γ and α phases are closely related cubic phases which satisfy the Landau criterion [7.9]. The intervening wurtzite phase [7.32] can be eliminated under pressure [7.33]. The specific heat in the γ phase shows a precursive behavior [7.32] suggestive of a lambda singularity. Similar precursive behavior has been observed in the ionic conductivity [7.34], the NMR relaxation [7.35], and in

Raman scattering [7.36]. It should be interesting to study CuI under pressure to test the suggestion that the transition can be made second order by elimination of the wurtzite phase.

7.3.2 Ionic Conductivity

The ionic conductivity is, of course, the property of primary interest in super-ionic conductors, and its behavior near the critical point is crucial to the understanding of the nature of the conduction process. The situation is complicated because of the geometrical requirement for the existence of "channels" without which the conductivity may be small even in the presence of a large hopping rate. An early attempt to treat the problem by SATO and KIKUCHI [7.37] used a single ion, mean field approach applied to Na β-alumina and β"-alumina. In the latter case, in which ordering sites are initially identical, a Landau transition occurs and was predicted to be accompanied by a sharp kink in the ionic conductivity. However, no phase transition has been reported for β"-alumina [7.38], and sharp kinks of the type predicted have not been observed in any material.

Ionic conductivity near the phase transitions in the lattice gas model has been calculated by MAHAN [7.39] and by LAM and BUNDE [7.21]. The determination of the dc conductivity is difficult, but some indication of a change in activation energy is found [7.39]. The lattice gas approach is reviewed in Chap.4 of this volume. Although not specifically applied to the phase transition problem, continuum or liquid-like models have been employed to calculate the ionic conductivity. These, too, have been most useful in explaining the frequency-dependent conductivity. A detailed review of such models is given in Chap.8 of this book.

A general treatment of the behavior of the ionic conductivity near an ordering transition has been given by VARGAS et al. [7.7]. Starting from the classical many-body theory of atomic diffusion [7.40], the authors divide the many-body potential energy function $V(\underline{s})$ into two parts, $V^0(\underline{s})$ which is the potential of a virtual crystal which would not order, and $V^i(\underline{s})$ which contains the ordering interactions. Both are functions of the 3N-dimensional, mass-weighted position vector $\underline{s}$. The virtual crystal potential has minima corresponding stable configurations while $V^i(\underline{s})$ is assumed to be a smooth function of $\underline{s}$, depending only on relative positions of the mobile ions. The result of the calculation is that the ionic conductivity is given by

$$\sigma = \sigma_0 \exp[(\langle V^i \rangle - \langle \hat{V}^i \rangle)/k_B T] \quad , \tag{7.22}$$

where σ_0 is the conductivity of the (nonordering) virtual crystal and $\langle V^i \rangle$ and $\langle \hat{V}^i \rangle$ are the configurational averages of the interaction potential at a minimum and a saddle point of the virtual crystal potential, respectively.

The quantity $<V^i>$ is just twice the interaction enthalpy per particle. The potential at the saddle point is somewhat more difficult to obtain, but is taken to be proportional to the enthalpy. Thus, the activation energy of the interacting system contains a contribution proportional to the enthalpy of interaction, so that the logarithmic derivative of the conductivity should obey

$$d(\ln\sigma)/dT = \kappa\Delta C_p/RT_c + \text{const.} \qquad (7.23)$$

where κ is a proportionality constant, ΔC_p is the critical part of the specific heat, and the constant arises from the smooth part of the conductivity.

This prediction has been tested in detail for the MAg_4I_5 family of ionic conductors [7.7]. The specific heat and ionic conductivity of single crystals were measured simultaneously using a temperature and current modulation scheme. This technique facilitates a direct test of (7.23). The results are shown in Fig.7.6 for KAg_4I_5. Similar curves are obtained for $RbAg_4I_5$ and $NH_4Ag_4I_5$ [7.7], and (7.23) is verified in detail.

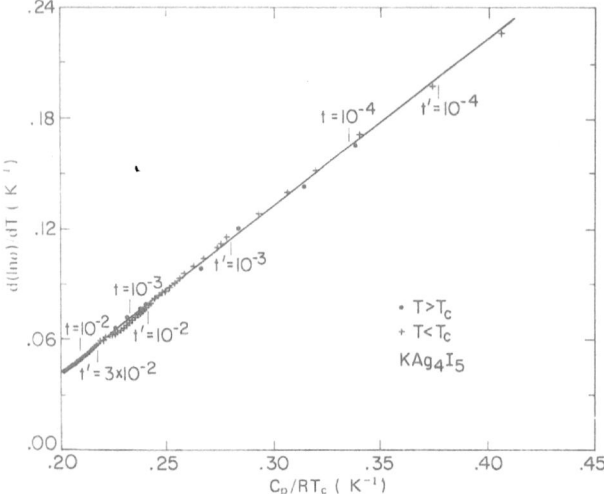

Fig.7.6. Direct comparison of $d(\ln\sigma)/dT$ with C_p. The reduced temperatures are $t = (T/T_c - 1)$ for $T > T_c$, and $t' = (1 - T/T_c)^p$ for $T < T_c$

Somewhat different results have been reported by BOROVKOV and IVANOV-SHITZ [7.41] for $RbAg_4I_5$. They claim to observe a step discontinuity in the ionic conductivity near 209.6 K rather than the smooth kink observed by VARGAS et al. The data of both groups are plotted together in Fig.7.7. The data appear to be consistent within experimental error, although there is a relative temperature shift of about 0.6 K. Whether the step observed by BOROVKOV and IVANOV-SHITZ is a manifes-

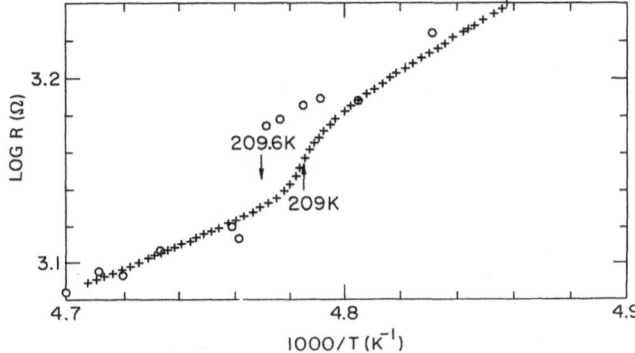

Fig.7.7. Ionic conductivity of $RbAg_4I_5$. Pluses denote data of [7.7]; circles, [7.41]

tation of the first-order transition predicted by the Landau theory in Sect.7.2 will require further testing. We note that the step was observed under experimental conditions in which the sample was clamped between electrodes, while in the other experiment, flexible wire leads were diffused into the sample. Given the sensitivity of structural transitions to external strain [7.42], this may explain the difference.

The pressure-dependent ionic conductivity of $RbAg_4I_5$ in the vicinity of the critical point has been measured by ALLEN and LAZARUS [7.43]. The critical temperature increases nearly quadratically with pressure to 6 kbar. The activation volume is found to be negative (in contrast to AgI) and is interpreted as further evidence that the saddle points are decoupled from ion-ion interactions and that increasing pressure *decreases* the depth of the potential wells at interstitial sites. They conclude that a significant fraction of the activation entropy is due to the ordering entropy.

The validity of (7.23) is not restricted to Landau transitions — it should hold for Class III transitions as well. Indeed, HIBMA's conductivity data for $PyAg_5I_6$ [7.6] shows a maximum slope near the peak in the specific heat. We have smoothed HIBMA's data numerically [7.44] and taken the derivative of log σ with respect to temperature. A small maximum is present, but it occurs near 295 K rather than at 310 K, where the specific heat is at a maximum, cf. Fig.7.5. The opposite is true in PbF_2, where the maximum in $d(\ln\sigma)/dT$ occurs at a higher temperature than the peak in C_p. As noted in Chap.5, anomalies in the optical properties coincide with the peak in $d(\ln\sigma)/dT$ and we suspect some error in the enthalpy measurement [7.30].

We conclude that the ionic conductivity near a critical point, or indeed, a smooth phase transition, is dominated by short-range order. However, the true correlated short-range order parameter must be taken into account for an accurate description. As with the specific heat, the mean-field calculations undertaken to date give a qualitative description, but a more general approach will be necessary to proceed beyond (7.23). Recent progress in this direction is discussed in Chaps.4 and 8 of this volume.

194

7.3.3 Acoustic Properties

The presence of an order parameter with the same symmetry as strain components leads to a renormalization or softening of the elastic constant for that strain. In the case of $RbAg_4I_5$, the order parameter couples to rhombohedral strains, and it is C_{44} which is affected. A calculation in the Landau approximation [7.24] gives, for $T > T_c$,

$$C_{44} = C_{44}^0[1 + A/(T - T_c)]^{-1} \quad ,$$

(7.24)

where C_{44}^0 is the elastic constant in the absence of coupling to the order parameter. The constant A can be estimated from the size of the rhombohedral distortion ($\sim 0.1°$) and the magnitude of the specific heat jump in a Landau model, with the result $A \approx 0.5$ K. Ultrasonic measurements by GRAHAM and CHANG [7.45] show quite clearly the softening of the C_{44} elastic constant. We have plotted these results for $C_{44}(T)/C_{44}(300 K)$ in Fig.7.8 along with the curve from (7.24) for A = 0.5 K and $C_{44}^0 = 1.02\ C_{44}(300 K)$, in agreement with the value extrapolated from Graham and Chang's data far from T_c. The mean-field result is seen to agree with the data sufficiently far from T_c; more precise data are required closer to T_c in order to make possible a detailed comparison with the Ising model.

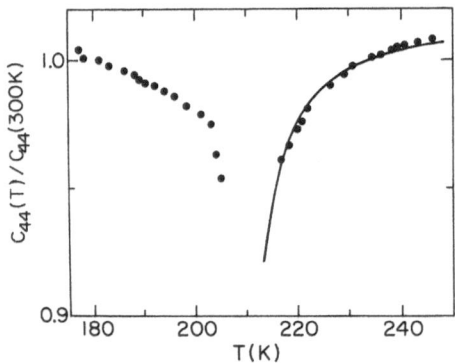

Fig.7.8. The ratio $C_{44}(T)/C_{44}(300 K)$ for RbAg4I5. The solid curve is the result of [7.24] with A = 0.5 K

An anomaly in the ultrasonic attenuation was also found in the vicinity of the 208 K transition of $RbAg_4I_5$ by GRAHAM and CHANG [7.45] for longitudinal and transverse waves along [110] which involve the C_{44} lattice constant, and by NAGAO and KANEDA [7.46], for longitudinal waves along [100], involving C_{11}. The latter authors determined that the attenuation increases as the square of the frequency and diverges as $(T - T_c)^{-1.7}$ above T_c.

Ultrasonic attenuation near critical points is a complicated property, and may arise from a variety of processes. In the case of the liquid-gas transition, to whose universality class $RbAg_4I_5$ belongs, the low frequency ultrasonic attenuation

α_s follows the law [7.47]

$$\alpha_s \sim \omega^2 (T - T_c)^{(\alpha/2)-2} \quad , \tag{7.25}$$

where α is the specific heat exponent. Both the frequency-squared behavior and the predicted exponent of 1.9 (see Table 1.7) agree within experimental uncertainty with the results of NAGAO and KANEDA [7.46]. This indicates that the sound wave in $RbAg_4I_5$ relaxes by coupling to thermal modes, rather than to other sound modes or viscous modes. A similar result has been obtained recently by LEUNG and HUBER [7.48].

The velocity of longitudinal sound waves in $HgAg_2I_4$ has been measured in the vicinity of the 324 K transition by BENGUIGUI and WEIL [7.49]. These waves show an anomalous decrease in polycrystalline samples, while the transverse waves do not. The effect is quite large; a decrease of more than 25% is observed. The authors obtain a result for the bulk modulus in a Landau model which is analogous to (7.24). A comparison of the prediction of this model with the data is shown in Fig.7.9 [7.49].

Finally, we note that an attempt to observe the softening of C_{44} directly by means of neutron diffraction was unsuccessful. This work [7.50], which is reported in more detail in Chap.3, found no shift in the energy of [100] phonons with C_{44} as the relevant elastic constant down to an energy of 0.27 meV. This may indicate either that the softening is too small to be resolved by neutron scattering, or that it only occurs at lower values of the wavevector and energy. This experiment did give strong evidence for a first-order transition in $RbAg_4I_5$ from the jump in the intensity of forbidden (600) and (700) reflections at T_c [7.50], as seen in Fig.7.10.

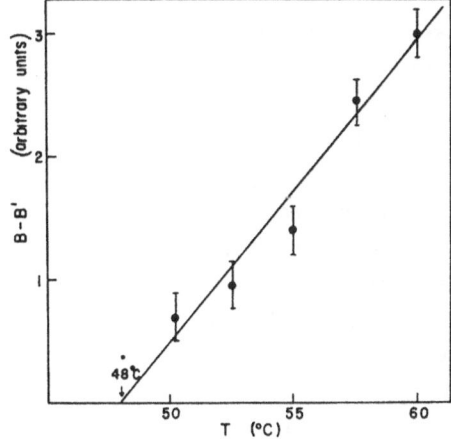

Fig.7.9. Change in bulk modulus for Ag_2HgI_4 [7.49]

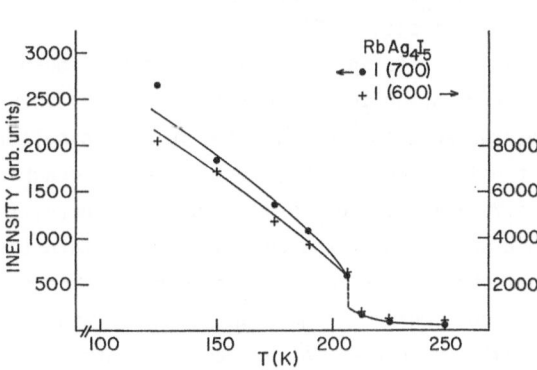

Fig.7.10. Neutron diffraction intensity of forbidden reflections

7.3.4 Other Properties

The full range of experimental techniques which have been applied to critical point
studies have not yet been brought to bear on the conducting-insulator transitions in
ionic salts. Optical methods are reviewed in Chap.5, but these have not, in general,
added to the understanding of the critical point.

In the case of $RbAg_4I_5$, however, optical data have proven crucial in the interpretation of the critical behavior [7.7]. Due to the structural aspects of this
transition, the β phase is rhombohedral and becomes birefringent, while the cubic
α phase is not. The onset of the spontaneous birefringence has been studied [7.7]
and the index of refraction is found to increase as $(T_c - T)^{0.35}$ below T_c. This behavior provides the crucial clue that the structural model presented in Sect.7.2
is valid, since it implies that the index of refraction, and hence the strain, increases linearly with the order parameter (cf. the value of β in Table 7.1). In
cases in which the coupling constant is derived from ion-ion repulsion alone, the
strain would depend on the square of the order parameter, in analogy with exchange
striction. In the present case, the analogy is with a cooperative Jahn-Teller model
[7.51] in which the interaction with phonon modes provides the major coupling between ions [7.24]. In this case, the critical point study is able to isolate the
dominant ion-ion interaction mechanism.

Even in the α phase, there is an effect on the transmission of polarized light
through $RbAg_4I_5$. This is due to fluctuations in the isotropy of the dielectric
tensor [7.52] which tend to depolarize the light. The degree of depolarization
can be shown to be related to the response function of the order parameter, and
thus to the exponent γ (see Table 7.1) through

$$1 - 2I/I_0 = \exp(ct^{-\gamma}) \quad , \tag{7.26}$$

where $2I/I_0$ is the fractional intensity transmitted, c is a constant and $t = (T/T_c-1)$.
For $RbAg_eI_5$, this gives a value of $\gamma = 1.25 \pm 0.1$, in agreement with Ising-model
values, as seen in Fig.7.11.

NMR results in ionic conductors have been reviewed in Chap.6, but several aspects relating to critical behavior are discussed here. BOYCE and HUBERMAN [7.35]
observe an anomalous increase in the spin-lattice relaxation rate of CuI and CuBr,
but not for CuCl [7.53]. This increase occurs in the γ phase as is shown in
Fig.7.12 and is due to quadrupolar effects. This has been interpreted as being due
to disordering of the Cu sublattice within the γ phase. It is interesting to note
that this occurs for CuI and CuBr which have highly conducting α phases, but not
for CuCl, which has no such phase. In contrast with these results, the spin-lattice
relaxation in PbF_2 follows the BPP form through its gradual transition, but with
a temperature-dependent relaxation rate which mirrors the behavior of the ionic
conductivity [7.54].

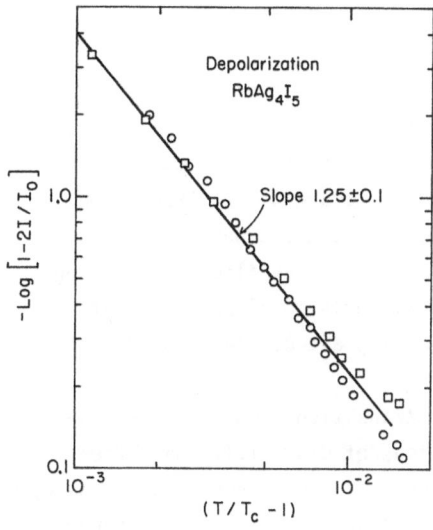

Fig.7.11. Depolarized intensity plotted to extract the exponent γ

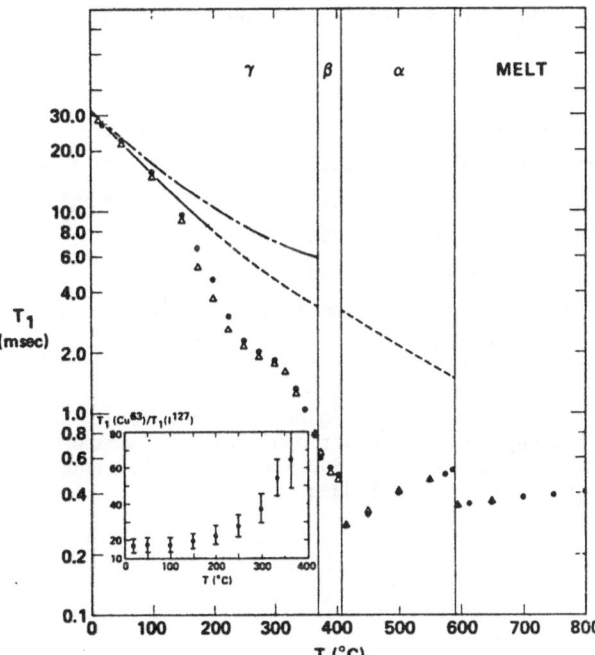

Fig.7.12. NMR spin-lattice relaxation time T_1 in CuI [7.35]

Other transport measurements, such as the Hall effect and thermopower, have not been carried out through the phase transitions in these ionic conductors. Since the various models presented for the superionic phases differ greatly in the details of the critical behavior, it is clear that studies of critical phenomena should be extended to more properties and in different materials.

7.4 Conclusions

The fact that the superionic phase is bounded by second-order or diffuse transitions is, in itself, strong evidence for the importance of collective effects in superionic conductivity. We have shown how a careful study of the critical points, where present, can be used as a tool in determining the nature of the transition, and thereby, the characteristics of the order parameter and the types of important ion-ion interactions. Thus, the great body of information now available concerning second-order transitions, and the powerful theoretical tools developed to treat them, can be employed in the analysis of this interesting and complex class of materials.

We have attempted here to unify the second-order transitions, which are a subset of O'Keefe's Class IIa, and the diffuse transitions of Class III. The latter were shown to be Landau transitions which are smeared by the presence of site-energy or site-degeneracy differences which remove the equivalence of sites in the superionic phase. The similarity of these contributions to an applied field in the Landau model should be useful in understanding the changes which accompany Class III transitions.

Dynamical aspects of these transitions remain largely unexplored. We have demonstrated the relationship between the ionic conductivity and the ordering entropy for a second-order transition, but this needs further study in the vicinity of Class III transitions. The interesting excess relaxation effects observed through NMR in CuI are also suggestive of critical dynamics associated with these transitions. The new theoretical methods for dynamical behavior discussed in Chaps.4 and 8 of this volume may provide the elements required for a treatment of critical dynamics at these superionic transitions.

Acknowledgements. The author has had useful discussions regarding this material with C.P. Flynn, R. Weil, and M. Delaney. The support of the National Science Foundation through the Illinois Materials Research Laboratory Grant No. NSF DMR77-23999 and through Grant No. NSF DMR76-10379 is gratefully acknowledged.

References

7.1 M.J. Rice, W.L. Roth: J. Solid State Chem. *4*, 294 (1972)
7.2 M. O'Keefe, B.G. Hyde: Phil. Mag. *33*, 219 (1976)
     M. O'Keefe: In *Superionic Conductors*, ed. by G.D. Mahan, W.L. Roth (eds.) (Plenum Press, New York 1976)
7.3 P.G. deGennes: *The Physics of Liquid Crystals* (Clarendon Press, Oxford 1974)
7.4 W.J. Pardee, G.D. Mahan: J. Solid State Chem. *15*, 310 (1975)
7.5 K.W. Browall, J.S. Kaspert: J. Solid State Chem. *15*, 54 (1975)
7.6 J. Hibma: Phys. Rev. B *15*, 5797 (1977)

7.7 R. Vargas, M.B. Salamon, C.P. Flynn: Phys. Rev. Lett. *37*, 1550 (1976); Phys. Rev. B *17*, 269 (1978)
7.8 L.D. Landau, E. Lifschitz: *Statistical Physics*, 2nd ed. (Pergamon Press, New York 1968)
7.9 G.Ya. Lyubarskii: *The Application of Group Theory in Physics* (Pergamon Press, Oxford 1960)
     N. Boccara: Ann. Phys. (NY) *47*, 40 (1968)
7.10 S. Geller: Phys. Rev. B *14*, 4345 (1976)
7.11 S.-K. Ma: *Modern Theory of Critical Phenomena* (Benjamin, Reading, Mass. 1976)
7.12 P. Pfeuty, G. Toulouse: *Introduction to the Renormalization Group and to Critical Phenomena* (Wiley and Sons, London 1977)
7.13 J.C. Slater: *Introduction to Chemical Physics* (Mc-Graw-Hill, New York 1939)
7.14 H.E. Stanley: *Phase Transitions and Critical Phenomena* (Oxford University Press, Oxford 1971)
7.15 K. Wilson, J. Kogut: Phys. Rep. *12C*, 75 (1974)
7.16 D. Mukamel, S. Krinsky: Phys. Rev. B *13*, 5078 (1976)
7.17 H. Wiedersich, W.V. Johnston: J. Phys. Chem. Solids *30*, 475 (1969)
7.18 M.J. Rice, S. Strässler, G.A. Toombs: Phys. Rev. Lett. *32*, 596 (1974)
7.19 B.A. Huberman: Phys. Rev. Lett. *32*, 1000 (1974)
7.20 D.O. Welch, G.J. Dienes: J. Phys. Chem. Solids *38*, 311 (1977)
7.21 L. Lam, A. Bunde: Z. Phys. B, *30*, 65 (1978)
7.22 S. Strässler, C. Kittel: Phys. Rev. *139*, A758 (1965)
7.23 S.M. Girvin, G.D. Mahan: Solid State Commun. *23*, 629 (1977)
7.24 M.B. Salamon: Phys. Rev. B *15*, 2236 (1977)
7.25 S. Geller: Science *157*, 310 (1967)
7.26 W.V. Johnston, H. Wiedersich, G.W. Linderg: J. Chem. Phys. *51*, 3739 (1969)
7.27 W.J. Pardee, G.D. Mahan: J. Chem. Phys. *61*, 2173 (1974)
7.28 F.L. Lederman, M.B. Salamon, H. Peisl: Solid State Commun. *19*, 147 (1976)
7.29 A.S. Dworkin, M.A. Bredig: J. Phys. Chem. *72*, 1277 (1968)
7.30 C.E. Derrington, A. Navrotsky, M. O'Keefe: Solid State Commun. *18*, 47 (1976)
7.31 W. van Gool, ed.: *Fast Ion Transport in Solids* (North Holland, Amsterdam 1973)
7.32 S. Miyake, S. Hoshino, T. Takenaka: J. Phys. Soc. Jpn. *7*, 19 (1952(
7.33 V. Meisalo, M. Kalliomäke: High Temp.-High Pressures *5*, 663 (1975)
7.34 J.B. Wagner, C.J. Wagner: J. Chem. Phys. *26*, 1597 (1957)
7.35 J.B. Boyce, B.A. Huberman: Solid State Commun. *21*, 31 (1977)
7.36 G. Burns, F.H. Dacol, M.W. Shafer: Solid State Commun. *24*, 753 (1977)
7.37 H. Sato, R. Kikuchi: J. Phys. Chem. *55*, 677 (1971)
7.38 W.L. Roth, F. Reidinger, S. La Placa: In *Superionic Conductors*, ed. by G.D. Mahan, W.L. Roth (Plenum Press, New York 1976) p.223
7.39 G.D. Mahan: Phys. Rev. B *14*, 780 (1976)
7.40 C.P. Flynn: *Point Defects and Diffusion* (Oxford University Press, Oxford 1972)
7.41 V.S. Borovkov, A.K. Ivanov-Shitz: Electrochim. Acta *22*, 713 (1977)
7.42 A. Aharony, K.A. Müller, W. Berlinger: Phys. Rev. Lett. *38*, 33 (1977)
7.43 P.C. Allen, D. Lazarus: Phys. Rev. B *17*, 1913 (1978)
7.44 J. Hibma: Private communication
7.45 L.J. Graham, R. Chang: J. Appl. Phys. *46*, 2433 (1975)
7.46 M. Nagao, T. Kaneda: Phys. Rev. B *11*, 2711 (1975)
7.47 D. Sette: *Proceedings of the International School of Physics, Course 51*, ed. by M.S. Green (Academic Press, New York 1971) p.508
7.48 K.M. Leung, D.L. Hubert: Phys. Rev. Lett. *42*, 452 (1979)
7.49 L. Benguigui, R. Weil: Phys. Rev. B *16*, 2569 (1977)
7.50 S.M. Shapiro, D. Semminsen, M.B. Salamon: In *Proceedings of the International Conference on Lattice Dynamics*, ed. by M. Balkansik (Flamarion, Paris 1978) p.538
7.51 G.A. Gehring, K.A. Gehring: Rep. Prog. Phys. *38*, 1 (1975)
7.52 M.B. Salamon: Bull. Am. Phys. Soc. *23*, 241 (1978)
     M.B. Salamon, C.-C. Huang: Phys. Rev. B
7.53 J.B. Boyce: Private communication
7.54 J.B. Boyce, J.C. Mikkelson, M.O' Keefe: Solid State Commun. *21*, 955 (1977)

8. Continuous Stochastic Models

T. Geisel

With 22 Figures

natura non facit saltus
[nature makes no jumps]
Comenius, Linné, Leibniz, Goethe, and others

This chapter concentrates on continuous models for the dynamics of superionic conductors as opposed to jump models and lattice gas models, which are considered in
Chap.4. By continuous models we mean that the diffusion of an ion is not represented
by instantaneous jumps from an equilibrium site to another one, but by a continuous
motion in between.

Superionic conductors are solids exhibiting a large mobility for the ions of
one sublattice (mobile ions), e.g., in α-AgI at $300°C$ the diffusion coefficient
for Ag^+ ions is $\sim 2 \times 10^{-5} cm^2/s$ and the conductivity is $\sim 2\Omega^{-1}cm^{-1}$. The diffusion coefficient of the I^- ions is negligible. In the context of the present chapter super-
ionic conductors may be generally characterized as follows: There is a cage of atoms
or ions which are bound to equilibrium sites (e.g., I^- ions) and there are mobile
ions (e.g., Ag^+ ions) which move in the (periodic) potential provided by the cage
(Fig.8.1a). There are several available potential minima per ion.

Fig.8.1a.
Ionic motion in a flat potential with potential barriers V_0

The large mobility of the mobile ions implies that they change position frequently and that the potential barriers V_0 separating these positions are relatively
small (~ 750 K in α-AgI [8.1]). Thus the potential seen by the ions is strongly
anharmonic and the ionic motion is rather complicated, involving oscillatory
features in the valleys of the potential correlated with motion over the potential
barriers. The continuous models attempt to treat both types of motion uniformly.
Because of the flatness of the potential the diffusive motion is not likely to be
performed by instantaneous jumps from one site to another, except at low temperature.

This qualitative conclusion was based on the large mobility of the ions. One may
ask whether there are more direct manifestations of deviations from the jump behavior, which underline the necessity of introducing the more complicated continuous
models. We quote some examples for α-AgI, part of which will be discussed in more
detail later.

a) EXAFS (see Chap.2): A measure for the continuous behavior is the ratio
τ_f/τ_r, where τ_f is a mean flight time and τ_r a mean residence time. EXAFS results
of BOYCE et al. [8.2] indicate that $\tau_f \sim 1/3 \ \tau_r$ at 198°C. As $\tau_f < \tau_r$, the authors
conclude that diffusion in α-AgI is more closely related to jump diffusion than to
free diffusion (i.e., without potential). However, since both times have the same
order of magnitude, rigorously neither of these limiting cases but rather an intermediate case is realistic. Pure jump diffusion would be encountered only if the
flight time were negligible. It was also concluded from quasi-elastic neutron scattering by ECKOLD et al. [8.3] that the flight time is comparable with the residence
time.

b) Neutron diffraction (see Chap.3): CAVA et al. [8.4] have derived the distribution of the Ag^+ scattering density. Their results indicate [Ref.8.4, Fig.2]
that at 160°C the probability of finding a mobile ion at an equilibrium site is
only 2.5 times larger than the probability of finding it at the potential barrier.
This demonstrates the importance of the motion in between the sites, which is skipped
in jump diffusion models.

c) Molecular dynamics (see Sect.8.3): The conclusion made in b) is supported
by molecular-dynamics calculations of VASHISHTA and RAHMAN [Ref.8.5, Fig.2] at
157°C. The probability of finding a mobile ion at an equilibrium site turns out to
be 2.7 times larger than the probability of finding it at the potential barrier.
Furthermore the authors found a "jump" frequency of $0.34 \times 10^{12} \ s^{-1}$. This is very
close to the frequency $0.5 \times 10^{12} \ s^{-1}$ of the low-lying optic phonon in β-AgI [8.6,7]
and means that a mobile ion moves to a neighboring site almost once in a period of
low-frequency phonons. We expect strong interference between diffusive and oscillatory motions.

d) Others: The continuous character of motion shows up less directly in some other
ways, e.g., in the ionic mean square displacements, in the lineshape of light scattering spectra (Sect.8.2.10), which deviates from the lineshape for jump diffusion,
etc.

The relatively free motion of the mobile ions makes the notion of phonons dubious.
Only for very short times is a mobile ion expected to participate in a collective
vibration with the cage lattice. Conventional perturbative approaches starting from
unperturbed harmonic phonons seem to be unrealistic because the perturbation by the
diffusive motion would be too strong and too fast, and would continuously change.
It was suggested that the mobile particles might behave like the particles of a
liquid and that one sublattice might be considered molten. This statement characterizes the situation; it should be understood in the sense that the mobile ions
are not bound to lattice sites but that they are free to move anywhere according to
their microscopic interactions. As in a liquid this motion comprises diffusion and
oscillation.

The ionic motion in superionic conductors is studied and visualized in a movie of the author, showing a mechanical molecular dynamics simulation. The ions and their interactions are represented by magnets of different color floating on an air table. A long-time photograph (time integration) displays the spatial probability distribution of the ions. In the black-and-white reproduction of Fig.8.1b, the cage and mobile ions appear black and gray, respectively. While the cage ions are centered around hexagonal lattice positions, the gray trace of the mobile ions is well interconnected. However, preferential positions of higher density in the centers of the triangles and positions of lower density can be distinguished. These correspond to potential minima and potential barriers, respectively, of the potential provided by the cage ions (Fig.8.1a). Although a snapshot would show that only half of the triangles are occupied at any instant, on the average all triangles are equally occupied. Besides, the movie shows different types of correlations, which cannot be reproduced in a single figure.

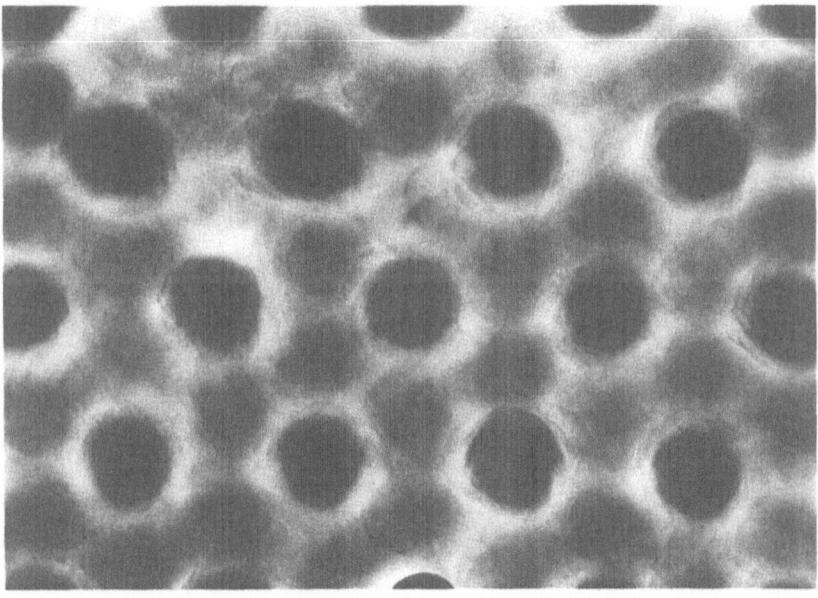

<u>Fig.8.1b.</u> Density distribution of cage and mobile ions obtained by a mechanical simulation. The photograph illustrates the notion of a molten sublattice

The methods presented in this chapter are steps towards a suitable dynamical description of superionic conductors. We shall discuss the justification of the models not only by comparison with specific experimental results but also by microscopic considerations, i.e., with respect to a general Hamiltonian. We attempt to avoid artifical restrictions for the motion of the mobile ions. Among other questions, we would like to learn which factors influence the dynamical properties, in particular the diffusion coefficient and the conductivity.

Most of the continuous models used up to now are stochastic models. The physical differences to other models are discussed microscopically in Sect.8.1. The correlation functions which are measured in scattering and absorption experiments are introduced. In Sect.8.2 a stochastic model is defined, and the correlation functions are calculated and compared with experiment. Extensions of the model and related models are also considered. Molecular-dynamics calculations are reviewed in Sect.8.3. This chapter deals primarily with effective one-particle models. Continuous theories covering the case of interacting mobile ions are still at an initial stage. First results obtained in this case are described in Sect.8.4.

8.1 Models for Superionic Conductors

8.1.1 The Hamiltonian

The models which are used to describe superionic conductors differ in whether certain interactions are neglected or considered. The lattice gas models stress the interactions among the mobile ions in the potential minima and usually neglect vibrations around the minima. Other models stress the liquidlike motion of the mobile ions and neglect interactions among these ions. In order to compare these models we start from a rather general Hamiltonian.

We only use the most general information we have about superionic conductors, that there are particles which form a lattice (species B) and others which do not (species A). The mobile particles A are allowed to take any position, whereas the cage particles B are restricted to small displacements from lattice sites. The possibility that species A may also form a lattice is of course included. The particles are treated as rigid atoms or rigid ions. Electronic degrees of freedom have not so far been considered in the dynamical models for diffusion. It was shown, however, that the electronic polarizability [8.1,8] and the quadrupolar deformability of the ions [8.9,10] may lower the potential barriers for diffusion.

The Hamiltonian is then one of interacting particles where only the position of species B can be described by displacements from reference positions $\underline{r}_\lambda^0$. For the sake of simplicity we write it down for the case of only two different kinds of atoms.

$$H = \sum_\ell \frac{1}{2m_A} \underline{p}_\ell^2 + \sum_\lambda \frac{1}{2m_B} \underline{p}_\lambda^2 + \phi(\underline{r}_\ell, \underline{r}_\lambda) \tag{8.1}$$

The index ℓ numbers the mobile particles (species A); the index λ numbers the cage particles (species B). The potential ϕ contains interactions among all particles. ϕ is now expanded with respect to displacements u_λ^α of the cage particles (α labels the cartesian components). If we assume centrosymmetric two-particle interactions we obtain, to second order in u_λ^α,

$$H = H_A + H_B + H' \quad , \tag{8.2}$$

$$H_A = \sum_\ell \frac{1}{2m_A} \, \underline{p}_\ell^2 + \sum_{\ell,\lambda} V^0(|\underline{r}_\ell - \underline{r}^0|) \tag{8.3}$$

$$- \underline{r}_\lambda^0|) + \frac{1}{2} \sum_{\ell \neq \ell'} V^1(|\underline{r}_\ell - \underline{r}_{\ell'}|) \quad ,$$

$$H_B = \sum_\lambda \frac{1}{2m_B} \, \underline{p}_\lambda^2 + \frac{1}{2} \sum_{\substack{\lambda\lambda' \\ \alpha\alpha'}} C_{\lambda\lambda'}^{\alpha\alpha'} u_\lambda^\alpha u_\lambda^{\alpha'} \quad , \tag{8.4}$$

$$H' = \sum_{\ell\lambda\alpha} V_\alpha^2(\underline{r}_\ell - \underline{r}_\lambda^0) u_\lambda^\alpha + \sum_{\substack{\ell\lambda \\ \alpha\alpha'}} V_{\alpha\alpha'}^3(\underline{r}_\ell - \underline{r}_\lambda^0) u_\lambda^\alpha u_\lambda^{\alpha'} \quad . \tag{8.5}$$

The coupling tensors can be easily expressed by derivatives of the two-particle potentials. V^0, V_α^2, and $V_{\alpha\alpha'}^3$ arise from mobile-particle-cage-particle interactions, V^1 is the interaction among the mobile particles and the force constant matrix $C_{\lambda\lambda'}^{\alpha\alpha'}$ arises from interactions among the cage particles.

This Hamiltonian is similar to that of an electron-phonon system. The interactions are illustrated in Fig.8.2. In the Hamiltonian H_A of the mobile ions, the contribution in V^0 may be regarded as an external potential (cage potential), which is periodic in the case of a crystalline cage. The cage Hamiltonian H_B does not represent the phonons of the entire solid but only the somewhat fictitious phonons of the cage particles. Anharmonic terms may be important, too. H' couples the mobile particles with the vibrations of the cage. V_α^2 and $V_{\alpha\alpha'}^3$ depend on the difference vector $\underline{r}_\ell - \underline{r}_\lambda^0$ whereas V^0 and V^1 only depend on its absolute value.

If we include the obvious anharmonic terms in u_λ^α, the Hamiltonian (8.2) is completely general in the framework of rigid ions and contains all relevant interactions. Different models for superionic conductors can now be compared on the basis of this Hamiltonian.

8.1.2 Comparison of the Models from Microscopic Considerations

Besides the continuous stochastic models we briefly discuss some other models which have been proposed for superionic conductors.

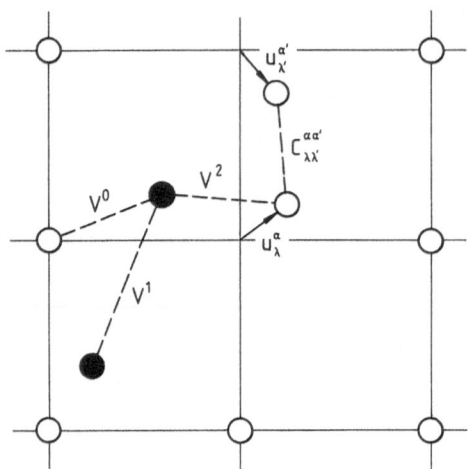

Fig.8.2. Illustration of the interactions in (8.3-5). The mobile particles (black) are allowed to take any position. The cage particles (white) are described by displacements u_λ^α from reference sites

a) Free Ion Model

One of the first attempts to treat the relatively free motion of the mobile ions was made by RICE and ROTH [8.11,12]. The motion in the cage potential V^0 is simulated by bound- and free-ion-like states which are separated by an energy gap. For a certain lifetime the ions can be thermally excited and propagate with the velocity of free ions. The model leads to a Drude behavior for the frequency-dependent ionic conductivity $\sigma(\omega)$. A phenomenological model of ECKOLD et al. [8.3, 13] for the interpretation of quasi-elastic neutron scattering in α-AgI is based on similar ideas.

b) Hopping and Lattice Gas Models [8.14-22]

The continuous motion of the mobile ions is replaced by discontinuous motion from site to site. In lattice gas models the potential V^0 and the interactions V^1 and V^2 are taken into account at these sites only, or they are meant as an average for the region around the sites. The potential barriers controlling diffusion are no longer incorporated in V^0. In order to get ionic transport a somewhat artificial transfer Hamiltonian is often introduced, which produces jumps from one site to another. The important question *how* transport occurs is thus excluded and part of the dynamics is anticipated by this ansatz. The problem is discussed in more detail in Chap.4.

Neutron and X-ray diffraction results for α-AgI and β-AgI [8.23-27] indicate that the mean square displacements of the mobile ions are larger than those of the cage ions. Lattice gas models, however, sometimes consider displacements of the cage ions but neglect such displacements for the mobile ions. One could do one more step by expanding V^0 with respect to small displacements of the mobile ions. Such a treatment would include vibrations of the mobile ions, but neglect that the diffusive motion of an ion is strongly correlated with its oscillatory motion. When the coupling V^2 between mobile ions and cage displacements is considered, it is often handled in terms of the small polaron concept [8.16-19]. Physically this allows for an accommodation of the cage to the site occupancy by mobile ions. Obviously the coupling is neglected for displaced positions of the mobile ions where it is expected, however, to be more important than at the sites. (For comparison: in continuous stochastic models the interaction V^2 is meant to produce the random forces, which are assumed not to be correlated with the position of the mobile ion.)

An advantage of the lattice gas models is that part of the interactions V^1 among the mobile particles (lattice gas interactions) can be treated more easily and accurately than in the continuous case. Interactions at displaced positions are again neglected. Therefore the lattice gas models have been used to describe the disordering at the phase transitions [8.28-33]. Recently LAM and BUNDE [8.34] have classified different lattice gas theories in relation to a general Hamiltonian which may be obtained by specializing (8.2-5) to the lattice gas case. They argue that also the models of HUBERMAN [8.35] and RICE et al. [8.36] may be formally obtained from this Hamiltonian in a limiting case.

The purpose of this discussion was to show the limitations of lattice gas models. The interactions of the mobile ions (V^0, V^1, V^2) are considered only at distinct sites and not at displaced positions. Lattice gas models may be useful to investigate static properties (e.g., static disorder). However, when we are interested in transport and dynamic properties we are concerned with motion over potential barriers separating the sites. This motion is generally expected to depend strongly on the interactions at the potential barriers, i.e., in between the sites. When we deal with dynamic properties we should therefore try to consider the interactions also at displaced positions, i.e., respect that the cage potential V^0, the ion-ion interaction V^1 and the interaction between mobile ions and cage vibrations V^2 are continuous functions of the ionic position.

c) Continuous Stochastic Models

They were introduced initially as phenomenological models [8.37-39]. Nevertheless it is possible to derive stochastic models microscopically from the Hamiltonian (8.2-5). This will be discussed in Sect.8.2.11. Here we quote only the underlying ideas. The mobile ions are assumed to move continuously in the cage potential V^0. This potential provides systematic forces, i.e., forces which depend on the positions of the mobile ions only (Fig.8.1a). The potential is strongly anharmonic. The restoring forces in the valleys of the potential lead to oscillatory types of motion, the potential barriers control the diffusive motion. The vibrations of the cage (H_B) are treated as a bath. Through the coupling H' they provide random forces and friction for the motion of the mobile ions. This is of course a simplification, dynamical coupling between mobile ions and cage degrees of freedom has been taken into account only roughly. Interactions among the mobile ions (V^1) are not treated explicitly, but are also considered to be incorporated in the bath. (Recent work overcoming this restriction is described in Sect.8.4). The aim of these models is to treat the oscillatory and diffusive motions simultaneously. As the problem is quite complex one can hope to find solutions for the simplest models only, e.g., by assuming white-noise random forces.

d) Continuum Models for the Hydrodynamic Region

In the long-wavelength and low-frequency region the mobile ions are described collectively by their density $n(\underline{r},t)$ and current $\underline{j}(\underline{r},t)$. They are coupled with the strain field of the crystalline cage. The theories [8.40-42] aim at a calculation of coupled liquid-cage fluctuations which are responsible for Brillouin scattering (see Sect.8.2.10). A rigorous microscopic derivation based on the Hamiltonian (8.2-5) is possible. Of course the continuum theories are valid for long wavelength fluctuations which do not involve individual motion of particular ions, such that the individual treatment can be abandoned in favour of a collective treatment.

8.1.3 Correlation Functions

In this section we specify what is meant by dynamical properties. We are interested in particular in the diffusion coefficient D, the dc and ac ionic conductivity $\sigma(0)$ and $\sigma(\omega)$ and the dynamic structure factor $S(q,\omega)$ of the mobile ions.

For noninteracting particles the diffusion coefficient is given by the Einstein relation [8.43]

$$D = k_B T \mu(0) \quad , \tag{8.6}$$

where $\mu(0)$ is the mobility at zero frequency. The conductivity is related to the dynamic mobility by

$$\sigma(\omega) = n(Ze)^2 \mu(\omega) \quad , \tag{8.7}$$

where n is the particle density and Ze the charge of the particle. Depending on the frequency range, $\sigma(\omega)$ is measured by infrared and microwave spectroscopy.

The quantities of interest can be calculated from velocity and density correlation functions. The fluctuation-dissipation theorem yields for the mobility [8.44]

$$\mu(\omega) = \frac{1}{3k_B T} \int_0^\infty dt \, \exp[i\omega t] <\underline{v}(0) \cdot \underline{v}(t)> \quad , \tag{8.8}$$

where $\underline{v}(t)$ is the velocity of the mobile ion at time t and the angular brackets denote a thermal average. The coherent and incoherent dynamic structure factors of the mobile ions are given by [8.45]

$$S_{coh}(\underline{q},\omega) = \frac{1}{2\pi} \int_{-\infty}^{+\infty} dt \, \exp[i\omega t] I_{coh}(\underline{q},t) \tag{8.9}$$

$$S_{inc}(\underline{q},\omega) = \frac{1}{2\pi} \int_{-\infty}^{+\infty} dt \, \exp[i\omega t] I_{inc}(\underline{q},t) \tag{8.10}$$

where the intermediate scattering functions I_{coh} and I_{inc} are defined as

$$I_{coh}(\underline{q},t) = \frac{1}{N} \sum_{\ell,\ell'} \, <\exp[i\underline{q}\cdot\underline{r}_\ell(t)]\exp[-i\underline{q}\cdot\underline{r}_{\ell'}(0)]> \tag{8.11}$$

$$I_{inc}(\underline{q},t) = \frac{1}{N} \sum_{\ell} \, <\exp[i\underline{q}\cdot\underline{r}_\ell(t)]\exp[-i\underline{q}\cdot\underline{r}_\ell(0)]> \quad . \tag{8.12}$$

The dynamic structure factors are obtained from neutron scattering experiments, q is the wave vector transfer of the scattered neutrons. Heterodyne quasi-elastic light scattering experiments can also measure the coherent structure factor. Due to the small wave vector transfer of the light, the frequencies of interest are smaller than 10^6 s^{-1} and light-beating techniques are required. Such an experiment is possible in principle but has not yet been reported for superionic conductors. It was shown recently [8.46a] that similar structure factors also enter the Raman scattering cross section (at higher frequencies) by a different scattering mechanism, however. In this case the wave vector q cannot be varied by the scattering geometry, but is imposed by the lattice periodicity (see Sect.8.2.10).

8.2 Continuous Models

As discussed earlier the aim of the continuous models for superionic conductors is to obtain a unified description of oscillatory and diffusive motions. From the theory of liquids numerous models are known which try to incorporate both types of motions, e.g., oscillatory diffusion models [8.47-49], oscillatory jump diffusion models [8.50], the itinerant oscillator model [8.51-55], etc. Similar attempts were made in models for the diffusion of hydrogen in metals [8.45]. These models usually consist in an assumption combining oscillatory and diffusive motion, which determines the form of the result for the dynamical structure factor. Thus a dynamical description in terms of oscillation and diffusion does not come out of the theory but is already adopted in the formulation of the model.

Our present interest, however, is to learn *how* oscillations and diffusion arise. We should therefore start from a microscopic model and try to avoid restrictive assumptions. The solutions should be as accurate as possible in order that the results be representative of the model, not of the method of solution. This is why simple one-dimensional one-particle models are studied—the goal is not to reproduce experimental results for a specific superionic conductor but to investigate some fundamental problems.

8.2.1 Langevin Equation

Continuing the discussion in Sect.8.1.2 we start from the Langevin equation for one-dimensional motion:

$$m\ddot{x} + m\gamma\dot{x} - K(x) = f(t) \quad , \tag{8.13}$$

where m is the mass of the mobile ion, x is its position and $K(x) = -\partial V^0/\partial x$ is the systematic force provided by interactions with the *rigid* cage. The friction constant γ and the random force $f(t)$ are due to interactions with *lattice vibrations* of the cage and with other mobile ions. The behavior of the random force is fixed by the condition [8.56]

$$<f(t)f(0)> = 2m\gamma k_B T\delta(t) \quad , \tag{8.14}$$

together with the assumption that the distribution of $f(t)$ be Gaussian with zero mean. According to (8.14) the random force at time t is uncorrelated with the random force at an earlier time. The Fourier transform of (8.14) gives a white noise spectrum. This model was first discussed in connection with superionic conductors by FULDE et al. [8.39].

In Sect.8.2.11 we discuss how generalized Langevin equations and generalized Fokker-Planck equations may be obtained microscopically from the Hamiltonian (8.2-5). Major simplifications inherent in (8.13,14) are the neglect of correlations among the mobile ions and of memory effects in the random forces. This restriction may be partially removed by introducing a set of Langevin equations for those variables (including cage variables) which are dynamically coupled and treating only the other variables as a bath. Such an approach, which treats part of the interaction with cage vibrations as a systematic and not as a random interaction, is described in Sect.8.2.6.

In the case of a crystalline cage the cage potential V^0 is a periodic function. For simplicity we often use a cosine potential

$$V^0(x) = -\frac{V_0}{2} \cos q_0 x \quad . \tag{8.15}$$

Here V_0 is the potential barrier height and $q_0 = 2\pi/a$ is the wave vector defining the reciprocal lattice (see Fig.8.1a). Other than sinusoidal potentials are investigated in Sect.8.2.4. FLYGARE and HUGGINS [8.1] and AJAYI et al. [8.8] have calculated the potential energy of ions along preferred paths in the α-AgI structure and in the rutile structure. An example is shown in Fig.8.3 for different mobile cation radii. The sinusoidal ansatz (8.15) reproduces such potential forms rather closely.

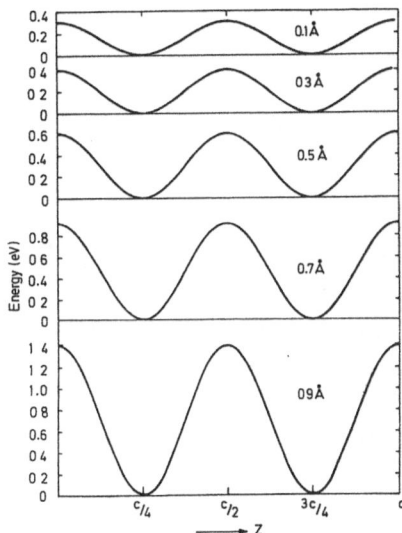

Fig.8.3. Potential energy vs distance along preferred paths in the rutile structure for different mobile cation radii [8.8]

8.2.2 Fokker-Planck Equation and Liouvillian

Our goal is to calculate correlation functions (Sect.8.1.3) in the model defined in the preceding section. In many cases these correlation functions cannot be obtained directly from the Langevin equation. It is then useful to work with the transition probability density $w(x',v',t|x,v)$ in the phase space (x,v) of the ion. This function determines the probability that an ion initially prepared at position x and velocity v be found at x' and v' after a time t. The correlation function for the mobility in (8.8) is written in terms of the transition probability density [8.57]

$$\langle v(0)v(t)\rangle = \int dx'dv'dxdvp(x,v)vw(x',v't|x,v)v' \quad , \tag{8.16}$$

where $p(x,v)$ is the stationary probability density. The correlation functions for the structure factors (8.11,12) are expressed analogously.

The transition probability density obeys partial differential equations, the coefficients (moments) of which are determined by the Langevin equation (8.13,14).

These partial differential equations are the Fokker-Planck equation [8.58-60]

$$\frac{\partial}{\partial t} w(x',v',t|x,v) = \hat{F}w(x',v',t|x,v) \quad , \tag{8.17}$$

with the Fokker-Planck operator

$$\hat{F} = - v' \frac{\partial}{\partial x'} - \frac{K(x')}{m} \frac{\partial}{\partial v'} + \gamma\left(1 + v' \frac{\partial}{\partial v'} + \frac{k_B T}{m} \frac{\partial^2}{\partial v'^2}\right) \quad ; \tag{8.18}$$

and the backward or Kolmogorov equation [8.60,61]

$$\frac{\partial}{\partial t} w(x',v',t|x,v) = \hat{L}w(x',v',t|x,v) \quad , \tag{8.19}$$

with the backward operator

$$\hat{L} = v \frac{\partial}{\partial x} + \frac{K(x)}{m} \frac{\partial}{\partial v} + \gamma\left(- v \frac{\partial}{\partial v} + \frac{k_B T}{m} \frac{\partial^2}{\partial v^2}\right) \quad . \tag{8.20}$$

The backward operator $\hat{L}$ acts on the initial variables x and v, whereas the Fokker-Planck operator $\hat{F}$ acts on the final variables x' and v'. The Fokker-Planck equation is therefore also called the forward equation.

The transition probability density w can be calculated as a solution of one of these differential equations with the initial condition

$$w(x',v',t = 0|x,v) = \delta(x - x')\delta(v - v') \quad . \tag{8.21}$$

The stationary solution p(x,v) is the Boltzmann factor

$$p(x,v) \propto \exp\left\{- \frac{1}{k_B T} \left[\frac{m}{2} v^2 + V^0(x)\right]\right\} \quad . \tag{8.22}$$

The formal solution of (8.19) can be written

$$w(x',v',t|x,v) = e^{\hat{L}t} w(x',v',0|x,v) = e^{\hat{L}t} \delta(x - x')\delta(v - v') \quad . \tag{8.23}$$

The backward operator $\hat{L}$ thus describes the time evolution of the system like the Liouvillian in classical mechanics. By analogy we shall also call it the Liouvillian. As a matter of fact the first two terms of this operator in (8.20) are identical with the Liouvillian of classical mechanics. The other two terms are dissipative; they are linear in γ and act on the initial velocity v.

In order to calculate the correlation functions, e.g., (8.16), we need not know the solutions w(x',v',t|x,v) explicitly. Using the formal solution (8.23) in (8.16) and integrating out the final variables x' and v' we obtain

$$<v(0)v(t)> = \int dx \; dv \; p(x,v)v \; e^{\hat{L}t}v \equiv <v \; e^{\hat{L}t}v> \quad . \tag{8.24}$$

The average can now be simply performed over the stationary solution $p(x,v)$ given by (8.22). Explicit knowledge of $w(x',v't|x,v)$ is thus not necessary if we can calculate the quantity $e^{\hat{L}t}$.

For the solutions of the problem, i.e., for the calculation of correlation functions like (8.24), several methods have been used [8.39,62-66]. Two of them are of more general interest and will be described in more detail. The first consists in a continued-fraction expansion for the Laplace transform of (8.24) (Sect.8.2.3). The other solves the differential equation (8.19) by an expansion of the solution in terms of a complete set of basis functions and will be described in connection with the dynamic structure factor (Sect.8.2.9.)

8.2.3 Continued-Fraction Expansion

For the calculation of correlation functions MORI [8.67,68] has introduced a continued-fraction expansion which is based on a projector technique for Hamiltonian systems. Our Liouvillian (8.20), however, contains dissipative terms and is non-Hermitian. A generalization of Mori's method to non-Hamiltonian systems has been worked out by SCHNEIDER [8.69] and was used by FULDE et al. [8.39] to calculate the dynamic mobility. More recently continued fractions were obtained directly from the moments of the Liouvillian [8.63] as described below.

When the operator $e^{\hat{L}t}$ is expressed explicitly (by its Taylor expansion) (8.24) becomes

$$<v(0)v(t)> = \sum_{n=0}^{\infty} \frac{1}{n!} <v\hat{L}^n v> t^n \quad . \tag{8.25}$$

With the Laplace transform of t^n

$$\int_0^{\infty} dt \; e^{-zt}t^n = \frac{n!}{z^{n+1}} \quad , \tag{8.26}$$

we obtain for the mobility (8.8)

$$\mu(\omega) = \frac{1}{k_B T} \sum_{n=0}^{\infty} \frac{c_n}{(-i\omega)^{n+1}} \quad , \tag{8.27}$$

with

$$c_n = <v\hat{L}^n v> \quad . \tag{8.28}$$

The moments c_n are calculated by applying the differential operator $\hat{L}$ n times to the velocity v and averaging with the Boltzmann factor (8.22). Equation (8.27) is

214

a power series in $(1/i\omega)$ for the mobility. By a well-known procedure [8.70] one can obtain a continued-fraction expansion (J-fraction) for the power series of the following form

$$\mu(\omega) = \frac{1}{m} \cfrac{a_0}{-i\omega+b_1 - \cfrac{a_1}{-i\omega+b_2-}} \\ \cfrac{}{\quad\quad\ddots \\ \quad\quad\quad - \cfrac{a_n}{-i\omega+R_{n+1}(\omega)}} \quad . \tag{8.29}$$

The coefficients a_i and b_i of this expansion are determined by the moments c_n and can be calculated by means of an algorithm which is described in [8.70]. This algorithm may also be performed by computer routines. The first few coefficients of the expansion are

$$a_0 = 1 \quad b_1 = \gamma \quad a_1 = \frac{<K'(x)>}{m}$$

where $K'(x)$ is the derivative of the periodic force. More coefficients are given in the appendix of [8.63].[1] The advantage of the continued-fraction expansion as compared with the power series (8.27) is that the divergence at $\omega = 0$ is avoided. The results obtained for $\mu(0)$ and $\mu(\omega)$ will be discussed in detail in the following sections.

8.2.4 Static Mobility, Diffusion Constant, dc Conductivity

Analytic results for the mobility $\mu(\omega = 0)$ are known only in certain limiting cases, i.e., for low temperatures [8.71], small friction [8.72] and large friction [8.73, 74]. With the continued-fraction expansion one can now obtain accurate results in the general case. The steps leading to the expansion (8.29) are performed by computer routines [8.62]. Details for the cosine potential are described in [8.63]; here we sketch the underlying ideas.

In order to calculate the moments c_n (8.28), the Liouvillian $\hat{L}$ (8.20) must operate repeatedly on the velocity v. In the case of a cosine potential $V^0 \propto \cos q_0 x$ the Liouvillian contains factors e^{iq_0x}. It is easy to see that the operation of $\hat{L}$ always leads to expressions of the form

$$\left(e^{iq_0x}\right)^j v^r \equiv M_{j,r} \quad ; \tag{8.30}$$

---

[1]In equation (A.1) of that paper, a_1 must be replaced by the expression given above.

for example

$$v = M_{0,1} \quad , \tag{8.31}$$

and

$$\hat{L}v = \hat{L}M_{0,1} = -\gamma v + \frac{q_0 V_0}{4im} \left(e^{iq_0 x} - e^{-iq_0 x} \right)$$

$$= -\gamma M_{0,1} + \frac{q_0 V_0}{4im} (M_{1,0} - M_{-1,0}) \quad . \tag{8.32}$$

The action of $\hat{L}$ on a general $M_{j,r}$ yields a recursion relation

$$\hat{L}M_{j,r} = -r\gamma M_{j,r} + \frac{\gamma kT}{m} r(r-1)M_{j,r-2}$$

$$+ ijq_0 M_{j,r+1} + \frac{q_0 V_0}{4im} r(M_{j+1,r-1} - M_{j-1,r-1}) \quad . \tag{8.33}$$

Then $\hat{L}^n v$ is also a linear combination in the functions $M_{j,r}$, and the moments $c_n = \langle v\hat{L}^n v \rangle$ are linear combinations of the averages $\langle M_{j,r} \rangle$. These averages can be performed analytically [8.63]. The actual combination of the $\langle M_{j,r} \rangle$ is generated recursively by a computer according to the steps (8.31,32) and the recursion relation (8.33). The moments c_n for the power series (8.27) for the mobility are thus obtained.

To get the continued fraction (8.29) one must determine the coefficients a_i and b_i from the moments c_n. This can also be done by a computer using the algorithm mentioned in the preceding section [8.70]. Continued fractions up to order $n = 30$ were calculated numerically in this way by DIETERICH et al. [8.63]. Convergence tests showed that the remainder $R_{n+1}(\omega)$ could be neglected for not too small friction.

Results for the static mobility $\mu(\omega = 0)$, which is connected with the diffusion constant and the dc conductivity according to (8.6) and (8.7) are shown in Fig.8.4. Here a normalized friction constant is used

$$\Gamma = \gamma \frac{2\pi}{\omega_0}$$

where ω_0 is the harmonic frequency for the potential V^0

$$\omega_0 = q_0 \sqrt{\frac{V_0}{2m}} \quad .$$

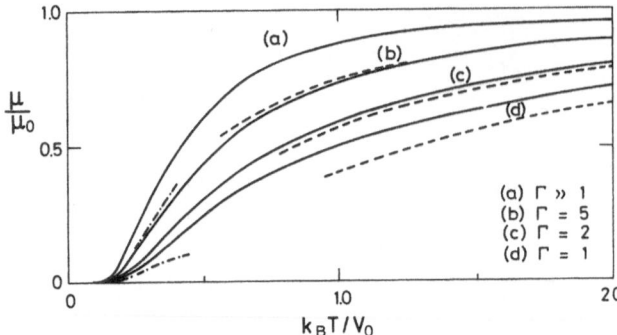

Fig.8.4. Static mobility $\mu(0)$ (dc conductivity) vs temperature in a cosine potential for different values of friction. Solid lines: continued fraction results; dot-dashed lines: escape rate theory; dashed lines: interpolation based on high temperature expansion. [8.63]

The mobility μ is normalized by

$$\mu_0 = \frac{1}{m\gamma} \quad .$$

At high temperatures μ saturates approaching the temperature-independent value μ_0.

It is interesting to compare these curves with low-temperature approximations. The dot-dashed lines are results obtained in the escape rate theory of KRAMERS [8.71], which yields

$$\mu(0) = \frac{2\pi}{q_0^2 k_B T} \left(\sqrt{\omega_0^2 + \gamma^2/4} - \gamma/2 \right) \exp\left(-\frac{V_0}{k_B T} \right) \quad . \tag{8.34}$$

This result implies an Arrhenius behavior for the diffusion constant. For $\Gamma = 1$ and $k_B T/V_0 = 0.3$ the escape rate value for $\mu(0)$ deviates by about 35% from the exact value. The deviation is less for larger friction. The assumption of thermally activated hopping in lattice gas models must be viewed in the light of these deviations. The dashed lines correspond to results of a perturbation expansion in V_0 for high temperatures. Details are described in [8.63].

As Fig.8.3 indicates, the cosine potential seems to be a realistic cage potential. Nevertheless it is instructive to study the influence of other forms of the potential. Again one might use the continued-fraction expansion. However, in the general case the derivatives in the Liouvillian cannot be performed by computer routines. Let us consider instead the large-friction limit, where one can obtain an analytical expression for the mobility [8.74] which is easy to compute for general forms of the potential

$$\mu(0) = \left(m\gamma \frac{1}{a} \int_0^a \exp\left[\frac{V(x)}{k_B T}\right] dx \; \frac{1}{a} \int_0^a \exp\left[\frac{-V(x)}{k_B T}\right] dx \right)^{-1} \quad . \tag{8.35}$$

Here $V(x)$ is a general periodic potential and a is its period. The results in Fig.8.5 were obtained using this expression and an interpolation formula [8.63]. The curves represent different periodic potentials with equal barrier heights V_0 but different curvatures (or harmonic frequencies). The highest static mobility $\mu(0)$ is found for the narrowest potential (a). The lowest mobility is found for an almost rectangular potential (c).

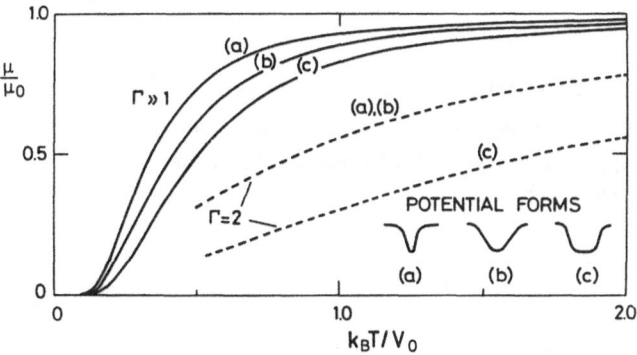

Fig.8.5. Static mobility $\mu(0)$ vs temperature for different periodic potentials having the same barrier heights [8.63]

8.2.5 Dynamic Mobility, ac Conductivity

The ac conductivity, which is proportional to the dynamic mobility $\mu(\omega)$ (8.7) can be calculated by a continued-fraction expansion in the same way as the static mobility [8.63]. Figure 8.6 shows the frequency dependence of the real part of $\mu(\omega)$ for a cosine potential in three representative cases. A comparison with experiment will be made in the next section. The frequency range is typical for far infrared experiments.

As expected at low temperature $k_B T/V_0 = 0.25$ a strong oscillatory peak shows up close to the harmonic frequency ω_0. With increasing temperatures this peak disappears and a peak at $\omega = 0$ emerges, which is due to increasing diffusive motion. At an intermediate temperature $k_B T/V_0 = 0.5$, both peaks are present with a minimum in between. This minimum is found only for small friction $\Gamma \sim 1$. For $\Gamma \gtrsim 2$ the peaks appear and disappear without the occurrence of a minimum at intermediate frequency.

Recently RISKEN and VOLLMER [8.66] used another method to calculate the mobility $\mu(\omega)$ for the same model. They solve the Fokker-Planck equation by a suitable eigen-

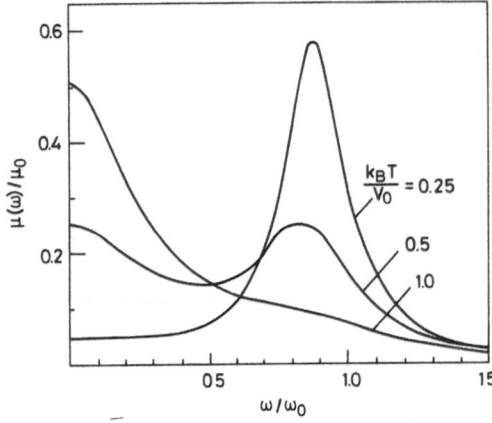

Fig.8.6. Real part of the dynamic mobility $\mu(\omega)$ for a cosine potential at three representative temperatures. The friction is $\Gamma = 1$. [8.63]

function expansion. Their results are similar to those of Fig.8.6. In certain cases the convergence of the procedure can be better than that of the continued-fraction expansion. The method was also applied to the calculation of the dynamic structure factor and will be described in Sect.8.2.9.

8.2.6 Approximate Solutions and Similar Models

One of the purposes of the accurate calculations in the simple model treated till now is to compare them with approximate results and to see under which circumstances the approximations are valid. More complicated situations may then be treated in these approximations. For the static mobility $\mu(0)$ the comparison was already made in Sect.8.2.4.

For an approximate expression for the dynamic mobility $\mu(\omega)$ one can truncate the continued fraction (8.29) after the second order

$$\mu(\omega) = \frac{1}{m} \frac{1}{-i\omega + \gamma - \dfrac{\langle K'(x) \rangle / m}{-i\omega + R_2}} \quad . \tag{8.36}$$

In this low order the remainder R_2 must not be neglected. It is often approximated by a value independent of ω which is determined from the static value $\mu(0)$. Equation (8.36) is then a two-pole formula, which reproduces oscillatory peaks at $\omega \sim \pm\omega_0$ at low temperatures and a diffusion peak at $\omega = 0$ at high temperatures much as in Fig.8.6. However, it cannot reproduce the simultaneous occurrence of both peaks as at $k_B T/V_0 = 0.5$ in Fig.8.6. Apart from this case, which shows up at low friction only, FULDE et al. [8.39] have found that (8.36) does not deviate significantly from higher-order continued fractions. We may consider (8.36) a reasonable approximation for $\Gamma \gtrsim 2$.

Apart from being an approximate solution of the well defined stochastic model, (8.36) has been established earlier in a similar form by a rather phenomenological reasoning [8.38]. This reasoning will give us some insight into the meaning of (8.36). If we look at the ion in a short time interval, it is likely to perform oscillations and may behave like a damped harmonic oscillator,

$$m\ddot{x} + m\gamma\dot{x} + m\bar{\omega}_0^2 x = f(t) \quad . \tag{8.37}$$

Unlike the periodic potential (8.15), the harmonic potential $m\bar{\omega}_0^2 x^2/2$ does not have finite potential barriers and the ion cannot escape. Since after a long time the ion has diffused, the long-time behavior cannot be described by (8.37) but rather by a free Langevin equation,

$$m\ddot{x} + m\gamma'\dot{x} = f(t) \quad . \tag{8.38}$$

Both equations are contained as limiting cases in a single generalized Langevin equation with a memory function $M(t - t')$:

$$m\ddot{x} + m\gamma\dot{x} + m\bar{\omega}_0^2 \int_0^t M(t - t')\dot{x}(t')dt' = f(t) \quad . \tag{8.39}$$

Oscillations and (8.37) are reproduced by the choice $M(t - t') = 1$, whereas (8.38) is reproduced by the choice $M(t - t') = 0$. In a superionic conductor, for intermediate times, there should be a continuous transition from oscillatory to free diffusive behavior, i.e., $M(t - t')$ should decay continuously from 1 to 0 with increasing time. The simplest choice for $M(t - t')$ which fulfils this condition is

$$M(t - t') = \exp[-(t - t')/\tau_c] \quad . \tag{8.40}$$

The mobility $\mu(\omega)$ which follows from (8.39,40) is [8.38,56]

$$\mu(\omega) = \frac{1}{m} \frac{1}{-i\omega + \gamma + \dfrac{\bar{\omega}_0^2}{-i\omega + 1/\tau_c}} \quad , \tag{8.41}$$

where the second fraction is due to the Laplace transform of the memory function. It is interesting to note that this physical reasoning leads to the same frequency dependence as in (8.36). There the second fraction is due to the presence of the periodic force $K(x)$. The memory function (8.40) thus simulates the motion in a periodic potential. In particular $1/\tau_c$ and R_2 reflect the nonlinearity of the problem; they do not occur for the linear (harmonic) case (8.37). Details of the shape of the potential (i.e., anharmonicities) also affect the coefficient $-\langle K'(x)\rangle/m$, which may be considered the square of an effective harmonic frequency for oscillations in the potential minima. Note that the function $\bar{\omega}_0^2 M(t - t')$ (although physically it accounts for the presence of a potential) may formally be considered a retarded friction in (8.39) and its Laplace transform a frequency-dependent friction.

The memory-function approach outlined above (i.e., continued fractions to second order) has been applied in several papers of BRÜESCH et al. [8.38,75-78]. These authors compared (8.41) with the conductivity $\sigma(\omega)$ of α-AgI derived from far-infrared reflectivity measurements. A fit of the data is shown in Fig.8.7. The model parameters are $\bar{\omega}_0 = 105$ cm^{-1}, $\gamma = 45$ cm^{-1}, $1/\tau_c = 53$ cm^{-1}. Note that the fit is good for $\omega \gtrsim 20$ cm^{-1}. In the experiment the conductivity drops off rapidly below 10 cm^{-1}. This was observed also by other authors [8.79] and in other super-ionic conductors [8.80-82]. The conductivity always seems to drop off close to 10 cm^{-1}.

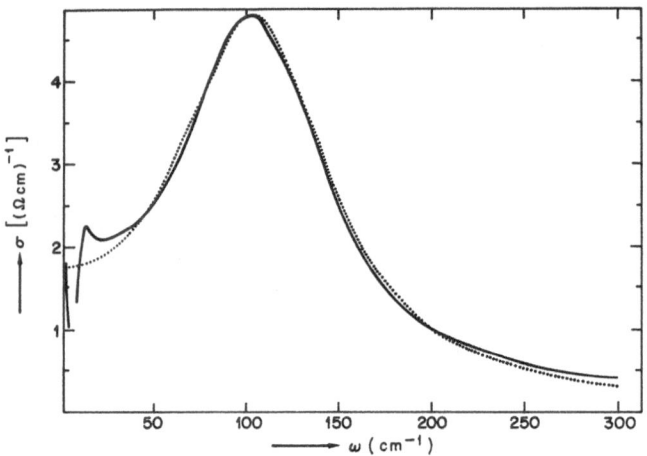

Fig.8.7. Conductivity $\sigma(\omega)$ of α-AgI at T = 453 K. Solid line: derived from FIR re-flectivity measurements; dotted line: fit based on a memory function approach (8.41) (second-order continued fraction). [8.78]

In order to account for this behavior BRÜESCH et al. [8.78] have extended the memory-function approach to the case of two coupled Langevin equations. This is based on the following ideas. In the stochastic models discussed so far, the cage vibrations are conceptually considered as a bath which exerts random forces. This picture is too rough, because there are interactions which produce coherent forces rather than random forces. Long-wavelength optical phonons, for example, are coupled motions of the mobile ions *and* the cage ions. More exactly, for a certain time interval, they perform coherent out-of-phase oscillations. This coupling should be treated explicitly outside the bath by coupled Langevin equations.

The problem reduces to a two-particle problem, i.e., to two coupled Langevin equations. The optical phonons contributing to $\sigma(\omega)$ have long wavelengths because the wavelength of the exiting electromagnetic wave is relatively large. These phonons can be treated by the relative motion of only two particles [8.83]. Again memory functions are constructed such that they reproduce these oscillatory motions. For short time intervals, a particular mobile particle performs a coupled oscillatory motion with a particular cage particle. With increasing time the mobile particle forgets about the cage particle and diffuses away, while the cage particle still performs oscillations in an effective medium, uncorrelated with the former oscillations.

Details of this theory including a discussion of local field corrections are described in [8.78]. Results for α-AgI are shown in Fig.8.8. Now the calculated conductivity also fits the observed structure near 10 cm^{-1}. The structure is attributed to the transition from dressed oscillations to bare diffusion. It should be mentioned that there is another possible explanation of this structure. A CPA calculation of lattice vibrations in the disordered α-AgI lattice also brings forth a peak at 15 cm^{-1}. For details see Chap.4.

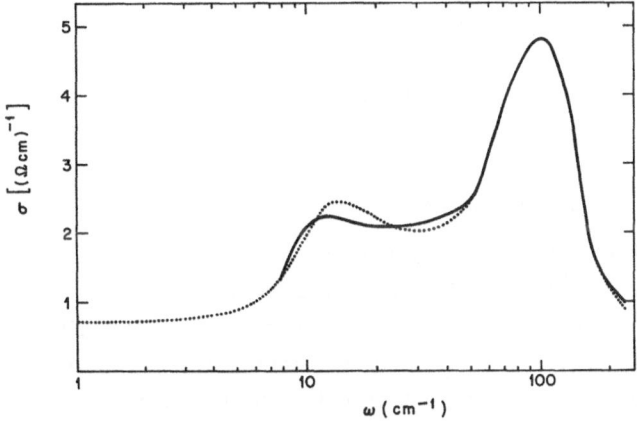

Fig.8.8. As Fig.8.7 for a memory function approach which considers coupled motions of mobile ions and cage ions [8.78]

An earlier work of HUBERMAN and SEN [8.37,84] is based on ideas similar to those which led to the introduction of a memory function in (8.39). They also argue that the response of the mobile ions is oscillatory for times shorter than the mean residence time and diffusive for longer times as described by the Langevin equations (8.37,38). The transition between the two regimes is not brought about by a memory function but by introducing a temperature dependent cutoff frequency ω_{co} which is determined by the mean residence time. The motion of the mobile ions is controlled by the free Langevin equation (8.38) for $\omega < \omega_{co}$, and by oscillator Langevin equations (8.37) for $\omega \geq \omega_{co}$. In addition to theories discussed so far the oscillatory response is not governed by a single oscillator $\bar{\omega}_0$ but by an ensemble of oscillators with a frequency distribution of the Debye type.

This theory, which is similar to a theory of RAHMAN et al. [8.85] for liquids, was applied to Na β-alumina. Comparison with the measured conductivity $\sigma(\omega)$ [8.86] is shown in Fig.8.9. The shape of the oscillatory peak is well reproduced by the theory. At high temperatures the theory yields a Lorentzian centered at $\omega = 0$ (curve b).

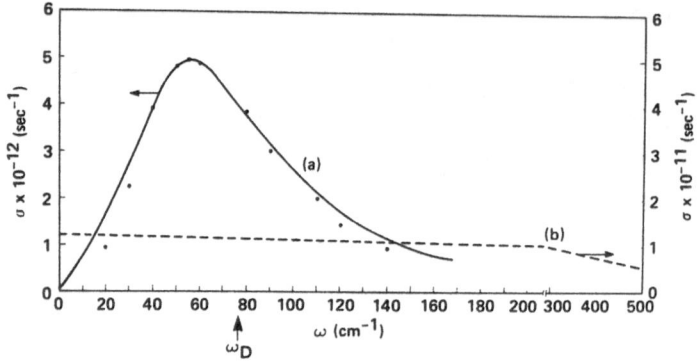

Fig.8.9. Conductivity $\sigma(\omega)$ of Na β-alumina at T = 300 K (circles). Curve a: stochastic theory with cutoff, curve b: theory at T = 2000 K. [8.37]

8.2.7 Dynamic Structure Factor for Jump Diffusion

We now turn to the calculation of the other correlation functions (8.9-12), the dynamic structure factors for the mobile ions. In models with noninteracting mobile ions (one-particle models), there is no difference between the coherent and incoherent case and we shall write $S(q,\omega)$ for the structure factor of the mobile ion. These models are more suited for the incoherent than for the coherent case.

For the sake of clarity we begin with the simplest model and then proceed to the more general cases. In the simple cases one obtains analytic results and can see how the improvements of the more general models come about.

In the simplest jump diffusion model for a solid [Ref.8.45, p.51] one assumes a periodic lattice of residence sites. Jumps between the sites happen instantaneously with the jump rate $1/\ell\tau$, where τ is a mean residence time and ℓ the number of available neighboring sites. It is clear that such a relaxation-time ansatz leads to correlation functions which decay exponentially in time, roughly $\sim\exp[-t/\tau]$; the Fourier transform gives a Lorentzian. The exact expression is [8.45]

$$S(\underline{q},\omega) = \frac{1}{\pi} \frac{\lambda(\underline{q})}{\omega^2 + \lambda^2(\underline{q})} \quad . \tag{8.42}$$

Here $\lambda(\underline{q})$ is a halfwidth depending on the scattering wave vector $\underline{q}$. In the one-dimensional case the dispersion $\lambda(q)$ is given by

$$\lambda(q) = \frac{1}{\tau} [1 - \cos(qa)] \quad , \tag{8.43}$$

where a is the distance between neighboring sites.

According to (8.42) the incoherent neutron scattering cross section consists of a Lorentzian quasi-elastic line, shown in Fig.8.10. Quasi-elastic means that it is centered at zero frequency. For small scattering vectors $q \ll 2\pi/a$ its half-width $\lambda(q)$ increases proportional to q^2. The increase of the halfwidth is connected with a decrease of the intensity $1/\pi\lambda(q)$. The function $\lambda(q)$ is a periodic function of q. For reciprocal lattice vectors $q = n2\pi/a$, the halfwidth $\lambda(q)$ vanishes. Thus in approaching reciprocal lattice vectors the quasi-elastic line sharpens and increases in intensity. The intensity of these Bragg peaks at reciprocal lattice vectors is the same for all reciprocal lattice vectors.

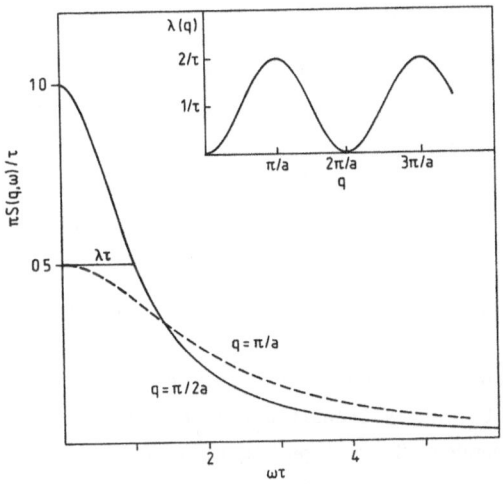

Fig.8.10. Structure factor $S(q,\omega)$ (8.42) and the dispersion of its halfwidth $\lambda(q)$ in a simple jump diffusion model

When there are inequivalent sites in the unit cell one obtains more than one relaxation mode for a given wave vector $\underline{q}$. This may be seen in analogy to the phonon branches in solids with more than one atom per unit cell. As in that case, the relaxation modes transform according to different representations of the space group and have different eigenvalues $\lambda_\alpha(\underline{q})$. The structure factor $S(\underline{q},\omega)$ becomes a super-position of Lorentzians (8.42) with halfwidths $\lambda_\alpha(\underline{q})$ all centered at zero frequency [8.87]. For the case of α-AgI with an tetrahedral interstitial lattice one obtains three relaxation modes at $\underline{q} = 0$ [8.65]

$$\Gamma_{1g} : \lambda_1 = 0 \quad \Gamma_{5u} : \lambda_{2,3,4} = \frac{1}{\tau} \quad \Gamma_{3g} : \lambda_{5,6} = \frac{3}{2\tau} \quad . \tag{8.44}$$

The representations Γ refer to the point group O_h. For $\underline{q} \to 0$ the relaxation frequency of the first mode is

$$\lambda_1 = Dq^2 \quad . \tag{8.45}$$

Only this frequency vanishes at $\underline{q} = 0$. The others $\lambda_{2,3,4}$ and $\lambda_{5,6}$ remain finite in (8.44). In practice this means that only one Lorentzian narrows and increases considerably as $\underline{q} \to 0$. The others retain a finite halfwidth; they reflect hopping motions among inequivalent sites within the unit cell. When the jump rate $1/\tau$ is large enough, which is the case in real superionic conductors like α-AgI, the Γ_{3g} mode can be observed in the Raman spectra [8.46a,88], whereas the halfwidth of the diffusive mode $\lambda_1 \lesssim 10^6$ s^{-1} is far below the resolution of Raman spectrometers. The Γ_{5u} mode is not Raman active.

8.2.8 Dynamic Structure Factor for Large Friction

We now return to the one-dimensional stochastic model defined by the Langevin equation in Sect.8.2.1 for general periodic potentials. It is instructive to consider first the large-friction regime.

When the friction γ is large, the equilibrium distribution of the velocity is attained rapidly (with a characteristic time $1/\gamma$). After this time the velocity distribution becomes Maxwellian and the transition probability w is relevant only for the positions x and x'. The Fokker-Planck equation (8.17) then reduces to the simpler Smoluchowski equation,

$$\frac{\partial}{\partial t} w(x't|x) = \frac{1}{m\gamma} \left[k_B T \frac{\partial^2}{\partial x'^2} - \frac{\partial}{\partial x'} K(x') \right] w(x't|x) \quad . \tag{8.46}$$

This equation can be derived from the Fokker-Planck equation by integrating out the velocity dependence [8.71]. Equation (8.46) can be easily transformed into a differential equation which has the form of a time-dependent Schrödinger equation for a periodic "potential" [8.63,65]

$$U(x') = \frac{K'(x')}{2k_B T} + \left[\frac{K(x')}{2k_B T} \right]^2 \quad . \tag{8.47}$$

Then the problem is very similar to the electron band problem. The eigenvalues $\lambda_n(q)$ and the eigenfunctions are labeled by a wave vector q and a band index n. For the dynamic structure factor, one obtains a superposition of Lorentzians

$$S(q,\omega) = \frac{1}{\pi} \sum_{n=0}^{\infty} |M_n(q)|^2 \frac{\lambda_n(q)}{\omega^2 + \lambda_n^2(q)} \quad . \tag{8.48}$$

Here $M_n(q)$ is a matrix element which determines the weight of the Lorentzians. At low temperatures we recover the jump diffusion result (8.42); only the lowest eigenvalue $\lambda_0(q)$ contributes, which has the same dispersion as (8.43)

$$\lambda_0(q) = \frac{2D}{a^2} [1 - \cos(qa)] \quad . \tag{8.49}$$

Also the higher eigenvalues $\lambda_n(q)$ are periodic functions of q. They must be clearly distinguished from the relaxation frequencies $\lambda_\alpha(\underline{q})$ for different relaxation modes α discussed in Sect.8.2.7. In the present case we deal with only one relaxation mode in the sense of Sect.8.2.7. The higher eigenvalues $\lambda_n(q)$ describe the deviations from the jump diffusion result $\lambda_0(q)$ of this relaxation mode. They belong to higher "bands" in the "potential" $U(x')$ (8.47).

Physically this means that diffusion in a periodic potential does not produce a Lorentzian quasi-elastic line, but a more complicated line shape, which in the large friction regime is given by a superposition of Lorentzians (8.48). This is shown for a cosine potential in Fig.8.11 [8.65]. For $k_B T/V_0 = 0.2$ one can recognize three Lorentzians. For higher temperatures the narrowest Lorentzian belonging to the lowest "band" looses intensity and at $k_B T/V_0 = 1.0$ the broad background belonging to higher "bands" dominates. This figure clearly demonstrates deviations from the hopping result (8.42,43). The corrections reflect the fine structure of the diffusive motion. The figure may be compared with quasi-elastic neutron scattering results in α-AgI [8.3] where similar line shapes were observed (see Chap.3). In that work the spectra were explained by a combination of a jump diffusion motion and a local random motion.

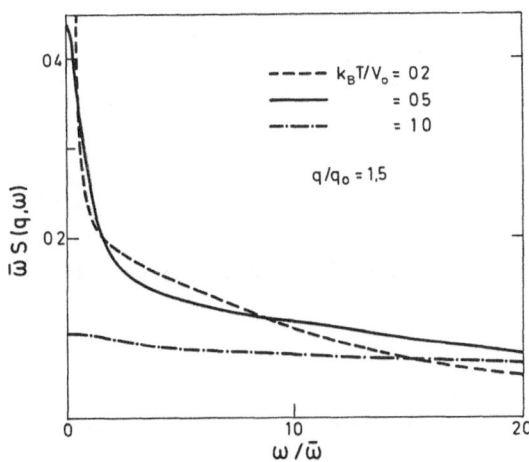

Fig.8.11. Dynamic structure factor for a cosine potential and large friction. $\bar{\omega} = \omega_0^2/\gamma$. [8.65]

8.2.9 Dynamic Structure Factor for General Friction

The dynamic structure factor was calculated also for smaller friction by DIETERICH et al. [8.65]. One can in principle derive a continued-fraction expansion analogous to (8.29). Better convergence, however, is obtained by the following method. The backward equation (8.19) is a partial differential equation. It is solved by representing the solution w in a basis of suitably chosen basis functions (a method which is frequently used in quantum mechanics). Here we use biorthogonal functions [8.89] because the differential operator $\hat{L}$ is not Hermitian. We consider the case of a cosine potential (8.15).

It is obvious that one can write down an equation analogous to (8.24) for the correlation function (8.12) entering the dynamic structure factor

$$\left\langle e^{-iqx(0)} e^{iqx(t)} \right\rangle = \left\langle e^{-iqx} e^{\hat{L}t} e^{iqx} \right\rangle = \left\langle g_q(x,v,t) \right\rangle \quad , \tag{8.50}$$

where we have defined the function

$$g_q(x,v,t) = e^{-iqx} e^{\hat{L}t} e^{iqx} \quad . \tag{8.51}$$

Knowledge of this function solves the problem. It is determined from an equation of motion which is easily obtained from (8.51)

$$\frac{\partial}{\partial t} [e^{iqx} g_q(x,v,t)] = \hat{L}[e^{iqx} g_q(x,v,t)] \quad , \tag{8.52}$$

together with the initial condition

$$g_q(x,v,t = 0) = 1 \quad . \tag{8.53}$$

Equations (8.52,53) are solved by expanding the unknown function $g_q(x,v,t)$ in terms of a complete set of biorthogonal functions. For the dependence on x we use a Fourier series, since we observe from (8.51) and (8.20) that $g_q(x,v,t)$ is a periodic function of x

$$g_q(x + a,v,t) = g_q(x,v,t) \quad . \tag{8.54}$$

For the dependence on v we use normalized Hermite polynomials, which may be written

$$\exp\left(\frac{u^2}{2}\right) \phi_\ell(u) \propto H_\ell(u) \tag{8.55}$$

where $\phi_\ell(u)$ are the normalized eigenfunctions for the quantum-mechanical harmonic oscillator and

$$u = v \sqrt{\frac{m}{2k_BT}} \tag{8.56}$$

is a normalized velocity, Hermite polynomials $H_\ell(u)$ are used because they are eigen-functions of the dissipative part of the Liouvillian [terms proportional to γ in (8.20)] as is easily verified. The expansion is then

$$g_q(x,v,t) = \exp\left(\frac{u^2}{2}\right) \sum_{\ell=0}^{\infty} \sum_{n=-\infty}^{+\infty} A_q^{\ell,n}(t)\, e^{inq_0x} \phi_\ell(u) \quad, \tag{8.57}$$

where $A_q^{\ell,n}(t)$ are the expansion coefficients which are still functions of time.

The problem now consists in the determination of the coefficients, which is performed in the usual way. We use this representation for $g_q(x,v,t)$ in the differential equation (8.52) and perform scalar products with the functions orthogonal to $\phi_\ell(u)$ and $\exp(inq_0x)$. We obtain a set of coupled ordinary linear differential equations for the coefficients $A_q^{\ell,n}(t)$. This set transforms into a set of linear equations for the Laplace transformed quantities $\hat{A}_q^{\ell,n}(z)$ [8.65]

$$\frac{z}{\gamma} \hat{A}_q^{\ell,n} - \delta_{\ell,0}\delta_{n,0}$$

$$= -\ell\hat{A}_q^{\ell,n} + \frac{1}{n}\left(n + \frac{q}{q_0}\right)(-1)^\ell\left(\sqrt{\ell + 1}\,\hat{A}_q^{\ell+1,n} + \sqrt{\ell}\,\hat{A}_q^{\ell-1,n}\right) \tag{8.58}$$

$$+ \frac{v_0}{4k_BT\eta}(-1)^\ell\sqrt{\ell + 1}\left(\hat{A}_q^{\ell+1,n+1} - \hat{A}_q^{\ell+1,n-1}\right)$$

where we have used the abbreviation $\eta = (\gamma/q_0)\sqrt{m/k_BT}$. To solve for $\hat{A}_q^{\ell,n}(z)$ the infinite system of equations is truncated, i.e., we set $\hat{A}_q^{\ell,n} = 0$ for $\ell > L$ or $|n| > N$. The remaining finite system of equations is solved numerically. This method can be used for any value of the friction constant γ. For small friction, however, L and N must be large numbers.

With the solutions $\hat{A}_q^{\ell,n}(z)$ the dynamic structure factor (8.10) follows from (8.12,50,57)

$$S(q,\omega) = \frac{1}{\pi\gamma} \mathrm{Re}\left\{\sum_{n=-\infty}^{+\infty} \hat{A}_q^{0,n}(-i\omega)\, \frac{I_n(V_0/2k_BT)}{I_0(V_0/2k_BT)}\right\} \quad, \tag{8.59}$$

where I_n are the modified Bessel functions, which result from the averages of $\exp(inq_0x)$ with the Boltzmann distribution (8.22). The dynamic structure factor was computed in this way for different wave vectors q, friction γ and temperatures T [8.65]. Figure 8.12 shows a strong quasi-elastic line and an inelastic peak due to diffusive and oscillatory motion, respectively. With increasing temperature the peaks broaden and finally merge.

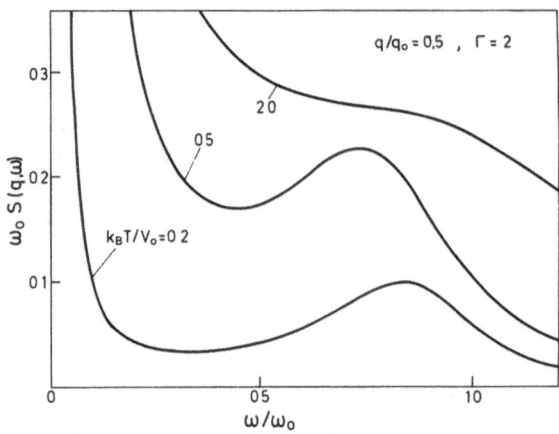

Fig.8.12.
Dynamic structure factor for
smaller friction Γ [8.65]

Deviations from jump diffusion behavior can be checked quantitatively by a plot of the halfwidth Δ of the quasi-elastic line. In Fig.8.13 its dependence on the wave vector q is compared with the jump diffusion result (8.43) or (8.49). Obviously the actual width is much smaller than the width predicted by the jump diffusion approximation. The differences, however, decrease with temperature, as expected, and they reduce to about 35% at $k_BT/V_0 = 0.1$ for $q/q_0 = 0.5$ and $\Gamma = 1$. The differences also decrease for larger friction; in the large-friction limit, the quasi-elastic line can be represented by a superposition of Lorentzians as discussed in the preceding section. These results show that for small friction, a jump diffusion description of the quasi-elastic line becomes valid only at very low temperatures. Equations (8.43,49) therefore should be used for estimates only and not for quantitative determinations of the "jump" rate, etc.

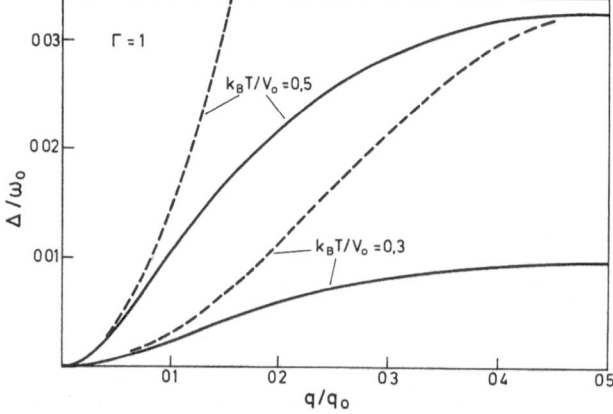

Fig.8.13. Halfwidth of the quasi-elastic line vs wave vector q for small friction.
The dashed lines show the jump diffusion result (8.43,49). [8.65]

Our Figs.8.11-13 should be compared in particular with *quasi-elastic* neutron
scattering measurements. Such measurements were performed by ECKOLD et al. [8.3]
on α-AgI (see Chap.3). The line shape of the quasi-elastic peak and its deviations
from a jump diffusion behavior were already discussed in the preceding section. An
oscillatory peak like that in Fig.8.12 cannot be seen in the measured spectrum due
to a different intensity scale. Most interesting results are expected from a comparison of Fig.8.13 with the measured halfwidth of the quasi-elastic peak. Its dependence on the wave vector q reflects the short-range diffusive motion and thus
contains information on the elementary diffusive step. At the moment, however, the
available experimental data are not detailed enough to allow such a comparison.

On the other hand it is a question whether Fig.8.12 should be compared with the
inelastic neutron scattering spectra. These spectra can probe different lattice vibrations, which are only schematically represented in a one-particle model. Note,
however, that similar line shapes were observed by SHAPIRO et al. [8.90] in $RbAg_4I_5$
(see Chap.3) and by DORNER et al. [8.91] in AgBr at elevated temperatures.

8.2.10 Light Scattering: Continuous and Continuum Models

The light scattering work on superionic conductors is reviewed in detail by Delaney
and Ushioda in Chap.5. In this section, light scattering is discussed in connection
with the continuous aspects of motion. As a specific system we refer to α-AgI.

From the properties of superionic conductors we expect that the light scattering
spectra might be rather complex. The disorder among the mobile ions can lead to
disorder-induced scattering reflecting a projected phonon density of states. The
lattice vibrations can be perturbed by the diffusive motion of the ions. The diffusive motion can lead to quasi-elastic scattering. The large mean square displacements of the ions can make higher-order scattering and anharmonic effects important. It is not possible to estimate a priori which one of these effects plays
a dominant role, since absolute intensities for the different scattering mechanisms
cannot be calculated reliably. Let us proceed empirically and first summarize some
of the experimental findings.

Raman spectra of α-AgI have been measured by several groups [8.88,92-96].
Figure 8.14 shows the results of HANSON et al. [8.92]. The polarized spectrum
(parallel polarization) exhibits a structureless Rayleigh wing (i.e., low-frequency
scattering with increasing intensity for $\omega \to 0$) and a shoulder close to 100 cm^{-1}.
The depolarized spectrum (crossed polarization) only exhibits a Rayleigh wing.
DELANEY and USHIODA [8.95] have found that the silver halide spectra at high
temperatures do not change appreciably upon melting. Similar spectra involving
Rayleigh wings are observed also in other superionic conductors, e.g., in Ag
β-Al_2O_3 [Ref.8.97, Fig.4, and Ref.8.98, Fig.6], in $RbAg_4I_5$ [Ref.8.99, Fig.4, and
Ref.8.100] and in PbF_2 [Ref.8.101, Fig.2].

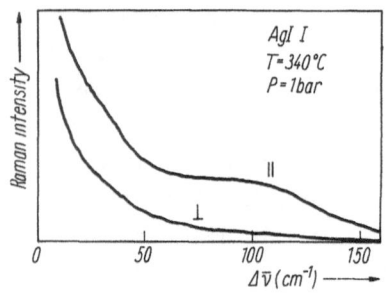

Fig.8.14. Raman spectrum of α-AgI for parallel and crossed polarization of the incident and scattered light [8.92]

A recent investigation of WINTERLING et al. [8.88] confirms that in α-AgI, the low-frequency scattering increases monotonically with decreasing frequency down to at least 1 cm^{-1}. The same is confirmed for $RbAg_4I_5$ by a very careful low-frequency investigation of FIELD et al. [8.100]. Using an iodine filter for the discrimination of the elastic peak, these authors have shown that there is real nonelastic low-frequency scattering (i.e., inelastic or quasi-elastic scattering). From the increase of the intensity with decreasing frequency we can conclude that the Rayleigh wings cannot be due to disorder-induced scattering because in that case the measured intensity would vanish like ω^2 as ω tends to zero [8.96], in contrast to what is observed. Some authors do not report the directly measured spectra but instead the spectra divided by [n(ω) + 1] where n(ω) is the Bose-Einstein population factor. In these reduced spectra, Rayleigh wings are no longer present, and the contrast to disorder-induced scattering is not as striking as in the directly measured spectra.

We shall explain the Rayleigh wings as quasi-elastic scattering due to the diffusive motion of the ions. This interpretation will be based on a calculation of light scattering in a continuous model, which predicts the width and relative intensity of the quasi-elastic peak. The other effects mentioned above are also expected more or less to show up in the spectra. Disorder-induced scattering, which can influence the inelastic part of the spectrum was considered by ALBEN and BURNS [8.96] and ELLIOTT et al. [8.101]. In these calculations the parameters determining the intensity must be adjusted to the experimental results, such that the theory cannot tell how much of the observed scattering is really due to disorder-induced scattering. It is thus a question, at which frequency quasi-elastic scattering ceases to be important and disorder-induced and also second-order scattering by phonons sets in.

Let us first analyze the quasi-elastic scattering in the jump diffusion model. It is now confirmed that in α-AgI the interstitial lattice is made up of tetrahedral sites and that the most probable diffusion path is between nearest-neighbor tetrahedral sites [8.2,4,25] (see Chap.2). As discussed in Sect.8.2.7, for this case the jump model yields three relaxation modes Γ_{1g}, Γ_{5u}, and Γ_{3g}, among which only the Γ_{3g} mode can be observed with Raman spectrometers. This mode should lead to a *single* Lorentzian at ω = 0 with halfwidth $\lambda_{5,6} = 3/2\tau$ which can be estimated from the diffusion coefficient to be ~3.5 cm^{-1}. The experiment, however, indicates that the quasi-elastic line rather consists of *two* components: a narrow component with halfwidth ~3.8 cm^{-1} and an additional broader component [8.88].

We can interpret the occurrence of the broader component as a deviation from the jump model. Such deviations are predicted by the continuous model. In the large friction regime (Sect.8.2.8) we obtain a superposition of several Lorentzians (see Fig.8.11) reflecting the fine structure of the diffusive motion. Quasi-elastic light scattering thus indicates deviations from the picture of thermally activated hopping and emphasizes the need of a continuous treatment. A similar conclusion is discussed by FIELD et al. [8.100] for RbAg$_4$I$_5$. Furthermore they note that one would need specific models for the change in polarizability during the flight of a silver ion. Such a continuous treatment of the scattering mechanism will be described below.

Generally, Raman scattering can be imagined to be generated by changes of the electronic polarizability due to the ionic motion. Hopping models only consider different polarizabilities at the site positions [8.87]. For Raman scattering by phonons one usually considers polarizability changes only for small displacements from equilibrium sites, leading to first- and second-order scattering. Of course, the motion in between the sites does not represent a small displacement. As mentioned at the beginning of this chapter there are experimental indications that the probability of finding an ion at the potential barrier in between the sites is only 2.5 times smaller than the probability of finding it at the sites [8.4,5] and that the flight time is only 3 times smaller than the residence time [8.2]. Therefore the change of the electronic polarizability during the "flight" of the ion must also be taken into account.

The following treatment considers polarizability changes for continuous motion of the ions [8.64]. The scattered intensity $I(q,\omega)$ at wave vector transfer q and frequency shift ω is determined by [8.102]

$$I(\underline{q},\omega) \propto \int dt d^3r d^3r' \exp[i\underline{q} \cdot (\underline{r} - \underline{r}') - i\omega t] \langle \alpha_{is}(\underline{r}',0) \alpha_{is}(\underline{r},t) \rangle \quad . \tag{8.60}$$

Here $\alpha_{is}(\underline{r},t)$ is a tensor element of the electronic polarizability density at the point $\underline{r}$. In perfect crystals the polarizability is expanded with respect to small displacements from equilibrium sites. Here we must allow the ions to move to any position. The electronic polarizability $\alpha_{is}(\underline{r})$ of a mobile ion and its neighborhood is a function of its position $\underline{r}$. As it moves in a periodic lattice this function is periodic and can be expanded in a Fourier series with Fourier components $\alpha_{is}(\underline{K})$, where $\underline{K}$ is a reciprocal lattice vector. With (8.60) the leading contribution of the mobile ions to the scattered intensity becomes [8.64]

$$I(\underline{q},\omega) \propto N \sum_{\underline{K},\underline{K}'} \alpha_{is}(\underline{K})\alpha_{is}(\underline{K}')S(\underline{K}' - \underline{q},\underline{K} + \underline{q},\omega) \quad , \tag{8.61}$$

where we have introduced a generalized dynamic structure factor

$$S(\underline{Q}',\underline{Q},\omega) = \frac{1}{2\pi N} \int dt \, \exp(-i\omega t) \sum_{\ell,\ell'} \langle \exp[i\underline{Q}'\underline{r}_{\ell'}(0)]\exp[i\underline{Q}\underline{r}_\ell(t)]\rangle \quad . \tag{8.62}$$

For $\underline{Q}' = -\underline{Q}$ (8.62) reduces to the conventional structure factor (8.9). The wave vectors $\underline{Q}'$ and $\underline{Q}$ are not determined by the scattering geometry as in neutron scattering but rather by the lattice periodicity (i.e., reciprocal lattice vectors $\underline{K}$).

Physically (8.61) and (8.62) mean that light scattering can probe *short* wavelength density fluctuations, because the polarizability passes one period when the ion moves from one unit cell to the next one. In neutron scattering the wavelength of the density fluctuations which are probed is determined by the momentum transfer of the neutrons. Besides (8.61) and (8.62) there is of course a contribution from lattice vibrations of the cage and there are other less prominent contributions from the mobile ions. These contributions might be written down formally. An explicit calculation then would contain the effects of disorder-induced and higher-order scattering. Due to the complexity of such a calculation this is still science fiction.

In order to study the role of the continuous ionic motion for the Raman spectra we use a simple and well-defined model. We calculate the Raman scattering due to an ion moving in a flat cosine potential according to (8.61) and (8.62). This calculation already brings forth most of the spectral features of Fig.8.14. The ionic motion is treated in the stochastic model of Sect.8.2.1. For the periodic variation of the polarizability $\alpha(x)$ we assume sinusoidal forms. Application to α-AgI is not straightforward. Difficulties arise because the stochastic model is one dimensional and does not have the required symmetry properties. We can only simulate the situation. The three-dimensional periodic potential is simulated in one dimension in the sense of Fig.8.15. Here the potential energy has been calculated along a preferred diffusion path connecting different tetrahedral sites in the unit cell and plotted as a function of the x coordinate [8.1]. The realistic case for Ag^+ ions is for the cation radius $r_i = 0.90$ Å. When the ion moves from one unit cell to the next one, it passes four potential minima. In order to simulate this behavior and to reproduce the curve in the middle of Fig.8.15, the period of the cosine potential must be $a/4$, where a is the real lattice parameter, i.e.,

$$V^0(x) = -\frac{V_0}{2} \cos(q_0 x) \quad \text{with} \quad q_0 = \frac{8\pi}{a} \quad . \tag{8.63}$$

We thus artificially enlarge the unit cell of the cosine potential to include a basis of four potential minima. q_0 is not the real reciprocal lattice vector, but only determines the period of the potential. The reciprocal lattice vector $K = 2\pi/a$ is four times smaller than q_0

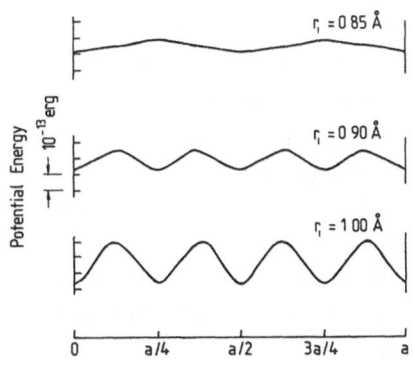

Fig.8.15. Variation of potential energy along preferred diffusion path within unit cell of α-AgI for cations of different size r_i. The realistic case is r_i = 0.90 Å. [8.1]

$$K = \frac{q_0}{4} \quad . \tag{8.64}$$

The polarizability, being a tensor, has different symmetry properties than the scalar V^0. The Fourier series of the periodic function $\alpha(x)$ starts with the reciprocal lattice vectors ±K. We only consider these terms, i.e., we assume that $\alpha(x)$ has a simple sinusoidal form

$$\alpha(x) = 2\alpha(K) \cos(Kx) \quad . \tag{8.65}$$

It would not make a difference to use a sine instead of a cosine for this particular choice of q_0 and K.

The model being defined, we can calculate the scattered intensity. The wavevector transfer q of the light can be neglected as compared with K. The scattered intensity is then determined by the dynamic structure factor (8.62) for $Q = -Q'$ $= K = q_0/4$. It was calculated using a continued-fraction expansion like that of Sect.8.2.3. The model parameters were chosen close to those which reproduced the dynamic mobility of α-AgI in Sect.8.2.6 [8.64]. The results are shown in Fig.8.16 for different temperatures.

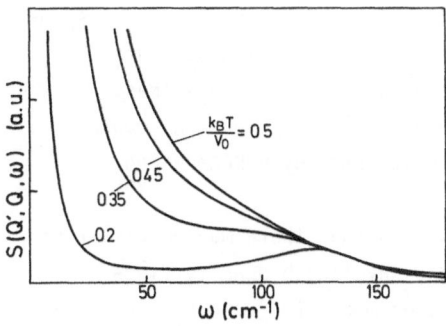

Fig.8.16. Raman scattering due to an ion moving continuously in a sinusoidal potential (8.63) for a variation of the polarizability given by (8.65). [8.64]

At low temperatures we find an inelastic peak due to oscillatory motion separated from a much stronger quasi-elastic peak due to diffusive motion. With increasing temperature the peaks broaden and finally merge. As the model is rather simplifying, we expect that it can reproduce the general features only. In this respect the spectrum for $k_B T/V_0$ = 0.35 agrees well with the measured polarized spectrum in Fig.8.14. The diffusion coefficient for this temperature is D = 1.23 × 10^{-5} cm^2/s. The width of the quasi-elastic line is thus not brought about by an unrealistically large value of D. Actually the halfwidth is ~4 cm^{-1}. It is due to the strong intensity of the quasi-elastic line (3 orders of magnitude more intense than the inelastic peak) that its wings are still observable at higher frequencies in Fig.8.16. A spectrum like that for $k_B T/V_0$ = 0.2 is not observed in the experiment because the transition to the β-phase of AgI changes the structure and the diffusion coefficient abruptly. The fact that the measured depolarized spectrum in Fig.8.14 does not exhibit an inelastic peak but only a quasi-elastic one can be understood by symmetry considerations.

The results of Fig.8.16 are representative for a certain variation of the polarizability α(x), given by (8.65). This choice simulates the behavior in α-AgI. Other possible forms of α(x) are investigated in general in [8.65].

Let us summarize what we can learn from this continuous treatment of Raman scattering in a simple model. The quasi-elastic line is strong and wide enough to show up in the Raman spectra. Thus, the measured Rayleigh wing can be explained by diffusive motions. In Fig.8.16 the Rayleigh wing interferes with the oscillatory peak. This demonstrates that the oscillations cannot be viewed separated from diffusive motions and underlines the significance of continuous treatments. In real superionic conductors additional peaks may occur due to lattice vibrations of the cage. Disorder-induced scattering and higher-order scattering from phonons may be superimposed altering the line shape or leading to additional structure in the spectrum.

Brillouin Scattering

The single particle picture can account for vibrations which behave like dispersionless Einstein oscillators. At frequencies below 1 cm^{-1} light is also scattered by long-wavelength acoustic phonons involving a large number of particles. These phonons have frequencies comparable with the "hopping" frequency and can couple with the relaxational motion. This coupling was investigated by HUBERMAN and MARTIN [8.40].

In the section on jump diffusion (Sect.8.2.7) we saw that there may be different relaxation modes: one mode (Γ_{1g}) connected with long wavelength density fluctuations [proper diffusive mode, see (8.45)] and other modes (Γ_{5u} and Γ_{3g}) connected with hopping motions within a unit cell. HUBERMAN and MARTIN treat the former by a diffusion equation for the density n($\underline{r}$,t) of the mobile particles, the latter as a relaxation process of a pseudospin variable s($\underline{r}$,t) describing the local site populations. Both are coupled to the crystalline strain e($\underline{r}$,t). From the coupled equations for these variables the dispersion relation of the coupled modes are obtained in three limiting cases. It is shown that the damping of long-wavelength acoustic modes is mainly due to coupling with pseudospin fluctuations.

In the frequency range of Brillouin scattering, the long-wavelength density fluctuations (Γ_{1g} mode in α-AgI) again do not show up ($Dq^2 \lesssim 10^6$ s^{-1}). Therefore the Brillouin spectra were calculated considering only acoustic phonons coupled with fluctuations in the local site populations. As shown in Fig.8.17 this leads to a central Lorentzian and to Lorentzian Brillouin peaks. The latter are damped and shifted in frequency due to the interactions of the acoustic phonons with the pseudospins [8.40]. The behavior of the central peak at an order-disorder phase transition was studied in another paper of HUBERMAN and MARTIN [8.103]. It is shown that close to the phase-transition temperature the relaxation frequencies can exhibit a dispersion in the form of a Z.

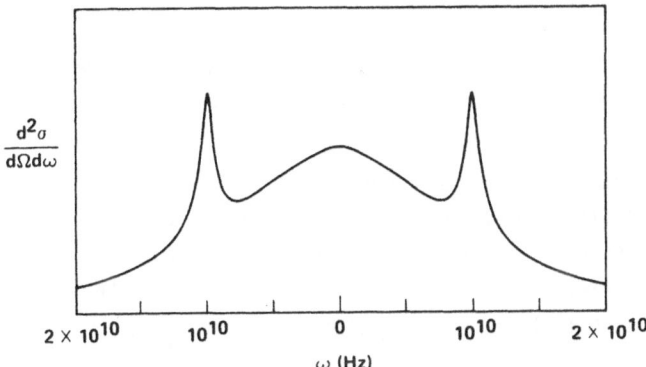

Fig.8.17. Light scattering in the Brillouin region showing a central component and inelastic Brillouin peaks [8.40]

Another approach to a continuum theory for the coupled liquid-cage system in the long-wavelength and low-frequency regime was made by SUBBASWAMY [8.41,42]. The coupled equations are derived from a Lagrangian ansatz. The first paper [8.41] predicted an extremely low frequency ($\sim$1 cm^{-1}) for longitudinal out-of-phase vibrations of the liquid and the cage. At low temperatures this vibration should be identical with a longitudinal optic phonon with much higher frequency. The predicted smallness of the frequency seems to be due to the neglect of restoring forces between liquid particles and cage particles (i.e., only interactions between the liquid and the cage *vibrations* are considered and not interactions between the liquid and the cage itself). This restriction is only partially removed in a second paper [8.42] which considers the effect of the macroscopic electric field. The frequency of the out-of-phase vibration is now shifted to the ionic plasma frequency.

One can use the Zwanzig-Mori method for a rigorous microscopic derivation of hydrodynamic fluctuations starting from the Hamiltonian (8.2-5). Such a theory by Zeyher is in progress. The ideas underlying this method are also sketched in the following section. Another continuum approach by Jäckle is in progress, too.

8.2.11 Microscopic Foundation

As mentioned earlier, the stochastic models were introduced phenomenologically. However, stochastic models can in principle be derived microscopically from the Hamiltonian (8.2-5). We shall discuss such attempts based on the Zwanzig-Mori formalism [8.67,68] for generalized Langevin equations, for a hydrodynamic description and for the Fokker-Planck equation. The formalism will be briefly sketched (for more details see, e.g., [8.104]).

For a given Hamiltonian H with f degrees of freedom the Liouvillian of classical mechanics is

$$\hat{L} = -i \sum_{j=1}^{f} \left(\frac{\partial H}{\partial p_j} \frac{\partial}{\partial q_j} - \frac{\partial H}{\partial q_j} \frac{\partial}{\partial p_j} \right) \quad , \tag{8.66}$$

where q_j are the generalized coordinates and p_j the conjugate momenta. The index j runs over all particle coordinates *including cage ions* besides the mobile ions. In this point it differs from the Liouville-like operator in the Fokker-Planck case used in preceding sections. In that case the motion of the cage ions is treated as a bath which is represented by a friction term. The Liouvillian describes the time evolution of dynamical variables A(t) [e.g., the particle velocity $v_j(t)$]

$$\frac{dA(t)}{dt} = i\hat{L}A(t) \quad . \tag{8.67}$$

Projection operators are used with the goal of projecting out from the time evolution of A the variables which are uncorrelated with A(0) and which will be considered to be bath variables. This leads to the following exact equation

$$\frac{dA}{dt} = i\Omega A(t) - \int_0^t M(\tau)A(t - \tau)d\tau + F(t) \quad . \tag{8.68}$$

Because of its similarity with (8.13) it is called the generalized Langevin equation. The frequency Ω is determined by the initial correlation of A with $\dot{A}$, F(t) is the random force obtained with the use of projection operators and the memory function $M(\tau)$ is given by the correlation of the random force analogous to (8.14). In contrast to this equation the random forces are in general not δ correlated.

If one were to apply this procedure to $A(t) = v_j(t)$, the velocity of a mobile particle, in order to get its generalized Langevin equation, the frequency Ω would vanish and all the information about friction and the cage potential $V^0(x)$ would be contained in the memory function $M(\tau)$. An explicit calculation of $M(\tau)$, however, is possible only in very simple cases. This is why BRÜESCH et al. [8.38] have assumed a simple ansatz (8.39,40) which simulates the friction and $V^0(x)$.

A microscopic calculation of the memory functions becomes more promising if one excludes such variables from the bath which are dynamically coupled with the variable of interest $v_j(t)$. This is achieved by considering a set of dynamical variables $A_i(t)$ which are expected to be dynamically coupled (relevant variables) and deriving a generalized Langevin equation like (8.68) in matrix form. If the set is suitably chosen, the random forces only contain irrelevant variables which merit treatment as a bath. Thus fewer requirements are made of the memory functions (i.e., correlations of the random forces) and their calculation becomes easier and more accurate.

This method was applied by ZEYHER [8.105] to a Hamiltonian like (8.2-5) but excluding interactions among the mobile ions. He assumes that the only relevant variables are the net ionic current j and its time derivative. The memory functions are calculated in lowest-order perturbation theory treating particles with energies above and below the potential barrier separately. This microscopic treatment leads to a spectrum for $\sigma(\omega)$ similar to Fig.8.6 exhibiting peaks at $\omega = 0$ and at the reststrahl frequency (corresponding to ω_0 in Fig.8.6). Numerical calculations were not performed.

The method can also be applied in the hydrodynamic regime. Consider conserved variables like the particle density, the momentum density and the energy density. The lifetime of fluctuations of the conserved variables becomes infinite for wave vectors $q \to 0$, i.e., they vary slowly as compared with other variables. These conserved variables can be used in a set of relevant variables for the procedure outlined above. Moreover in the hydrodynamic regime the calculation of the memory-function matrices, etc., can be simplified considerably by expanding them with respect to small wave vectors q and frequencies ω. The correlation functions can be derived rigorously in this way [8.104]. This method is applied to superionic conductors by ZEYHER. First results are under way.

Outside the hydrodynamic regime the generalized Langevin equations discussed above have some drawback. The cage potential $V^0(x)$ does not occur explicitly as in the phenomenological Langevin equation (8.13) [by the force K(x)]; at least its anharmonicities must be simulated by the memory functions (cf. also Sect.8.2.6). Considering that the potential barriers, i.e., the extreme anharmonicities of $V^0(x)$ and nonlinearities of K(x) are the important features, the generalized Langevin equations are less satisfactory than the phenomenological Langevin equation (8.13). Since (8.13) is equivalent to the Fokker-Planck equation (8.17,18), we can

238

improve upon this point if we can derive a Fokker-Planck equation microscopically. This was done by BECKER [8.106] based on a work of MORI et al. [8.107,108] [see also Ref.8.104, Chap.6.4] and will be briefly illustrated.

In the following the transition probability in the phase space of the mobile ions is abbreviated for simplicity by $w(x_1t|x_0)$ from a point x_0 in phase space to a point x_1. It can be written

$$w(x_1t|x_0) = <\delta[x(t) - x_1]\delta[x(0) - x_0]> \frac{1}{p(x_0)} \quad , \tag{8.69}$$

where $p(x_0) = <\delta[x(0) - x_0]>$ is the stationary distribution. An equation of motion for the transition probability (Fokker-Planck equation) can be obtained in the same way as the generalized Langevin equation (8.68) using projection operators and setting

$$A(t) = \delta[x(t) - x_1] \quad . \tag{8.70}$$

Multiplying the equation of motion (8.68) for this $A(t)$ with $\delta[x(0) - x_0]/p(x_0)$ and averaging yields the desired Fokker-Planck equation for w (8.69). This equation now contains the cage potential explicitly in the same way as in (8.18). However, one obtains a *retarded* friction which in general depends not only on the time t' but also on the whole history of $x(t')$ in between 0 and t.

The generalized Fokker-Planck equation is thus obtained microscopically. However, there seems to be no hope of finding its solution in the general case because of the complicated form of the friction. The only approximation which seems tractable at present is to neglect the retardation of the friction. This leads to a Fokker-Planck equation similar to (8.17,18), which was solved and investigated in previous sections. There the friction constant γ was used as an independent model parameter. The derivation of BECKER now yields a microscopic expression for γ which also contains a temperature dependence.

8.3 Computer Simulations

This section reviews molecular-dynamics (MD) calculations on CaF_2 and α-AgI. These calculations consider a limited number of anions and cations interacting with each other. The dynamical evolution is calculated stepwise according to Newton's laws. As all the interactions are included, MD calculations can give us information which cannot as yet be obtained by analytical means.

Calculations on CaF_2 were performed by RAHMAN and JACUCCI [8.109,110] using repulsive potentials tabulated by KIM and GORDON [8.111]. A system of 108 Ca^{2+} ions
and 216 F^- ions was studied at 1590 K. The Ca^{2+} ions are found to form a fcc lattice
with rms displacements of about 0.3 Å. The mean square displacements of both types
of ions are shown as a function of time in Fig.8.18. Note that the calculations
reproduce the fact that the F^- ions diffuse away (D = 2.6 × 10^{-5} cm^2/s) whereas the
Ca^{2+} ions are bound to their sites. In order to follow the F^- motion statistically
the unit cell is divided into subcells such that there are 8 subcells available per
F^- ion. Among 216 F^- ions on the average 131.3 are found in subcells belonging to
regular sites and 84.7 are found in other subcells. The "jump" rate is 0.21 × 10^{12} s^{-1}.

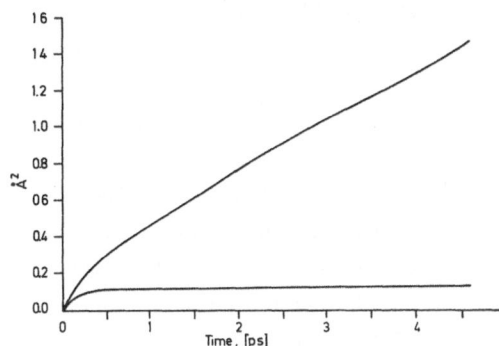

Fig.8.18. Mean square displacements (MD)
of F^- ions (upper curve) and Ca^{2+} ions
(lower curve) in CaF_2 at 1590 K [8.109]

MD calculations on α-AgI were performed by SCHOMMERS [8.112,8.113] for 256 Ag^+
and I^- ions at 563 K. When only Coulomb and repulsive Born-Mayer potentials are
assumed, the I^- ions do not form a stable lattice but leave their sites. In order
to improve on this behavior a restoring harmonic potential for the I^- ions is included.

The mean square displacements of the mobile Ag^+ ions are shown by the solid
curve in Fig.8.19. The role of the cage vibrations was investigated by a reference
calculation where the I^- ions were not allowed to be displaced from their sites.
The mean square displacements obtained in this case are shown by the dashed curve
in Fig.8.19. Note that the diffusion coefficient (given by the slope of the curves
for large times) is much smaller when cage ion displacements are inhibited. The
question *how* the cage ions support the diffusion of the mobile ions is not answered
at present.

The role of dynamical coupling between mobile ions is studied in Fig.8.20. The
dashed line shows the velocity autocorrelation function of an Ag^+ ion as a function
of time. The solid line shows the current-current correlation function, which involves all ions. The figure demonstrates that the autocorrelation function decays
much faster than the current correlations, i.e., there exist relatively long-lived
correlations between the velocities of pairs of ions in α-AgI.

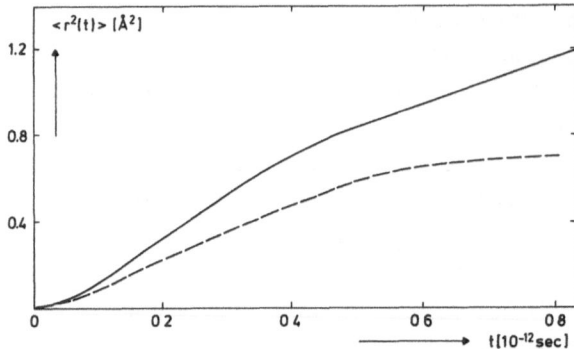

<u>Fig.8.19.</u> Mean square displacements (MD) of Ag^+ ions in α-AgI at 563 K (solid line). Dashed line: MD result for a rigid cage. [8.112]

In a recent MD work of VASHISHTA and RAHMAN [8.5] the use of more refined effective pair potentials effectuates stability of the I^- sublattice. The authors have made up the density distribution of I^- and Ag^+ ions on the (100) faces of the bcc cage (Fig.8.21). Different positions are marked in the inset of the figure. Such a face contains 4 tetrahedral sites (t), which are the preferential positions of the Ag^+ ions. The solid circles show the MD results, the solid line is the only experimental result in this figure (deduced from neutron diffraction); all other lines are guides for the eye. In accordance with experiment [8.4], Fig.8.21 displays the following results: the only local maxima are at the t sites, i.e., these are the only residence sites. The point S is a saddle point with a minimum along t-S-t and a maximum along I-S-C. The potential is likely to behave the other way with a saddle point at S, which offers the lowest potential barrier. Thus, it is not astonishing that diffusion happens between the tetrahedral sites, predominantly (82%) along the line t-S-t. Considering that S is not a residence site but a potential barrier, the density at this point is remarkably high. This fact is incompatible with the jump diffusion picture.

Recently, a molecular dynamics simulation was also performed by mechanical means by the author and has been recorded in a movie. One of the scenes is reproduced in Fig.8.1b showing the distribution of mobile ions and cage ions.

8.4 Correlations Among the Mobile Ions

While the effective one-particle model has been studied extensively, continuous theories investigating correlations among the mobile ions are still at an initial

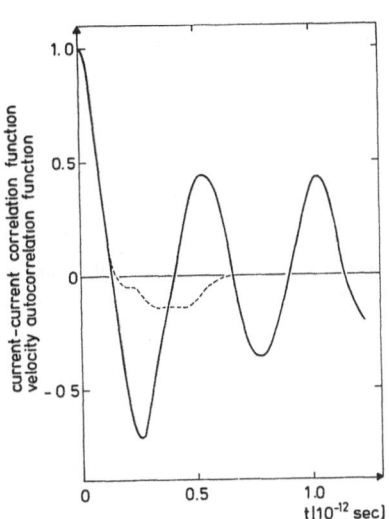

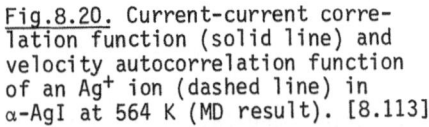

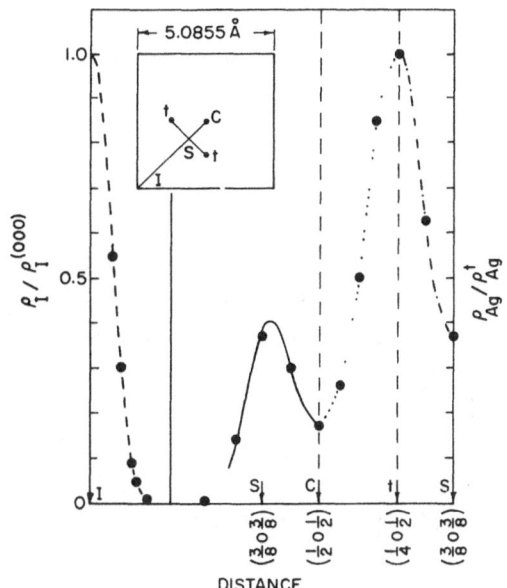

Fig.8.20. Current-current correlation function (solid line) and
velocity autocorrelation function
of an Ag$^+$ ion (dashed line) in
α-AgI at 564 K (MD result). [8.113]

Fig.8.21. Density distribution (MD) of Ag$^+$
and I$^-$ in α-AgI at 157°C. I: iodine site;
t: tetrahedral site; S: saddle point; C:
octahedral site. Diffusion occurs between
tetrahedral sites, predominantly along
t-S-t. [8.5]

stage. This section outlines a method for the many-particle case and discusses a
result for which correlated motion and a continuous treatment are of crucial importance.

It has been noted very recently that under some limiting conditions for the
many-particle Langevin equation, the sine-Gordon equation and soliton solutions
are recovered [8.114-116]. (For a review on solitons, see [8.117].) The method
described below does not depend on the existence of soliton solutions and applies
also to the physically important case where the average spacing between mobile ions
differs from the period of the cage potential.

The model is defined by many-particle Langevin equations analogous to (8.13,14)

$$m\ddot{x}_j + m\gamma\dot{x}_j + \frac{\partial}{\partial x_j} H_A = f_j(t) \quad , \tag{8.71}$$

where the Hamiltonian H_A (8.3) gives rise to systematic forces due to the cage potential V^0 and interactions V^1. The one-particle model treated the interactions V^1
as part of the bath (friction and random forces). Here, the bath is due only to
fluctuations of the cage.

As before, a Fokker-Planck equation (8.17), its backward analogue (8.19), and a "Liouvillian" $\hat{L}$ (8.20) can be derived. Let us concentrate on the frequency-dependent conductivity in one dimension,

$$\sigma(\omega) = \frac{(Ze)^2}{k_B T} \sum_{\ell,m} \int_0^\infty dt\, e^{i\omega t} <\dot{x}_\ell(0)\dot{x}_m(t)> \quad , \tag{8.72}$$

which can be calculated by a continued fraction expansion (8.29) as in Sect.8.2.3. In the present context, however, it is preferable to derive the continued fraction by the Zwanzig-Mori projector technique [8.67,68], which gives an expression for the remainder R_{n+1} in (8.29). The parameters defined as in (8.29) for $\sigma(\omega)$ are

$$a_0 = N(Ze)^2 \quad b_1 = \gamma \quad b_2 = 0$$

$$a_1 = -(1/Nmk_B T) \sum_{\ell,m} <K_0(x_\ell)K_0(x_m)> \quad , \tag{8.73}$$

$$a_2 = (1/a_1 Nm^2) \sum_{\ell} [<K_0'(x_\ell)K_0'(x_\ell)> - <K_0'(x_\ell)>^2] \quad , \tag{8.74}$$

$$R_3 = \gamma \quad , \tag{8.75}$$

where $K_0(x_\ell) = -\partial V^0/\partial x_\ell$ is the force due to the cage potential, and (8.75) is valid in the large friction regime. The accuracy of this termination of the continued fraction can be checked by comparison with the analytic expression (8.35) in the one-particle case. For a sinusoidal potential and $k_B T/V_0 \geq 0.5$ we find deviations of only 1.5%.

Note that the interaction V^1 only occurs in the averages of (8.73,74). This method can be used to calculate the frequency-dependent conductivity for any one-particle potential V^0 and interaction potential V^1 in the large friction regime. In the following, the static conductivity will be calculated for a sinusoidal cage potential V^0 (8.15). The interaction potential V^1 should be realistic for superionic conductors but simple enough so as to permit a reliable calculation of the averages (8.73,74). However, the calculation of (8.73) is as complicated as that of the static structure factor for a dense liquid at short wavelength in the presence of a periodic potential. Under these circumstances, a harmonic nearest-neighbor interaction appears to be the most appropriate, because (8.73,74) can be calculated accurately and because it is fairly suited to the physical situation in superionic conductors. (For hard-core interactions, see [8.118].)

It is assumed that for a given density n of mobile ions, their mutual repulsion tends to keep them separated by an average distance b = 1/n. Deviations from this distance lead to restoring forces which to second order are described by the interaction potential

$$V^1 = \frac{\alpha}{2} \sum_{\ell} (x_{\ell+1} - x_\ell - b)^2 \quad .$$

(8.76)

The averages (8.73,74) can now be computed using the transfer integral operator. For high temperatures, this yields

$$\sigma(0) = \frac{N(Ze)^2/m\gamma}{1 + \frac{1}{2}\left(\dfrac{V_0}{2k_BT}\right)\dfrac{[1 - \exp(-k_BTq_0^2/\alpha)]^2}{[\exp(-k_BTq_0^2/\alpha) + 1 - 2\cos(q_0b)\exp(-k_BTq_0^2/2\alpha)]^2}} \quad .$$

(8.77)

(For lower temperatures, and further details, see [8.119].) Here V_0 is the potential barrier and $q_0 = 2\pi/a$ where a is the distance between the potential minima. This expression is displayed in Fig.8.22. Note that the ratio $a/b = a \cdot n$ is proportional to the density of mobile particles. One thus can obtain expressions for $\sigma(0)$ covering the whole density range, including the conventionally considered limiting cases of vacancy ($n \lesssim 1/a$), interstitialcy ($n \gtrsim 1/a$), and single-ion ($n \ll 1/a$) conduction. The figure shows that maximum conductivity is obtained in neither of these limiting cases, but rather for $n \approx 0.7/a$, where the potentials V^0 and V^1 compete. For lower temperatures or other parameters the conductivity changes may be much more drastic. A comparison with specific superionic conductors must await more detailed investigations. It should be recalled, however, that in many superionic conductors the highest conductivity is found for nonstoichiometric densities n.

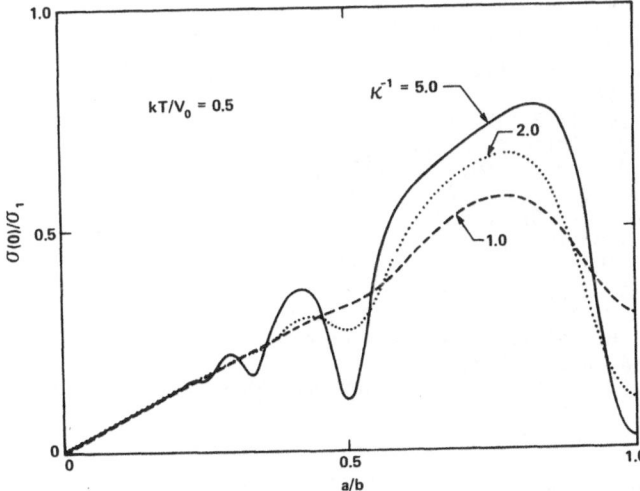

Fig.8.22. Static conductivity versus concentration of mobile ions for different correlation lengths $\kappa^{-1} = \alpha b^2/2\pi^2 k_BT$. The normalization is $\sigma_1 = (Ze)^2L/am\gamma$ where L is the length of the sample

Note that this result, which involves correlated motion of mobile ions, is not dependent on the existence of soliton solutions. However, this result requires a finite spacing b ≠ a between the mobile ions and, of course, a continuous treatment.

Acknowledgement. I wish to thank J. Keller and K.W. Becker for many helpful discussions and critical reading of the manuscript.

References

8.1 W.H. Flygare, R.A. Huggins: J. Phys. Chem. Solids *34*, 1199 (1973)
8.2 J.B. Boyce, T.M. Hayes, W. Stutius, J.C. Mikkelsen, Jr.: Phys. Rev. Lett. *38*, 1362 (1977)
8.3 G. Eckold, K. Funke, J. Kalus, R.E. Lechner: J. Phys. Chem. Solids *37*, 1097 (1976)
8.4 R.J. Cava, F. Reidinger, B.J. Wuensch: Solid State Commun. *24*, 411 (1977)
8.5 P. Vashishta, A. Rahman: Phys. Rev. Lett. *40*, 1337 (1978)
8.6 W. Bührer, P. Brüesch: Solid State Commun. *16*, 155 (1975)
8.7 G.L. Bottger, C.V. Damsgard: J. Chem. Phys. *57*, 1215 (1972)
8.8 O.B. Ajayi, L.E. Nagel, I.D. Raistrick, R.A. Huggins: J. Phys. Chem. Solids *37*, 167 (1976)
8.9 W.G. Kleppmann, H. Bilz: Commun. Phys. *1*, 105 (1976)
8.10 W.G. Kleppmann: J. Phys. C. *9*, 2285 (1976)
8.11 M.J. Rice, W.L. Roth: J. Solid State Chem. *4*, 294 (1972)
8.12 C.W. Haas: J. Solid State Chem. *7*, 155 (1973)
8.13 G. Eckold, K. Funke, J. Kalus, R.E. Lechner: Phys. Lett. *55* A, 125 (1975)
8.14 H. Sato, R. Kikuchi: J. Chem. Phys. *55*, 677 (1971)
8.15 R. Kikuchi, H. Sato: J. Chem. Phys. *55*, 702 (1971)
8.16 F.G. Mahan, W.J. Pardee: Phys. Lett. *49* A, 325 (1974)
8.17 W.J. Pardee, G.D. Mahan: J. Solid State Chem. *15*, 310 (1975)
8.18 H. Hinkelmann, B.A. Huberman: Solid State Commun. *19*, 365 (1976)
8.19 H. Hinkelmann: Solid State Commun. *21*, 975 (1977)
8.20 R. Vargas, M.B. Salamon, C.P. Flynn: Phys. Rev.Lett. *37*, 1550 (1976)
8.21 G.D. Mahan: Phys. Rev. B *14*, 780 (1976)
8.22 P.M. Richards: Phys. Rev. B *16*, 1393 (1977)
8.23 W. Bührer, W. Hälg: Helv. Phys. Acta *47*, 27 (1974)
8.24 A.F. Wright, B.E.F. Fender: J. Phys. C *10*, 2261 (1977)
8.25 S. Hoshino, T. Sakuma, Y. Fujii: Solid State Commun. *22*, 763 (1977)
8.26 S. Hoshino: J. Phys. Soc. Jpn. *12*, 315 (1957)
8.27 G. Burley: J. Phys. Chem. Solids *25*, 629 (1964)
8.28 K.R. Subbaswamy, G.D. Mahan: Phys. Rev. Lett. *37*, 642 (1976)
8.29 F. Claro, G.D. Mahan: J. Phys. C *10*, L73 (1977)
8.30 M. Girvin, G.D. Mahan: Solid State Commun. *23*, 629 (1977)
8.31 F.L. Lederman, M.B. Salamon, H. Peisl: Solid State Commun. *19*, 147 (1976)
8.32 M.B. Salamon: Phys. Rev. B *15*, 2236 (1977)
8.33 H.U. Beyeler, L. Pietronero, S. Strässler, H.J. Wiesmann: Phys. Rev. Lett. *38*, 1532 (1977)
8.34 L. Lam, A. Bunde: Z. Phys. *30*, 65 (1978)
8.35 B.A. Huberman: Phys. Rev. Lett. *32*, 1000 (1974)
8.36 M.J. Rice, S. Strässler, G.A. Toombs: Phys. Rev. Lett. *32*, 596 (1974)
8.37 B.A. Huberman, P.N. Sen: Phys. Rev. Lett. *33*, 1379 (1974)
8.38 P. Brüesch, S. Strässler, H.R. Zeller: Phys. Status Solidi (a) *31*, 217 (1975)
8.39 P. Fulde, L. Pietronero, W.R. Schneider, S. Strässler: Phys. Rev. Lett. *35*, 1776 (1975)

8.40 B.A. Huberman, R.M. Martin: Phys. Rev. B *13*, 1498 (1976)
8.41 K.R. Subbaswamy: Solid State Commun. *19*, 1157 (1976)
8.42 K.R. Subbaswamy: Solid State Commun. *21*, 371 (1977)
8.43 G.H. Wannier: *Statistical Physics* (Wiley and Sons, New York 1966)
8.44 R. Kubo: J. Phys. Soc. Jpn. *12*, 570 (1957)
     R. Kubo: Rep. Progr. Phys. *29*, 255 (1966)
8.45 e.g. T. Springer: *Quasielastic Neutron Scattering for the Investigation of Diffusive Motions in Solids and Liquids*, Springer Tracts in Modern Physics, Vol. 64 (Springer, Berlin, Heidelberg, New York 1972)
8.46 M. Balkanski (ed.): *Lattice Dynamics*, International Conference Paris 1977 (Flammarion, Paris 1978)
8.46a T. Geisel: In Ref.8.46, p.549
8.47 P.A. Egelstaff, P. Schofield: Nucl. Sci. Eng. *12*, 260 (1962)
8.48 K.S. Singwi, A. Sjölander: Phys. Rev. *119*, 863 (1960)
8.49 V.S. Oskotskii: Sov. Phys.-Solid State *5*, 789 (1963)
8.50 C.T. Chudley, R.J. Elliott: Proc. Phys. Soc. (London) *77*, 353 (1961)
8.51 V.F. Sears: Proc. Phys. Soc. (London) *86*, 953 (1965)
8.52 Y. Nakahara, H. Takahashi: Proc. Phys. Soc. (London) *89*, 749 (1966)
8.53 P.S. Damle, A. Sjölander, K.S. Singwi: Phys. Rev. *165*, 277 (1968)
8.54 Caroli, G. Jannink, D. Saint-James: J. Phys. C *4*, 545 (1971)
8.55 G. Wyllie: J. Phys. C *4*, 564 (1971)
8.56 R. Kubo: Rep. Progr. Phys. *29*, 255 (1966)
8.57 e.g. H. Haken: *Synergetics, an Introduction*, 2nd ed. (Springer, Berlin, Heidelberg, New York 1978) p.86
8.58 For an elementary introduction see F. Reif: *Fundamentals of statistical and thermal physics* (McGraw-Hill, New York 1965) Chap.15.11
8.59 R.L. Stratonovich: *Topics in the Theory of Random Noise*, Vol.I (Gordon and Breach, New York 1963)
8.60 H. Haken: Rev. Mod. Phys. *47*, 67 (1975)
8.61 A.T. Bharucha-Reid: *Elements of the Theory of Markov Processes and Their Applications* (McGraw-Hill, New York 1960) Chap.3
8.62 W.R. Schneider, S. Strässler: Z. Phys. B *27*, 357 (1977)
8.63 W. Dieterich, I. Peschel, W.R. Schneider: Z. Phys. B *27*, 177 (1977)
8.64 T. Geisel: Solid State Commun. *24*, 155 (1977)
8.65 W. Dieterich, T. Geisel, I. Peschel: Z. Phys. B *29*, 5 (1978)
8.66 H. Risken, H.D. Vollmer: Preprint (1978), submitted to Z. Phys.
8.67 H. Mori: Progr. Theor. Phys. *33*, 423 (1965)
8.68 H. Mori: Progr. Theor. Phys. *34*, 399 (1965)
8.69 W.R. Schneider: Z. Phys. B *24*, 135 (1976)
8.70 H.S. Wall: *Analytic Theory of Continued Fractions* (Chelsa, New York 1948) p.197f
8.71 H.A. Kramers: Physica *7*, 284 (1940)
8.72 C.A. Wert: Phys. Rev. *79*, 601 (1950)
8.73 Y.M. Ivanchenko, L.A. Zil'berman: Sov. Phys.-JETP *28*, 1272 (1969)
8.74 V. Ambegaokar, B.I. Halperin: Phys. Rev. Lett. *22*, 1364 (1969)
8.75 H.R. Zeller, P. Brüesch, L. Pietronero, S. Strässler: In *Superionic Conductors*, ed. by G.D. Mahan, W.L. Roth (Plenum Press, New York 1976) p.201
8.76 P. Brüesch, L. Pietronero, H.R. Zeller: J. Phys. C *9*, 3977 (1976)
8.77 P. Brüesch, L. Pietronero, S. Strässler, H.R. Zeller: Electrochim. Acta *22*, 717 (1977)
8.78 P. Brüesch, L. Pietronero, S. Strässler, H.R. Zeller: Phys. Rev. B *15*, 4631 (1977)
8.79 W. Jost, K. Funke, A. Jost: Z. Naturforsch. *25* a, 983 (1970)
8.80 U. Strom, P.C. Taylor, S.G. Bishop, T.L. Reinecke, K.L. Ngai: Phys. Rev. B *13*, 3329 (1976)
8.81 G. Eckold, K. Funke: Z. Naturforsch. *28* a, 1042 (1973)
8.82 S.J. Allen, Jr., A.S. Cooper, F. DeRosa, J.P. Remeika, S.K. Ulasi: Phys. Rev. B *17*, 4031 (1978)
8.83 See e.g., C. Kittel: *Introduction to Solid State Physics*, 4th ed. (Wiley and Sons, New York 1971) p.181

8.84 P.N. Sen, B.A. Huberman: Phys. Rev. Lett. *34*, 1059 (1975)
8.85 A. Rahman, K.S. Singwi, A. Sjölander: Phys. Rev. *126*, 997 (1962)
8.86 S.J. Allen, Jr., J.P. Remeika: Phys. Rev. Lett. *33*, 1478 (1974)
8.87 See e.g., M.V. Klein: In *Light Scattering in Solids*, ed. by M. Balkanski, R.C.C. Leite, S.P.S. Porto (Flammarion, Paris 1976) p.351
8.88 G. Winterling, W. Senn, M. Grimsditch, R. Katiyar: In Ref.8.46, p.553
8.89 P.M. Morse, H. Feshbach: *Methods of Theoretical Physics* (McGraw-Hill, New York 1953) p.884
8.90 S.M. Shapiro, D. Semmingsen, M.B. Salamon: In Ref.8.46, p.538
8.91 B. Dorner, J. Windscheif, W. von der Osten: In Ref.8.46, p.535
8.92 R.C. Hanson, T.A. Fjeldly, H.D. Hochheimer: Phys. Status Solidi (b) *70*, 567 (1975)
8.93 G. Burns, F.H. Dacol, M.W. Shafer: Solid State Commun. *19*, 291 (1976)
8.94 M.J. Delaney, S. Ushioda: Solid State Commun. *19*, 297 (1976)
8.95 M.J. Delaney, S. Ushioda: Phys. Rev. B *16*, 1410 (1977)
8.96 R. Alben, G. Burns: Phys. Rev. B *16*, 3746 (1977)
8.97 C.H. Hao, L.L. Chase, G.D. Mahan: Phys. Rev. B *13*, 4306 (1976)
8.98 L.G. Chase: In Ref.8.75, p.299
8.99 D. Gallagher, M.V. Klein: J. Phys. C *9*, L 687 (1976)
8.100 R.A. Field, D.A. Gallagher, M.V. Klein: Preprint 1978, submitted to Phys. Rev. B
8.101 R.J. Elliott, W. Hayes, W.G. Kleppmann, A.J. Rushworth, J.F. Ryan: Proc. R. Soc. (London) A *360*, 317 (1978)
8.102 e.g. B.J. Berne, R. Pecora: *Dynamic Light Scattering* (Wiley and Sons, New York 1976)
8.103 B.A. Huberman, R.M. Martin: Phys. Rev. Lett. *39*, 478 (1977)
8.104 D. Forster: *Hydrodynamic Fluctuations, Broken Symmetry, and Correlation Functions* (Benjamin, New York 1975)
8.105 R. Zeyher: In Ref.8.46, p.532
8.106 K.W. Becker: Unpublished
8.107 H. Mori, H. Fujisaka: Progr. Theor. Phys. *49*, 764 (1973)
8.108 H. Mori, H. Fujisaka, H. Shigematsu: Progr. Theor. Phys. *51*, 109 (1974)
8.109 A. Rahman: J. Chem. Phys. *65*, 4845 (1976)
8.110 G. Jacucci, A. Rahman: Preprint 1978
8.111 Y.S. Kim, R.G. Gordon: J. Chem. Phys. *60*, 4332 (1974)
8.112 W. Schommers: Phys. Rev. Lett. *38*, 1536 (1977)
8.113 W. Schommers: Phys. Rev. B *17*, 2057 (1978)
8.114 M. Remoissenet: Solid State Commun. *27*, 681 (1978)
8.115 A.R. Bishop: J. Phys. C *11*, L329 (1978)
8.116 L. Pietronero, S. Strässler: Solid State Commun. *27*, 1041 (1978)
8.117 A.R. Bishop, T. Schneider (eds.): *Solitons and Condensed Matter Physics*, Springer Series in Solid-State Sciences,Vol.8 (Springer, Berlin, Heidelberg, New York 1978)
8.118 W. Dieterich, I. Peschel: To be published
8.119 T. Geisel: To be published

Additional References with Titles

Chapter 2

P. Vashishta, A. Rahman: Ionic motion in α-AgI. Phys. Rev. Lett. *40*, 1137 (1978)
G. Jacucci, A. Rahman: Diffusion of F^- ions in CaF_2. J. Chem. Phys. *69*, 4117 (1978)
M.J. Gillan, D.D. Richardson: Disorder in superionic fluorites. J. Phys. C *12*, L61 (1979)
T. Sakuma: Comment on 'The Structure of α-Ag_2S'. J. Phys. C *11*, L747 (1978)
Y. Tsuchiya, Y. Waseda, S. Tamakai: A reply to "Comment on 'The Structure of α-Ag_2S'". J. Phys. C *12*, L93 (1979)
S. Hoshino, T. Sakuma, Y. Fujii: The existence of the order phase in superionic conductor Ag_3SI. J. Phys. Soc. Jpn. *45*, 705 (1978)
H.U. Beyeler, P. Brüesch, T. Hibma, W. Bührer: Phonon-induced diffuse X-ray scattering in β-AgI. Phys. Rev. B *18*, 4570 (1978)

Chapter 3

J.P. Boilot, Ph. Colomban, R. Collongues, G. Collin, R. Comès: X-ray scattering evidence for sublattice phase transition in stoichiometric silver β-alumina. Phys. Rev. Lett. *42*, 785 (1979)
R.J. Cava, F. Reidinger, B.J. Wuensch: Single crystal neutron diffraction study of the fast-ion conductor β-Ag_2S between 186° and 325°C. To be published
M.H. Dickens, W. Hayes, M.T. Hutchings, C. Smith: Investigation of the structure of strontium chloride at high temperatures using neutron diffraction. J. Phys. C*12*, L97 (1979)
M.H. Dickens, M.T. Hutchings, J.K. Kjems, R.E. Lechner: Quasielastic neutron scattering by superionic strontium chloride. J. Phys. C*11*, L583 (1978)

Chapter 4

J. Bernasconi, H.U. Beyeler, S. Strässler, S. Alexander: Anomalous frequency. Dependent conductivity in disordered one-dimensional systems. Phys. Rev. Lett., (March 1979)
H.U. Beyeler, L. Pietronero: About the conductivity anomlay in (N_a,K)-Ga_2O_3 mixed crystals. Solid State Commun. To appear (1979)
H.U. Beyeler, S. Strässler: The state of order in α-AgI. Submitted to Phys. Rev. B
J.C. Kimball, L.W. Adams, Jr.: Hopping conduction and superionic conductors. Phys. Rev. B *18*, 5851 (1978)
L. Pietronero: Stability and melting of simple ionic systems. Phys. Rev. B *17*, 3946 (1978)
L. Pietronero, S. Strässler: Conductivity of a polarizable ion. Phys. Rev. B *18*, 2016 (1978)
L. Pietronero, S. Strässler: Anomalous specific heat of a one-dimensional disordered solid. Phys. Rev. Lett. *42*, 188 (1979)
L. Pietronero, S. Strässler: Conductivity of a generalized hopping system with internal dynamics. Submitted to Phys. Rev. B (1979)
L. Pietronero, S. Strässler, H.R. Zeller: The microwave conductivity of solid electrolytes. Solid State Commun. To appear (1979)

P.M. Richards: Hopping conductivity in a quasi-one-dimensional lattice gas with three-dimensional ordering. Phys. Rev. B *18*, 945 (1978)

J.B. Sokoloff, W. Widom: Theory of one-dimensional ionic and solitary wave conduction in potassium hollandite. Phys. Rev. B *18*, 2824 (1978)

R.A. Vargas, M.B. Salamon, C.P. Flynn: Ionic Conductivity and heat capacity of the solid electrolytes $M Ag_1 I_5$ near T_c. Phys. Rev. B *17*, 269 (1978)

P. Vashista, A. Rahman: Ionic motion in α-AgI. Phys. Rev. Lett. *40*, 1337 (1978)

Chapter 5

S.J. Allen, Jr., A.S. Cooper, F. DeRosen, J.P. Remeika, S.K. Ulasi: Far infrared absorption and ionic conductivity of Na, K, Rb, Ag and Tl-alumina. Phys. Rev. B*17*, 4031 (1978)

C.R.A. Catlow, J.D. Comins, F.A. Germano, R.T. Harley, W. Hayes: Brillouin scattering and theoretical studies of high temperature disorder in fluorite crystals. J. Phys. C*11*, 3197 (1978)

H.R. Chandrasekhar, G. Burns, G.V. Chandrasekhar: Infrared spectra of superionic conductors Na, K, Rb, Ag and Tl-alumina. Solid State Commun. *27*, 829 (1978)

W. Hayes, A.J. Rushworth, J.F. Ryan: "Effect on anharmonicity and disorder on the Raman spectrum in fluorite crystals at high temperatures", in *Lattice Dynamics*, Int. Conf. Paris 1977, ed. by M. Balkanski (Flammarion, Paris 1977) p. 519

T. Kaneda, J.B. Bates, J.C. Wang: Raman scattering from $^6Li^+$ and $^7Li^+$ in lithium-alumina. Solid State Commun. *28*, 469 (1978)

R.J. Nemanich, J.C. Mikkelsen, Jr.: "Raman scattering from the copper halides CuI, CuBr and CuCl in the high temperature phases", in Proc. of the 14th Int. Conf. on Semiconductors, Edinburgh, 1978)

R.J. Nemanich, R.M. Martin, J.C. Mikkelsen, Jr.: Light scattering from correlated ion fluctuations in ionic conductors. Submitted to Solid State Commun.

Chapter 7

J.P. Boilot, Ph. Colomban, R. Collongues, G. Collin, R. Comes: X-ray scattering evidence for sublattice phase transition in stoichiometric silver β-alumina. Phys. Rev. Lett. *42*, 785 (1979)

J.B. Boyce, B.A. Huberman: Superionic conductors: Transitions, structures, and dynamics. Phys. Rep. (to be published)

D. Brinkman, W. Freudenreich: Ion dynamics and transition to the superionic state in α-AgI. Solid State Commun. *25*, 625 (1978)

D. Brinkman, W. Freudenreich, H. Arend, J. Roos: Evidence for a first-order transitions at 209 K in the superionic conductor $RbAg_4I_5$. Solid State Commun. *27*, 133 (1978)

Yu.Ya. Gurevich, Y.I. Kharkats: Transitions in superionic crystals. J. Phys. Chem. Sol. *39*, 751 (1978)

S. Hoshino, T. Sakuma, Y. Fujii: The existence of the order phase in superionic conductor Ag_3SI. J. Phys. Soc. Jpn. *45*, 705 (1978)

Chapter 8

A.K. Das, P. Schwendiman: Fokker-Planck equation for a periodic potential. Physica *89*A, 605 (1977)

R. Festa, E.G. d'Agliano: Diffusion coefficient for a Brownian particle in a periodic field of force. Physica *90*A, 229 (1978)

R.A. Field, D.A. Gallagher, M.V. Klein: Rayleigh-Brillouin spectra of the solid electrolyte $RbAg_4I_5$. Phys. Rev. B *18*, 2995 (1978)

J. Jäckle: Long-wavelength density fluctuations in superionic conductors. Z. Phys. B *30*, 255 (1978)

I. Peschel, W. Dieterich: The diffusion of interacting classical particles as a many boson problem. Z. Phys. B *31*, 195 (1978)

H. Risken, H.D. Vollmer: Correlation functions for the diffusive motion of particles in a periodic potential. Z. Phys. B *31*, 209 (1978)

R. Zeyher: Hydrodynamics of superionic conductors. Z. Phys. B *31*, 127 (1978)

Subject Index

T_1 minimum 151,154,156ff,171

T_1/T_2 ratio 151,166ff
    quadrupole effect 152

$TaS_2(NH_3)$, resonance 161

Thermal bath 207

Threshold energy 10

$TiH_{1.7}$, T_1 158,159

TO optical mode 68

Tracer conductivity 88

Transit time 2,78

Transition probability 211
   temperature, fluorites 68

Translational diffusion 63

Trapping time 78,88

Two-dimensional superionic conductor
  154

Two-fluid case 91

Ultrasonic attenuation 194

Universality 180

Vacancy 80

Vacancy + distortion 89

Variance-coveriance matrix 47ff

$VD_{0.79}$ 54,56

Velocity correlation matrix 81

Virtual crystal approximation 86,95
    potential 191

White line 10

Window function 8,12

Wurtzite structure 34

X-ray diffraction 6,14

Zincblende structure 34

Zwanzig-Mori method 236

H. Haken
Synergetics
An Introduction

Nonequilibrium Phase Transitions and Self-Organization in Physics, Chemistry and Biology

2nd, enlarged edition 1978. 152 figures, 4 tables. XII, 355 pages
ISBN 3-540-08866-0

Contents: Goal. – Probability. – Information. – Chance. – Necessity. – Chance and Necessity. – Self-Organization. – Physical Systems. – Chemical and Biochemical Systems. – Applications to Biology. – Sociology: A Stochastic Model for the Formation of Public Opinion. – Chaos. – Some Historical Remarks and Outlook.

Synergetics, deals with profound and striking analogies recently discovered between the self-organized behavior of seemingly quite different systems in physics, chemistry, biology, sociology and other fields. The cooperation of many subsystems such as atoms, molecules, cells, animals, or humans may produce spatial, temporal or functional structures. Their spontaneous formation out of chaos is often strongly reminiscent of phase transitions.

This book, written by the founder of synergetics, provides an elementary introduction into its basic concepts and mathematical tools. Numerous exercises, figures and simple examples greatly facilitate the understanding. The profound analogies are demonstrated by various realistic examples from fluid dynamics, laser physics, mechanical engineering, chemical and biochemical systems, ecology, sociology and theories of evolution and morphogenesis.
The second edition differs from the first by an additional chapter on chaotic motion, a rapidly growing field, and by new sections on laser pulses and on morphogenesis.

C.P. Slichter
Principles of Magnetic Resonance

2nd, revised and expanded edition 1978. 115 figures. X, 397 pages
(Springer Series in Solid State Sciences, Volume 1)
ISBN 3-540-08476-2

Contents: Elements of Resonance. – Basic Theory. – Magnetic Dipolar Broadening of Rigid Lattices. – Magnetic Interactions of Nuclei with Electrons. – Spin-Lattice Relaxation and Motional Narrowing of Resonance Lines. – Spin Temperature in Magnetism and in Magnetic Resonance. – Double Resonance. – Advanced Concepts in Pulsed Magnetic Resonance. – Electric Quadrupole Effects. – Electron Spin Resonance. – Summary. – Problems. – Appendices.

Principles of Magnetic Resonance is a textbook for graduate students or others beginning research in magnetic resonance who wish to learn either nuclear magnetic resonance or electron spin resonance. It is intended for physicists, chemists, applied scientists, or others who have had a one year graduate course in quantum mechanics from one of the standard textbooks.
The text aims at developing a physical understanding of magnetic resonance and familiarity with the principal theoretical techniques needed to read resonance articles in scientific journals. The author seeks to develop depth of understanding of the most important topics rather than giving a comprehensive account of all of resonance.
The new edition differs from the original one, principally, in the addition of chapters on spin temperature in magnetic resonance, double resonance and on techniques for line-narrowing in solids (the Waugh-Mansfield-approach). There are two new appendices.

O. Madelung
Introduction to Solid-State Theory

Translated from the German by B. C. Taylor
1978. 144 figures. XI, 486 pages
(Springer Series in Solid-State Sciences, Volume 2)
ISBN 3-540-08516-5

Contents: Fundamentals. – The One-Electron Approximation. – Elementary Excitations. – Electron-Phonon Interaction: Transport Phenomena. – Electron-Electron Interaction by Exchange of Virtual Phonons: Superconductivity. – Interaction with Photons: Optics. – Phonon-Phonon Interaction: Thermal Properties. – Local Description of Solid-State Properties. – Localized States. – Disorder. – Appendix: The Occupation Number Representation.

This is a textbook in solid-state theory for graduate students of physics and material science. In addition, it provides the theoretical background needed by physicists doing research in both pure solid-state physics and solid-state physics as applied to electrical engineering.
The fundamentals of solid-state theory are developed starting from one unifying point of view: from the description by delocalized and localized states and – within the concept of delocalized states – by elementary excitations. The development of solid-state theory within the last ten years has shown that by a systematic introduction of these concepts, large parts of the theory can be described in a unified way. At the same time, this form of description gives a "pictorial" formulation of many elementary processes in solids, facilitating their understanding. The book is a revised and partly rewritten version of a German textbook published a few years ago.

Dynamics of Solids and Liquids by Neutron Scattering

Editors: S. W. Lovesey, T. Springer

1977. 156 figures, 15 tables. XI, 379 pages
(Topics in Current Physics, Volume 3)
ISBN 3-540-08156-9

Contents:
S. W. Lovesey: Introduction. – *H. G. Smith,*
N. Wakabayashi: Phonons. – *B. Dorner, R. Comès:*
Phonons and Structural Phase Transformations. –
J. W. White: Dynamics of Molecular Crystals,
Polymers, and Absorbed Species. – *T. Springer:*
Molecular Rotations and Diffusion in Solids, in
Particular Hydrogen in Metals. – *R. D. Mountain:*
Collective Modes in Classical Monoatomic
Liquids. – *S. W. Lovesey, J. M. Loveluck:* Magnetic
Scattering.

Neutron Diffraction

Editor: H. Dachs

1978. 138 figures, 32 tables. XIII, 357 pages
(Topics in Current Physics, Volume 6)
ISBN 3-540-08710-9

Contents:
H. Dachs: Principles of Neutron Diffraction. –
J. B. Hayter: Polarized Neutron. – *P. Coppens:*
Combining X-Ray and Neutron Diffraction: The
Study of Charge Density Distributions in Solids. –
W. Prandl: The Determination of Magnetic Struc-
tures. – *W. Schmatz:* Disordered Structures. –
P.-A. Lindgård: Phase Transitions and Critical
Phenomena. – *G. Zaccai:* Application of Neutron
Diffraction to Biological Problems. – *P. Chieux:*
Liquid Structure Investigation by Neutron
Scattering. – *H. Rauch, D. Petrascheck:* Dynamical
Diffraction on Perfect Crystals and its Application
in Neutron Physics.

Neutron Physics

1977. 40 figures, 11 tables. VII, 135 pages
(Springer Tracts in Modern Physics, Volume 80)
ISBN 3-540-08022-8

Contents:
I. Koester: Neutron Scattering Lengths and Funda-
mental Neutron Interactions. – *A. Steyerl:* Very Low
Energy Neutrons.

T. Springer

Quasielastic Neutron Scattering for the Investigation of Diffusive Motions in Solids and Liquids

1972. 36 figures. II, 100 pages
(Springer Tracts in Modern Physics, Volume 64)
ISBN 3-540-05808-7

Contents:
Scattering Theory. – Methodical and Experimental
Aspects. – Monoatomic Liquids with Continuous
Diffusion. – Jump Diffusion in Liquids. – Diffusion
of Hydrogen in Metals. – Rotational Diffusion in
Molecular Solids. – Molecular Liquids. – Polymers
and other Complicated Systems. – Effects of
Coherent Scattering. – Quasielastic Scattering and
other Methods.

Solid Electrolytes

Editor: S. Geller

1977. 85 figures, 24 tables. XI, 229 pages
(Topics in Applied Physics, Volume 21)
ISBN 3-540-08338-3

Contents:
S. Geller: Introduction. – *H. Sato:* Some Theoretical
Aspects of Solid Electrolytes. – *S. Geller:* Halogenide
Solid Electrolytes. – *B. B. Owens, J. E. Oxley,*
A. F. Sammells: Applications of Halogenide Solid
Electrolytes. – *J. H. Kennedy:* The Beta-Aluminas. –
W. L. Worrel: Oxide Solid Electrolytes. – *L. Heyne:*
Electrochemistry of Mixed Ionic-Electronic
Conductors.

Springer-Verlag
Berlin
Heidelberg
New York